Synthesis Lectures on Mathematics & Statistics

Series Editor

Steven G. Krantz, Department of Mathematics, Washington University, Saint Louis, USA

This series includes titles in applied mathematics and statistics for cross-disciplinary STEM professionals, educators, researchers, and students. The series focuses on new and traditional techniques to develop mathematical knowledge and skills, an understanding of core mathematical reasoning, and the ability to utilize data in specific applications.

Vladyslav Babenko · Volodymyr Kofanov ·
Peter Kogut · Oleg Kovalenko ·
Nataliia Parfinovych

Extremal Problems of Analysis and Applications

Vladyslav Babenko
Department of Mathematical Analysis
and Optimization
Oles Honchar Dnipro National University
Dnipro, Ukraine

Peter Kogut
Department of Mathematical Analysis
and Optimization
Oles Honchar Dnipro National University
Dnipro, Ukraine

Nataliia Parfinovych
Department of Mathematical Analysis
and Optimization
Oles Honchar Dnipro National University
Dnipro, Ukraine

Volodymyr Kofanov
Department of Mathematical Analysis
and Optimization
Oles Honchar Dnipro National University
Dnipro, Ukraine

Oleg Kovalenko
Department of Mathematical Analysis
and Optimization
Oles Honchar Dnipro National University
Dnipro, Ukraine

ISSN 1938-1743   ISSN 1938-1751 (electronic)
Synthesis Lectures on Mathematics & Statistics
ISBN 978-3-031-94731-5   ISBN 978-3-031-94732-2 (eBook)
https://doi.org/10.1007/978-3-031-94732-2

© The Editor(s) (if applicable) and The Author(s), under exclusive license to Springer Nature Switzerland AG 2025

This work is subject to copyright. All rights are solely and exclusively licensed by the Publisher, whether the whole or part of the material is concerned, specifically the rights of translation, reprinting, reuse of illustrations, recitation, broadcasting, reproduction on microfilms or in any other physical way, and transmission or information storage and retrieval, electronic adaptation, computer software, or by similar or dissimilar methodology now known or hereafter developed.
The use of general descriptive names, registered names, trademarks, service marks, etc. in this publication does not imply, even in the absence of a specific statement, that such names are exempt from the relevant protective laws and regulations and therefore free for general use.
The publisher, the authors and the editors are safe to assume that the advice and information in this book are believed to be true and accurate at the date of publication. Neither the publisher nor the authors or the editors give a warranty, expressed or implied, with respect to the material contained herein or for any errors or omissions that may have been made. The publisher remains neutral with regard to jurisdictional claims in published maps and institutional affiliations.

This Springer imprint is published by the registered company Springer Nature Switzerland AG
The registered company address is: Gewerbestrasse 11, 6330 Cham, Switzerland

If disposing of this product, please recycle the paper.

Introduction

Inequalities that estimate the deviation of a value of a function at some point from its mean value using some characteristics of the function, are sometimes called Ostrowski-type inequalities. The first result of this kind was obtained by Ostrowski in 1938 and such inequalities were heavily studied since then. Ostrowski-type inequalities can be viewed as a partial case of the problem to find the deviation between operators, or as a simplest form of the problem of optimization of cubature formulae. At the same time, it appears that such inequalities are often a key step in solutions of other important extremal problems in approximation theory, including the problems of optimization of cubature formulae, the Stechkin problem about approximation of unbounded operators by bounded ones, inequalities for derivatives of Landau–Kolmogorov type and of Nagy type, and others.

In Chap. 1 we discuss a general approach to some extremal problems of approximation theory, which in particular allows us to obtain many Ostrowski-type inequalities for various classes of functions. We show how Ostrowski-type inequalities can be applied to obtain sharp inequalities for derivatives and related problems. Using Ostrowski-type inequalities as a primary tool, we obtain solutions for problems of optimization of cubature formulae.

It is well known that the problem to find a sharp constant in a Kolmogorov type inequality for functions defined on the real axis, is equivalent to the extremal Kolmogorov problem to find the exact upper bound of the norm of an intermediate derivative of a function on the class of functions with restrictions on the norms of the function and its higher derivative. Despite a large number of works devoted to Kolmogorov-type inequalities, sharp constants for derivatives of arbitrary order are known only in a few cases. Therefore, the modification of the Kolmogorov problem considered by Boyanov and Naidyonov is interesting. In this modification, the norm of the intermediate derivative on the entire line is substituted by its norm on an arbitrary finite segment.

In Chap. 2, the Boyanov-Naidyonov problem is solved on classes of functions with a given comparison function for norms of the positive and negative parts of the intermediate derivative of the function. In particular, this problem is solved on the Sobolev classes and on the spaces of trigonometric polynomials and polynomial splines. In addition, a solution to an analogue of the Erdos problem is obtained; we characterize a polynomial (spline) with a given uniform norm that has maximal possible total length of the arcs of the graph of its positive (negative) part on a given segment.

Remez-type inequalities play an important role in approximation theory. This topic was initiated in the work of Remez in 1936, in which he found a sharp constant in an inequality of this type for algebraic polynomials. At the end of the 20th century, a new surge of works on this topic was observed. The efforts of many mathematicians aimed at finding the sharp constant in the Remez-type inequality for trigonometric polynomials. Only in 2019 this problem was solved in the work of Tikhonov and Yuditski.

In the author's works, Remez-type inequalities were extended to wider classes of functions. In Chap. 3, sharp Remez-type inequalities for functions with a given comparison function are obtained in various metrics. As a result, such type of inequalities were proved for functions from the Sobolev classes, for trigonometric polynomials and polynomial splines with a given ratio of norms of their positive and negative parts.

In Chap. 4, we focus on the development of a variational approach for simultaneous contrast enhancement of color images and their denoising. With that in mind we propose a new variational model in Sobolev-Orlicz spaces with non-standard growth conditions of the objective functional and discuss its applications to the simultaneous fusion and denoising of each spectral channel for an input color images. The characteristic feature of the proposed model is the fact that we deal with a constrained minimization problem with a special objective functional that lives in variable Sobolev-Orlicz spaces. This functional contains a spatially variable exponent characterizing the growth conditions and it can be seen as a replacement for the standard 1-norm in TV regularization. We show that the proposed model allows to synthesize at a high level of accuracy noise- and blur-free color images, which were captured in extremely low light conditions.

The main purpose of Chap. 5 is to describe a robust approach for the simultaneous fusion and denoising of non-smooth multispectral images defined on grids with different resolution using for that a special extremal problem with nonstandard growth of the energy functional. In fact, we use the L^1-norm of the noise in the minimization function and a special form of anisotropic diffusion tensor for the regularization term. Following this approach, we increase the noise robustness of the proposed model albeit it makes such variational problem completely non-smooth, non-convex, and, hence, significantly more difficult from a minimization point of view. The principle characteristic feature of the proposed model is that we consider the energy functional with nonstandard growth for each spectral channel separately. The second point that should be emphasized is the fact that we do not predefine the variable exponents a priori using for that the original noisy images, but instead we associate these characteristics with each feasible solution.

Contents

1 On Ostrowski-Type Inequalities and Their Applications 1
- 1.1 Introduction 1
- 1.2 Ostrowski-Type Inequalities 4
  - 1.2.1 Abstract Distance Spaces 4
  - 1.2.2 Some Notes About Estimates for the Deviation Between Operators 12
  - 1.2.3 Classes of Smooth Functions Defined on a Segment 14
  - 1.2.4 Some Auxiliary Results 20
  - 1.2.5 Classes of Functions with Given Majorant of Modulus of Continuity 28
  - 1.2.6 Classes of Functions with a Restriction on Their Gradient .. 29
  - 1.2.7 Classes of Functions with Bounded Norms of Their Laplacian ... 31
  - 1.2.8 Classes of Random Processes 35
  - 1.2.9 Sets and Functions of Bounded Variation 39
- 1.3 Inequalities for Derivatives 47
  - 1.3.1 Hypersingular Integral Operator 47
  - 1.3.2 Kolmogorov-Type Inequalities for Classes $H_K^\omega(C)$ 50
  - 1.3.3 Kolmogorov-Type Inequalities for Classes $L_{\infty,p}^\Omega(C)$ 51
  - 1.3.4 Kolmogorov-Type Inequalities for Classes $L_{\infty,p}^\Delta(\mathbb{R}^d)$ 52
  - 1.3.5 Nagy-Type Inequalities 55
  - 1.3.6 Kolmogorov-Type Inequalities for Charges 60
  - 1.3.7 Inequalities for a Mixed Derivative of a Function 61
  - 1.3.8 Approximation of Unbounded Operators by Bounded Ones and Related Problems 64

	1.4	Optimization of Cubature Formulae	68
		1.4.1 Optimization of Cubature Formulae on Multivariate Sobolev Classes	68
		1.4.2 Optimization of Cubature Formulae on Classes of Random Processes	71
	References		79

2 The Bojanov–Naidenov Problem for Differentiable Functions and the Erdös Problem for Polynomials and Splines ... 85

2.1	Introduction	85
2.2	Auxiliary Statements	89
2.3	The Bojanov–Naidenov Problem for the Classes of Functions with a Given Comparison Function	95
2.4	The Bojanov–Naidenov Problem for Sobolev Classes	99
2.5	The Bojanov–Naidenov Problem for Trigonometric Polynomials	100
2.6	The Bojanov–Naidenov Problem for Splines	102
2.7	The Erdös Problem for Spaces of Trigonometric Polynomials and Splines	103
References		104

3 Remez-Type Inequalities ... 105

3.1	Introduction	105
3.2	Classes $S_\varphi(\omega)$	107
3.3	Classes $L^r_\infty(I_{2\pi})$	111
3.4	Classes of Trigonometric Polynomials	114
3.5	Classes of Splines	117
References		121

4 Restoration of the Noise Corrupted Optical Images with Their Simultaneous Contrast Enhancement ... 123

4.1	Preliminaries	126
	4.1.1 Functional Spaces	127
	4.1.2 Basic Facts on the Lebesgue and Sobolev Spaces with Variable Exponents	128
	4.1.3 On the Dual Sobolev Space $H^{-1}(\Omega)$	130
	4.1.4 Level Sets, Directional Gradients, and Texture Indexes	131
4.2	Statement of the Problem	132
4.3	Optimality Conditions	136
4.4	Existence Issues and Regularization of the Original Optimization Problem	143
4.5	Numerical Results	152
4.6	Conclusions	161
References		162

5 Variational Approach to Simultaneous Fusion and Denoising of the Color Images with Different Spatial Resolution 165
   5.1 Preliminaries 168
   5.2 Existence Result 171
   5.3 On Relaxation of the Minimization Problem (5.7)–(5.8) 175
   5.4 Proximal Alternating Minimization Algorithm and Its Modification 182
   5.5 Optimality Conditions 190
   5.6 Numerical Scheme and Settings 192
   5.7 Numerical Results 195
   References 203

On Ostrowski-Type Inequalities and Their Applications

Abstract

Inequalities that estimate the deviation of a value of a function at some point from its mean value using some characteristics of the function, are sometimes called Ostrowski-type inequalities. The first result of this kind was obtained by Ostrowski in 1938 and such inequalities were heavily studied since then. Ostrowski-type inequalities can be viewed as a partial case of the problem to find the deviation between operators, or as a simplest form of the problem of optimization of cubature formulae. At the same time, it appears that such inequalities are often a key step in solutions of other important extremal problems in approximation theory, including the problems of optimization of cubature formulae, the Stechkin problem about approximation of unbounded operators by bounded ones, inequalities for derivatives of Landau–Kolmogorov type and of Nagy type, and others. In this chapter we discuss a general approach to some extremal problems of approximation theory, which in particular allows us to obtain many Ostrowski-type inequalities for various classes of functions. We show how Ostrowski-type inequalities can be applied to obtain sharp inequalities for derivatives and related problems. Using Ostrowski-type inequalities as a primary tool, we obtain solutions for problems of optimization of cubature formulae.

1.1 Introduction

Let two operators Λ and I defined on a set $\mathcal{A}$ of functions f be given, and h measure distance in the range of the operators Λ and I. The quantity

$$U(\Lambda, I; \mathcal{A}) := \sup_{f \in \mathcal{A}} h(\Lambda f, If) \qquad (1.1)$$

defines the deviation between operators Λ and I on the set $\mathcal{A}$. The problem to find such deviation occurs in many questions of approximation theory and numerical analysis. For example, if $\mathcal{A}$ is a set of continuous functions $f\colon T \to \mathbb{R}$, $\Lambda f = \int_T f(s)ds$, $If = \sum_{k=1}^n c_k f(x_k)$ ($c_k \in \mathbb{R}$, $x_k \in T$, $k = 1, \ldots, n$) and h is the usual metric in $\mathbb{R}$, then (1.1) gives the worst-case error of the cubature formula I on the class $\mathcal{A}$; if Λf is the identity operator, h is some metric, and I is the operator of the best approximation by elements from a set $\mathcal{B}$ i.e., $If = \arg\min_{g \in \mathcal{B}} h(f, g)$, then quantity (1.1) becomes the best approximation of the set $\mathcal{A}$ by the set $\mathcal{B}$ in metric h that is given by the formula $\sup_{f \in \mathcal{A}} \inf_{g \in \mathcal{B}} h(f, g)$, provided the minimum in the definition of I exists for each $f \in \mathcal{A}$.

In 1938 Ostrowski [66] proved the following theorem.

Theorem 1.1 *Let $f\colon [-1, 1] \to \mathbb{R}$ be a differentiable function and let for all $t \in (-1, 1)$, $|f'(t)| \leq 1$. Then for all $x \in [-1, 1]$ the following inequality holds*

$$\left| \frac{1}{2} \int_{-1}^{1} f(t)dt - f(x) \right| \leq \frac{1}{2} \int_{-1}^{1} |t - x|dt = \frac{1 + x^2}{2}. \tag{1.2}$$

The inequality is sharp in the sense that for each fixed $x \in [-1, 1]$, the upper bound $\frac{1+x^2}{2}$ cannot be reduced.

Inequalities that estimate the deviation of a value of a function at some point from its mean value using some characteristics of the function, are sometimes called Ostrowski-type inequalities. Such inequalities were intensively studied, see for example [40, 41, 63].

Observe that this theorem can be rewritten as

$$\sup_{\sup_{t \in (-1,1)} |f'(t)| \leq 1} \left| \frac{1}{2} \int_{-1}^{1} f(t)dt - f(x) \right| = \frac{1 + x^2}{2},$$

and hence this kind of inequalities are in fact a partial case of the problem to find the deviation between operators.

At the same time, it appears that such inequalities are often a key step in solutions of other important extremal problems in approximation theory, including the problems of optimal recovery of operators and functionals (in particular, the problems of optimization of cubature formulae) and the questions of inequalities for derivatives.

Let X and Y be linear spaces equipped with a seminorm $\|\cdot\|_X$ and a norm $\|\cdot\|_Y$ respectively. A linear operator $A\colon X \to Y$ is called bounded if

$$\|A\| = \|A\|_{X \to Y} := \sup_{\|x\|_X \leq 1} \|Ax\|_Y < \infty.$$

Otherwise, the operator A is called unbounded. By $\mathcal{L}(X, Y)$ we denote the space of all linear bounded operators $S\colon X \to Y$.

1.1 Introduction

Quantity (1.1) was stated using an abstract notion of a distance. It will be discussed in such a general form in Sects. 1.2.1 and 1.2.2, but more often is considered in metric or normed spaces. If $\Lambda, I\colon X \to Y$ and $\mathcal{A} \subset X$, then quantity (1.1) becomes

$$U(\Lambda, I; \mathcal{A}) := \sup\{\|\Lambda x - Ix\|_Y : x \in \mathcal{A}\}.$$

The Stechkin problem of approximation of a generally speaking unbounded operator A by linear bounded operators on $\mathcal{A}$ is stated as follows. For a given number $N > 0$ find the quantity

$$E_N(A, \mathcal{A}) := \inf\{U(A, S; \mathcal{A}) : S \in \mathcal{L}(X, Y), \|S\| \leq N\}, \tag{1.3}$$

and an operator S on which the infimum is attained, if such an operator exists. The statement of this problem, first important results, and solutions to this problem for differential operators of small orders were presented in [71]. For a survey of further results on this problem see [5].

Inequalities of the form

$$\|f^{(k)}\|_{L_q(G)} \leq K \|f\|_{L_p(G)}^{\mu} \|f^{(r)}\|_{L_s(G)}^{\lambda},$$

where $0 < k < r$, $1 \leq q, p, s \leq \infty$, $\mu = 1 - \lambda$, $\lambda = \frac{k-1/q+1/p}{r-1/s+1/p}$, and G is some domain of definition are called Kolmogorov-type (or Landau–Kolmogorov-type) inequalities in the multiplicative form. The first results of this kind were obtained by Hardy and Littlewood [47] in 1912. The first sharp (i.e., with the smallest possible constant) inequalities for derivatives were obtained by Landau [61] and Hadamard [46]. In these articles for functions defined on $\mathbb{R}$ or $\mathbb{R}_+$ inequalities that estimate the uniform norm of the derivative of a function via its uniform norm and the uniform norm of its second derivative were obtained. Kolmogorov [54] proved sharp inequalities for all natural r and k in the case $G = \mathbb{R}$, $q = p = s = \infty$. Many results in this topic for univariate functions can be found in [27]. The multivariate situation is substantially harder to study and only results for classes of low smoothness are known. In Sect. 1.3 we discuss inequalities for derivatives mainly for multivariate functions.

Quantity (1.1) also appears in problems of optimal recovery of operators and functionals. Let a metric space (X, h_X), sets Y and W, and mappings $\Lambda\colon W \to X$ and $I\colon W \to Y$ be given. An arbitrary function $\Phi\colon Y \to X$ is called a method of recovery of the mapping Λ on the class W using the information given by the mapping I. The error of recovery of the operator Λ on the class W by the method Φ using the information given by the mapping I is defined by the formula

$$\mathcal{E}(\Lambda, W, I, \Phi, X) = U(\Lambda, \Phi \circ I; W) = \sup_{w \in W} h_X(\Lambda(w), \Phi(I(w))).$$

The quantity

$$\mathcal{E}(\Lambda, W, I, X) = \inf_{\Phi} \mathcal{E}(\Lambda, W, I, \Phi, X) \tag{1.4}$$

is called the optimal error of recovery of the operator Λ on the class W, using the information given by the operator I. The problem of optimal recovery is to find quantity (1.4) and a method Φ^* (if it exists), on which the infimum on the right-hand side of (1.4) is attained. If $\mathcal{I}$ is some class of informational operators, then it is interesting to find the quantity

$$\mathcal{E}(\Lambda, W, \mathcal{I}, X) = \inf_{I \in \mathcal{I}} \mathcal{E}(\Lambda, W, I, X)$$

and an optimal information operator (if it exists), or a sequence of asymptotically optimal operators, if an optimal operator does not exists, or it is hard to find one.

In Sect. 1.4 we consider problems of optimization of cubature formulae i.e., problems of optimal recovery of the integral operator $\Lambda w = \int_T w(x) dx$ for some classes of functions w. It is well known that under rather mild conditions on the classes of functions, there exist linear methods among the optimal methods of recovery. If the informational operator I gives the values of a function at n points $x_1, \ldots, x_n \in T$, then the error of recovery (1.4) becomes

$$\inf_{c_k} \sup_{w \in W} \left| \int_T w(s) ds - \sum_{k=1}^n c_k w(x_k) \right|.$$

Observe that for $n = 1$ the problem of optimization of the cubature formula essentially becomes a problem to find a sharp Ostrowski-type inequality. Thus Ostrowski-type inequalities might be also viewed as a partial case of the problem of optimal recovery. On the other hand, for classes of functions of low smoothness, it is sometimes possible to obtain a solution to the problem of optimal cubature formulae for arbitrary n, from the corresponding sharp Ostrowski-type inequality. Some applications of this kind will be given in Sect. 1.4.

This chapter is organized as follows. In Sect. 1.2 we discuss a general approach that allows to compute quantity (1.1), and using this approach deduce many Ostrowski-type inequalities for various classes of functions. In Sect. 1.3 we show how Ostrowski-type inequalities can be applied to obtain sharp inequalities for derivatives and related problems. Finally, in Sect. 1.4 we apply some Ostrowski-type inequalities to problems of optimization of cubature formulae.

1.2 Ostrowski-Type Inequalities

1.2.1 Abstract Distance Spaces

1.2.1.1 Notations and Definitions

The results of this section are contained in [15]. The notion of a distance (in particular, a metric) plays an important role in many branches of mathematics. Definitions of numeric-valued distances or metrics and a detailed discussion of these notions can be found e.g., in monograph [52]. We refer to [12, 13, 48, 77] for metrics that take value in more general sets. We consider a rather general definition for this notion.

A set M with a reflexive, antisymmetric and transitive relation $\leq$ is called *partially ordered*.

1.2 Ostrowski-Type Inequalities

Let X be an arbitrary set and M be a partially ordered set that has a smallest element, which we denote by θ (i.e., $\theta \leq m$ for any $m \in M$). A function $h_X \colon X \times X \to M$ is called an *M-distance* in X, if for arbitrary $x, y \in X$

1. $h_X(x, x) = \theta$,
2. $h_X(x, y) = h_X(y, x)$.

The pair (X, h_X) will be called an *M-distance space*.

In [12, 13] the notion of M-distance was introduced for the case when M is a partially ordered monoid.

Let two M-distance spaces (X, h_X) and (Y, h_Y) be given. The class $H(X, Y)$ of mappings $f \colon X \to Y$ that satisfy the Lipschitz condition can be defined in a standard way:

$$H(X, Y) = \{f \colon X \to Y \colon h_Y(f(x_1), f(x_2)) \leq h_X(x_1, x_2) \, \forall x_1, x_2 \in X\}.$$

In this section, speaking of a partially ordered set M, we assume that some M-distance h_M is defined in M.

We say that an M-distance h_X in X *agrees* with an M-distance h_M in M, if

$$h_M(h_X(x, x_1), h_X(x, x_2)) \leq h_X(x_1, x_2) \, \forall \, x, x_1, x_2 \in X. \tag{1.5}$$

Note that inequality (1.5) holds (and is equivalent to the triangle inequality) if $M = \mathbb{R}_+$ with the usual metric, and (X, h_X) is a pseudo metric space (for a definition of a pseudo metric and a pseudo metric space see, for example [51, Chapter 4]). That is why we introduce the following definition.

An M-distance h_X on a set X will be called an *M-pseudo metric*, if it agrees with M-distance h_M i.e., inequality (1.5) holds. In this case the pair (X, h_X) will be called an *M-pseudo metric space*.

In Lemma 1.2 we will give a general sufficient condition that an M-metric h (see the definition prior to Lemma 1.2) agrees with h_M.

We need the following lemma.

Lemma 1.1 *Let T, X, Y be M-distance spaces. Then*

1. *If $f \in H(T, X)$ and $g \in H(X, Y)$, then $g \circ f \in H(T, Y)$.*
2. *If $f \in H(T, X)$ and h_X is an M-pseudo metric, then $h_X(f(\cdot), f(t)) \in H(T, M)$ for any fixed $t \in T$. In particular, if $T = X$, then $h_T(\cdot, t) \in H(T, M)$.*

Proof The first statement of the lemma is obvious.

If $f \in H(T, X)$ and h_X is an M-pseudo metric, then for arbitrary $t_1, t_2 \in T$,

$$h_M(h_X[f(t_1), f(t)], h_X[f(t_2), f(t)]) \stackrel{(1.5)}{\leq} h_X(f(t_1), f(t_2)) \stackrel{f \in H(T,X)}{\leq} h_T(t_1, t_2).$$

Therefore $h_X(f(\cdot), f(t)) \in H(T, M)$. If $T = X$ and $f(\tau) = \tau$, $\tau \in T$, we obtain that $h_T(\cdot, t) \in H(T, M)$.

1.2.1.2 Classes of Operators and an Ostrowski-Type Inequality

For an M-distance space X, an operator $\lambda \colon H(X, M) \to M$ will be called *monotone*, if for arbitrary $u, v \in H(X, M)$

$$(\forall x \in X \; u(x) \leq v(x)) \implies (\lambda(u) \leq \lambda(v)).$$

Let T, Y be M-distance spaces, X be an M-pseudo metric space, and $t \in T$ be fixed. We say that an operator $\Lambda \colon H(T, X) \to Y$ and a monotone operator $\lambda \colon H(T, M) \to M$ *agree*, if $\forall f \in H(T, X)$

$$h_Y(\Lambda f(\cdot), \Lambda f(t)) \leq \lambda(h_X(f(\cdot), f(t))). \tag{1.6}$$

Here and below $\Lambda f(t)$ means the value of the operator Λ on the constant function $\tau \mapsto f(t)$, $\tau \in T$ (the same notation will be used for other operators whose arguments are functions).

Theorem 1.2 *Let (T, h_T) and (X, h_X) be M-pseudo metric spaces, (Y, h_Y) be an M-distance space, and $t \in T$ be fixed. Assume that an operator $\Lambda \colon H(T, X) \to Y$ and a monotone operator $\lambda \colon H(T, M) \to M$ agree. Then for arbitrary function $f \in H(T, X)$ the following Ostrowski-type inequality holds:*

$$h_Y(\Lambda f(\cdot), \Lambda f(t)) \leq \lambda(h_T(\cdot, t)). \tag{1.7}$$

If

$$\lambda(\theta) = \theta, \tag{1.8}$$

and there exists an operator $\phi_X \colon H(T, M) \to H(T, X)$ and $\phi_Y \in H(M, Y)$ with the following property

$$h_Y(\phi_Y(m), \phi_Y(\theta)) = m, \text{ if } m = \lambda(h_T(\cdot, t)), \tag{1.9}$$

such that the diagram

$$\begin{array}{ccc} H(T, X) & \xrightarrow{\Lambda} & Y \\ \phi_X \uparrow & & \uparrow \phi_Y \\ H(T, M) & \xrightarrow{\lambda} & M \end{array}$$

is commutative i.e.,

$$\Lambda \circ \phi_X = \phi_Y \circ \lambda, \tag{1.10}$$

then inequality (1.7) is sharp and becomes equality on the function

$$f_t(\cdot) = \phi_X(h_T(\cdot, t)). \tag{1.11}$$

1.2 Ostrowski-Type Inequalities

Proof Let $f \in H(T, X)$. Since (T, h_T) and (X, h_X) are M-pseudo metric spaces, we have due to Lemma 1.1 that $h_X(f(\cdot), f(t)) \in H(T, M)$ and $h_T(\cdot, t) \in H(T, M)$. So both of these functions belong to the domain of λ.

Since the operators Λ and λ agree, and the operator λ is monotone, for each $f \in H(T, X)$ one has

$$h_Y(\Lambda f(\cdot), \Lambda f(t)) \leq \lambda(h_X(f(\cdot), f(t))) \leq \lambda(h_T(\cdot, t)),$$

and inequality (1.7) is proved.

The function from (1.11) belongs to the class $H(T, X)$, since $h_T(\cdot, t) \in H(T, M)$ and $\phi_X : H(T, M) \to H(T, X)$. Using condition (1.10), one has

$$h_Y(\Lambda f_t(\cdot), \Lambda f_t(t)) = h_Y(\Lambda(\phi_X(h_T(\cdot, t))), \Lambda(\phi_X(h_T(t, t))))$$

$$= h_Y((\Lambda \circ \phi_X)(h_T(\cdot, t)), (\Lambda \circ \phi_X)(h_T(t, t)))$$

$$\stackrel{(1.10)}{=} h_Y((\phi_Y \circ \lambda)(h_T(\cdot, t)), (\phi_Y \circ \lambda)(\theta))$$

$$= h_Y(\phi_Y(\lambda(h_T(\cdot, t))), \phi_Y(\lambda(\theta))) \stackrel{(1.9),(1.8)}{=} \lambda(h_T(\cdot, t)).$$

The theorem is proved.

Note that classes of operators that satisfy the properties analogous to properties (1.6), (1.9) and (1.10) were considered in [20]. In order to explain the nature of these properties we give the following example. Condition (1.6) is a relaxed version of the following condition:

$$h_Y(\Lambda f, \Lambda g) \leq \lambda(h_X(f(\cdot), g(\cdot)))$$

for all f, g. If $M = \mathbb{R}_+$, $X = Y$ is a Banach space, $\Lambda f = \int_{-1}^{1} f(t)dt$ is the Bochner integral of f, and λ is the Lebesgue integral on $[-1, 1]$, then this condition becomes

$$\left\| \int_{-1}^{1} f(t)dt - \int_{-1}^{1} g(t)dt \right\| \leq \int_{-1}^{1} \|f(t) - g(t)\|dt.$$

Moreover, if α is an integrable real-valued function and $x \in X$, then

$$\int_{-1}^{1} \alpha(t) \cdot x dt = \left(\int_{-1}^{1} \alpha(t)dt \right) \cdot x$$

i.e., condition (1.10) is satisfied with ϕ_X and ϕ_Y being multiplication by a fixed element $x \in X$. If the element x is such that $\|x\| = 1$, then the operator ϕ_X preserves the Lipschitz property, and condition (1.9) holds for arbitrary $m \in \mathbb{R}_+$. The function $h_T(\cdot, t) = |\cdot - t|$ is extremal in inequality (1.2) for the real-valued functions from $H([-1, 1], \mathbb{R})$. Therefore for any x such that $\|x\| = 1$ the function $f_t(\cdot) = |\cdot - t| \cdot x$ is extremal in the Ostrowski-type inequality for Banach space-valued functions.

In the majority results that we know (see e.g. [3, 4, 14, 20]) for extremal problems on classes of non-numeric-valued functions $f : T \to X$, extremal functions are built based on the real-valued extremal function for the extremal problem on the corresponding class of real-valued functions $f: T \to \mathbb{R}$: if $f_e: T \to \mathbb{R}$ is an extremal function in the real-valued case, then the function $f_e \cdot x: T \to X$ usually becomes an extremal function in the non-numeric-valued situation for some specially chosen element $x \in X$. This corresponds to the described above approach in the case, when ϕ_X and ϕ_Y are operators of multiplication by some elements.

More generally, if for the operators $\lambda: H(T, M) \to M$ and $\Lambda = \lambda$, fixed $t \in T$ and $f \in H(T, M)$ inequality (1.6) holds, then

$$h_M(\lambda f(\cdot), \lambda f(t)) \leq \lambda(h_T(\cdot, t)).$$

If in addition, property (1.9) holds with $Y = M$ and ϕ_Y being the identity function, then the latter inequality become equality on the function $h_T(f(\cdot, t))$. Function (1.11) is obtained from this function as a result of applying to it the operator $\phi_X: H(T, M) \to H(T, X)$.

Recall that an operator $\phi: H(T, M) \to X$ can be considered as an operator $\phi: M \to X$ (if $m \in M$, then $\phi(m) := \phi(f)$, where $f(\cdot) \equiv m$ on T).

Corollary 1.1 *Assume that operators Λ, λ and ϕ_X and $\phi_Y = \phi_X$ satisfy the conditions of Theorem 1.2 with $X = Y$. Let also there exist an operator $P \in H(X, X)$ such that*

1. $\Lambda f = (\Lambda \circ P)f = (P \circ \Lambda)f \quad \forall f \in H(T, X);$
2. $\Lambda f(t) = (\Lambda \circ P)f(t) = Pf(t) \quad \forall t \in T \, \forall f \in H(T, X);$
3. *For any $t \in T$*

$$h_X((P \circ \phi_X)(m), (P \circ \phi_X)(\theta)) = m, \text{ if } m = \lambda(h_T(\cdot, t)). \tag{1.12}$$

Then for arbitrary $t \in T$ the following sharp inequality holds:

$$h_X(\Lambda f(\cdot), Pf(t)) \leq \lambda(h_T(\cdot, t)). \tag{1.13}$$

The inequality becomes equality for the function

$$\tilde{f}_t(\cdot) = (P \circ \phi_X)(h_X(\cdot, t)).$$

If $P = \text{Id}$ (the identity operator) satisfies the above conditions, then inequality (1.13) has the form

$$h_X(\Lambda f(\cdot), f(t)) \leq \lambda(h_T(\cdot, t)).$$

and becomes equality for function (1.11).

Proof It is easy to check that operators Λ, λ and $\widetilde{\phi}_X = P \circ \phi$ and $\widetilde{\phi}_Y = \widetilde{\phi}_X$ instead of ϕ_X and ϕ_Y satisfy the conditions of Theorem 1.2 with $X = Y$. Therefore for $h_X(\Lambda f(\cdot), \Lambda f(t))$ we obtain

$$h_X(\Lambda f(\cdot), \Lambda f(t)) \leq \lambda(h_T(\cdot, t)).$$

Due to the properties of P

$$h_X(\Lambda f(\cdot), Pf(t)) = h_X(\Lambda f(\cdot), \Lambda f(t)) \leq \lambda(h_T(\cdot, t)),$$

and the inequality (1.13) is proved.

For the function $\widetilde{f}_t(\cdot) = (P \circ \phi_X)(h_X(\cdot, t))$ we have

$$h_X(\Lambda \widetilde{f}_t(\cdot), \Lambda \widetilde{f}_t(t)) = h_X((\Lambda \circ (P \circ \phi_X))(h_T(\cdot, t)), (\Lambda \circ (P \circ \phi_X))(h_T(t, t)))$$

$$\stackrel{(1.10)}{=} h_X(((P \circ \phi_X) \circ \lambda)(h_T(\cdot, t)), ((P \circ \phi_X) \circ \lambda)(\theta))$$

$$= h_X((P \circ \phi_X)(\lambda(h_T(\cdot, t)), (P \circ \phi_X)(\lambda(\theta))) \stackrel{(1.12)}{=} \lambda(h_T(\cdot, t)).$$

Therefore inequality (1.13) becomes equality for the function $\widetilde{f}_t(\cdot)$.

The last statement of the Corollary is obvious.

The case when Λ is the integral operator and P is the convexifying operator for multi-valued (see e.g. [43]), L-space-valued (see e.g. [20, 75]), or quasilinear-space-valued functions (see e.g. [7]), is an important example of the operators that satisfy the conditions of Corollary 1.1. The case, when Λ is the integral operator, and P is the identity operator occurs in the case of real-valued functions and functions with values in Banach spaces. Thus in the case $M = \mathbb{R}_+$ many known Ostrowski-type inequalities for real-valued, multi-valued and fuzzy-valued functions, as well as for functions with values in Banach spaces (in particular, random processes) and in L-spaces follow from Theorem 1.2 and Corollary 1.1 with appropriately chosen spaces T, X and Y and operators Λ, λ, P, ϕ_X and ϕ_Y.

The only result that we know, where related questions were considered for $M \neq \mathbb{R}_+$, is article [59].

1.2.1.3 Classes $H^\omega(T, X)$ and Ostrowski-Type Inequalities

Let h_M be an M-distance in a set M. A function $\omega: M \to M$ is called a *modulus of continuity*, if it satisfies the following properties:

1. $\omega(\theta) = \theta$;
2. ω is non-decreasing i.e., $\omega(m_1) \leq \omega(m_2)$, whenever $m_1 \leq m_2$;
3. ω is semi-additive in the following sense: for all $m_1, m_2 \in M$

$$h_M(\omega(m_1), \omega(m_2)) \leq \omega(h_M(m_1, m_2)).$$

In the case of $M = \mathbb{R}_+$ a modulus of continuity as an independent notion was introduced by Nikolsky [65].

Let a modulus of continuity ω and two M-distance spaces (T, h_T), (X, h_X) be given. We consider the classes

$$H^\omega(T, X) = \{f: T \to X : h_X(f(t_1), f(t_2)) \leq \omega(h_T(t_1, t_2)) \ \forall t_1, t_2 \in X\}.$$

Classes $H^\omega(T, X)$ play an important role in approximation theory. Many papers are devoted to solutions of different extremal problems for these classes. Some results for real-valued functions can be found e.g., in [33, 55, 73]. Some results regarding extremal problems for classes $H^\omega(T, X)$ of functions with non-numeric values can be found in [14, 16, 20, 21, 42, 58].

Observe that the class $H(T, X)$ is a partial case of the class $H^\omega(T, X)$ in the case, when $\omega = \mathrm{Id}$, where $\mathrm{Id}: M \to M$ is the identity mapping. On the other hand, as it is easy to see, the function $h_T^\omega: T \times T \to M$, given by the formula

$$h_T^\omega(t_1, t_2) = \omega(h_T(t_1, t_2))$$

is a new M-distance in T, which becomes a M-pseudo metric, if h_T is an M-pseudo metric. Consideration of the classes $H^\omega(T, X)$ with different ω and fixed distance h_T in T allows to appreciate the properties of the functions $f: T \to X$ in a more detailed manner. This makes the classes $H^\omega(T, X)$ important for approximation theory.

If in Theorem 1.2 the M-pseudo metric in T is understood as h_T^ω, then we obtain the following

Corollary 1.2 *For arbitrary modulus of continuity ω and arbitrary function $f \in H^\omega(T, X)$ the following inequality holds:*

$$h_Y(\Lambda f(\cdot), \Lambda f(t)) \leq \lambda(\omega(h_T(\cdot, t))),$$

which is sharp under the corresponding conditions and becomes equality on the function

$$f_{\omega,t}(\cdot) = \phi_X(\omega(h_T(\cdot, t))).$$

1.2.1.4 On Agreement of M-Distances

A partially ordered set M with a smallest element θ will be called a *partially ordered monoid*, if an associative binary operation $+$ is defined in M and the following properties hold:

1. For all $m \in M$, $\theta + m = m = m + \theta$.
2. If $m, n \in M$ are such that $m \leq n$, then $m + p \leq n + p$ for all $p \in M$.

1.2 Ostrowski-Type Inequalities

An element s in a partially ordered set M is called a *supremum* of two elements $m, n \in M$, if the following two conditions are satisfied

1. $s \geq m$ and $s \geq n$;
2. If $u \geq m$ and $u \geq n$, then $u \geq s$.

If a supremum of $m, n \in M$ exists, then it is unique and we denote it by $\sup\{m, n\}$.

A mapping $h_X \colon X \times X \to M$ is called an *M-metric*, if the following conditions hold:

1. For all $x, y \in X$, $x = y$ if and only if $h_X(x, y) = \theta$;
2. For all $x, y \in X$, $h_X(x, y) = h_X(y, x)$;
3. For all $x, y, z \in X$, $h_X(x, y) \leq h_X(x, z) + h_X(z, y)$.

Next we give a sufficient condition on an M-metric h_M in a partially ordered monoid M to agree with an arbitrary M-metric h_X on a set X. Before doing so, we note that generally speaking M-metric h_X need not agree with h_M. For example, if $M = \mathbb{R}_+$, h_X is a metric such that $0 < h_X(\alpha, \beta) < 1$ for some $\alpha, \beta \in X$, and h_M is the discrete metric on $\mathbb{R}_+$ (i.e., $h_M(a, b) = 0$, if $a = b$ and $h_M(a, b) = 1$ for all $a \neq b$), then inequality (1.5) does not hold for $x = x_1 = \alpha$ and $x_2 = \beta$. Moreover, an M-metric h_M does not necessarily agree with itself. Consider for example $M = \mathbb{R}_+$, and let

$$h_M(a, b) = \begin{cases} 0, & a = b = 0, \\ \frac{3}{4}, & \text{exactly one of } a, b \text{ is } 0, \\ \min\left\{1, \left|\ln \frac{a}{b}\right|\right\}, & a \neq 0 \text{ and } b \neq 0. \end{cases}$$

It is easy to verify that it is actually a metric on M (the fact that this function satisfies the property $h_M(ra, rb) = h_M(a, b)$ for all $a, b \geq 0$ and $r > 0$ allows to reduce the number of different cases to consider during verification of the triangle inequality). Since $\ln 2 < \frac{3}{4}$, for $x = x_1 = 1$, $x_2 = 2$ inequality (1.5) with h_X substituted by h_M does not hold.

Lemma 1.2 *Let M be a partially ordered monoid and assume there is a function $e \colon M \times M \to M$ such that for all $x, y, z \in M$ the following properties hold:*

$$x \leq y \iff e(x, y) = \theta;$$
$$e(x, \theta) \leq x;$$
$$e(x, y) \leq e(x, z) + e(z, y);$$
$$e(z + x, z + y) \leq e(x, y).$$

If for arbitrary $x, y \in M$ the supremum $\sup\{x, y\}$ exists, then

$$h_M(x, y) = \sup\{e(x, y), e(y, x)\}$$

is an M-metric. Moreover, arbitrary M-metric h agrees with h_M.

For example, if $M = \mathbb{R}_+$, the function $e(x, y) = \max\{x - y, 0\}$, $x, y \in \mathbb{R}_+$, satisfies the conditions of Lemma 1.2. In this case $h_M(x, y) = |x - y|$.

Proof We prove that h_M is an M-metric first. If $x \in M$, then

$$h_M(x, x) = \sup\{e(x, x), e(x, x)\} = \sup\{\theta, \theta\} = \theta.$$

Moreover, if $h_M(x, y) = \theta$, then $e(x, y) = e(y, x) = \theta$, hence $x \le y$ and $y \le x$, thus $x = y$. Since $\sup\{a, b\} = \sup\{b, a\}$ for all $a, b \in M$, we obtain that $h_M(x, y) = h_M(y, x)$ for all $x, y \in M$. Finally, for all $x, y, z \in X$,

$$h_M(x, y) = \sup\{e(x, y), e(y, x)\} \le \sup\{e(x, z) + e(z, y), e(y, z) + e(z, x)\}$$
$$\le \sup\{e(x, z), e(z, x)\} + \sup\{e(z, y), e(y, z)\} = h_M(x, z) + h_M(z, y).$$

Let $x \le y$ and $z \in M$. Then $e(x, y) = \theta$, and

$$e(y, z) = e(x, y) + e(y, z) \ge e(x, z)$$

i.e., the function e is non-decreasing in its first variable.

Finally, if h is an M-metric on a set T, then for arbitrary $t, t_1, t_2 \in T$,

$$h_M(h(t, t_1), h(t, t_2)) = \sup\{e[h(t, t_1), h(t, t_2)], e[h(t, t_2), h(t, t_1)]\}$$
$$\le \sup\{e[h(t, t_2) + h(t_2, t_1), h(t, t_2)], e[h(t, t_1) + h(t_1, t_2), h(t, t_1)]\}$$
$$\le \sup\{e[h(t_2, t_1), \theta], e[h(t_1, t_2), \theta]\} = e(h(t_1, t_2), \theta) \le h(t_1, t_2).$$

The idea to use an order-defining function e (with properties similar to the ones stated in the lemma) as a tool to define partially ordered metric spaces was introduced in [11].

1.2.2 Some Notes About Estimates for the Deviation Between Operators

We start with some remarks that apply to an arbitrary partially ordered set.

A sequence $\{x_n\} \subset N$ will be called *extremal* for a set N in a partially ordered space M, if $y \in M$ and $y \ge x_n$ for all $n \in \mathbb{N}$ implies that $y \ge x$ for all $x \in N$.

Note that if $\sup N$ exists and belongs to N, then the constant sequence $\{\sup N\}$ is extremal for the set N.

The following observation holds.

Lemma 1.3 *Let a set $\mathcal{A}$, a partially ordered set M, and two functions $\phi, \Phi : \mathcal{A} \to M$ be given. Assume the following properties hold.*

(a) *For all $f \in \mathcal{A}$ one has $\phi(f) \le \Phi(f)$.*
(b) *There exists a subset $\mathcal{B} \subset \mathcal{A}$ such that $\phi(f) = \Phi(f)$ for each $f \in \mathcal{B}$.*
(c) *For some sequence $\{f_n\} \subset \mathcal{B}$ the sequence $\{\Phi(f_n)\}$ is extremal for the set $\Phi(\mathcal{A})$.*

1.2 Ostrowski-Type Inequalities

Then the sequence $\{\phi(f_n)\}$ *is extremal for the set* $\phi(\mathcal{A})$. *If there exists* $\sup \Phi(\mathcal{A})$, *then* $\sup \phi(\mathcal{A})$ *exists and the following equality holds*

$$\sup \phi(\mathcal{A}) = \sup \Phi(\mathcal{A}). \tag{1.14}$$

Proof Let $\{f_n\} \subset \mathcal{B}$ be as in property (c). Assume that $m \in M$ is such that $m \geq \phi(f_n)$ for all $n \in \mathbb{N}$. Since $\{f_n\} \subset \mathcal{B}$, we obtain that $m \geq \Phi(f_n)$ for all $n \in \mathbb{N}$, due to condition (b). Hence $m \geq \Phi(x) \geq \phi(x)$ for all $x \in \mathcal{A}$, due to conditions (c) and (a), which implies the first statement of the lemma.

Suppose that $s := \sup \Phi(\mathcal{A})$ exists. Then due to condition (a) for any $f \in \mathcal{A}$ one has $s \geq \phi(f)$. Moreover, if $u \in M$ is such that $u \geq \phi(f)$ for any $f \in \mathcal{A}$, then for the sequence $\{f_n\}$ from condition (c), due to condition (b) we obtain $u \geq \phi(f_n) = \Phi(f_n)$ for all $n \in \mathbb{N}$. Therefore $u \geq \Phi(f)$ for any $f \in \mathcal{A}$, and hence by the definition of supremum, $u = s$. This implies (1.14). The lemma is proved.

It appears that in many known situations, the solution of the problem to find quantity (1.1) uses the scheme of Lemma 1.3. We restate it in Lemma 1.4 in slightly different notations, which are closer to the ones used below. The motivation for these notations is as follows. Information about many classes of functions $\mathcal{A}$ is given in terms of the value λf, $f \in \mathcal{A}$, of some operator λ. For example, the Sobolev classes $W_q^r(a, b)$, $r \in \mathbb{N}$, $q \in [1, \infty]$, $(a, b) \subset \mathbb{R}$, are determined by the condition $\|f^{(r)}\|_{L_q(a,b)} \leq 1$ on the function $\lambda f = f^{(r)}$. More examples will be considered below.

For two sets A and B, by B^A we denote the set of all functions $f : A \to B$.

Lemma 1.4 *Let M be a partially ordered set, S, T, X, Y, Z be some sets and $h_Y : Y \times Y \to M$ be some function. Assume also $\mathcal{A} \subset X^T$, $\Lambda, I : \mathcal{A} \to Y$, $\lambda : \mathcal{A} \to Z^S$, and $\varphi : \lambda(\mathcal{A}) \to M$ be such that the following properties hold.*

(a) *For all $f \in \mathcal{A}$ one has*

$$h_Y(\Lambda f, I f) \leq \varphi \circ \lambda(f). \tag{1.15}$$

(b) *There exists a subset $\mathcal{B} \subset \mathcal{A}$ such that inequality (1.15) becomes equality for each $f \in \mathcal{B}$.*

(c) *For some sequence $\{f_n\} \subset \mathcal{B}$ the sequence $\{\varphi \circ \lambda(f_n)\}$ is extremal for the set $\varphi \circ \lambda(\mathcal{A})$.*

Then the sequence $\{h_Y(\Lambda f_n, I f_n)\}$ is extremal for the set $\{h_Y(\Lambda f, I f) : f \in \mathcal{A}\}$.

Proof It is enough to apply Lemma 1.3 to $\phi(f) = h_Y(\Lambda f, I f)$, and $\Phi = \varphi \circ \lambda$.

The statement of Lemma 1.4 can be rephrased as follows. Under the conditions of Lemma 1.4 the inequality

$$h_Y(\Lambda f, If) \leq \sup \varphi \circ \lambda(\mathcal{A}) \tag{1.16}$$

holds for all $f \in \mathcal{A}$, provided the supremum on the right-hand side of the inequality exists. Inequalities (1.15) and (1.16) are sharp.

There are two key steps in the application of Lemma 1.4: the first one is to obtain inequality (1.15) that becomes equality for an enough wide family of functions; the second one is to solve extremal problem

$$\varphi(g) \to \sup \text{ over } g \in \lambda(\mathcal{A}).$$

Each of these steps can be non-trivial, but in some known in the literature results both of them are either trivial, or already known. In particular we show that many results on the Ostrowski type inequalities in fact can be obtained using standard arguments as an application of Lemma 1.4. For example, a majority of results from survey [41] in fact follow from Theorem 1.5 with $n = 1$ and $n = 2$.

1.2.3 Classes of Smooth Functions Defined on a Segment

The results of this section are mainly from [59].

1.2.3.1 Auxiliary Results

Using the following technical lemmas, we give several applications of Lemma 1.4. For a given integrable on $[a, b]$ function w and $x \in [a, b]$ set

$$r_x(s) = r_x(w; s) = \begin{cases} -\int_a^s w(t)dt, & s \leq x, \\ \int_s^b w(t)dt, & s \geq x; \end{cases} \tag{1.17}$$

Lemma 1.5 *Let $f, w \colon [a, b] \to \mathbb{R}$ be absolutely continuous functions, w be positive on $[a, b]$ and $p \colon [a, b] \to \mathbb{R}$ be an integrable on $[a, b]$ function. Then for each $x \in [a, b]$*

$$\int_a^b p(t)f(t)dt - \left(\int_a^b p(t)w(t)dt\right)\frac{f(x)}{w(x)} = \int_a^b r_x(pw; s)Df(s)ds,$$

where $Df = \left(\frac{1}{w}f\right)'$.

Proof

$$\int_a^b p(t)f(t)dt - \left(\int_a^b p(t)w(t)dt\right)\frac{f(x)}{w(x)}$$
$$= \int_a^b p(t)w(t)\left(\frac{f(t)}{w(t)} - \frac{f(x)}{w(x)}\right)dt$$

1.2 Ostrowski-Type Inequalities

$$= \int_a^x p(t)w(t)\left(\frac{f(t)}{w(t)} - \frac{f(x)}{w(x)}\right) dt + \int_x^b p(t)w(t)\left(\frac{f(t)}{w(t)} - \frac{f(x)}{w(x)}\right) dt$$

$$= -\int_a^x p(t)w(t) \int_t^x Df(s) ds\, dt + \int_x^b p(t)w(t) \int_x^t Df(s) ds\, dt$$

$$= \int_a^x Df(s)\left(-\int_a^s p(t)w(t) dt\right) ds + \int_x^b Df(s)\left(\int_s^b p(t)w(t) dt\right) ds$$

$$= \int_a^b r_x(s) Df(s) ds,$$

as required.

Assume that $n \in \mathbb{N}$ positive on $[a,b]$ functions $w_1, \ldots, w_n$ such that $w_k^{(n-k)}$ is absolutely continuous, $k = 1, \ldots, n$, are given. Consider differential operators

$$D_0 f = f, \quad D_k f = \left(\frac{1}{w_k} D_{k-1} f\right)', \quad k = 1, \ldots, n. \tag{1.18}$$

Such type of operators were studied in [50, Chapter 6]. Starting with an integrable function $p : [a,b] \to \mathbb{R}$ and a point $x \in [a,b]$, we define a sequence of functions $r_x^k : [a,b] \to \mathbb{R}$ by the formula

$$r_x^0 = p, \text{ and } r_x^k = r_x(w_k r_x^{k-1}), \; k = 1, \ldots, n, \tag{1.19}$$

where the function r_x is defined by (1.17). The following representation holds.

Lemma 1.6

$$\int_a^b p(t)f(t)dt - \sum_{k=0}^{n-1}\left(\int_a^b r_x^k(t)w_{k+1}(t)dt\right)\frac{D_k f(x)}{w_{k+1}(x)} = \int_a^b r_x^n(t) D_n f(t) dt.$$

Proof We proceed by induction on n. The case $n = 1$ immediately follows from Lemma 1.5. Assume the lemma is true for $n = s$. Then

$$\int_a^b p(t)f(t)dt - \sum_{k=0}^{s}\left(\int_a^b r_x^k(t)w_{k+1}(t)dt\right)\frac{D_k f(x)}{w_{k+1}(x)}$$

$$= \int_a^b r_x^s(t) D_s f(t) dt - \left(\int_a^b r_x^s(t) w_{s+1}(t) dt\right)\frac{D_s f(x)}{w_{s+1}(x)}.$$

Applying Lemma 1.5 with $p = r_x^s$ and $w = w_{s+1}$, we obtain the required.

In the case, when $w_k \equiv 1$ for all $k = 1, \ldots, n$, the previous lemma can be rewritten in a more explicit way.

Lemma 1.7 *If* $n \in \mathbb{N}$, f *has* $n-1$ *derivatives on* $[a,b]$ *and* $f^{(n-1)}$ *is absolutely continuous on* $[a,b]$, *then*

$$\int_a^b p(t)f(t)dt - \sum_{k=0}^{n-1} \frac{1}{k!}\left(\int_a^b p(t)(t-x)^k dt\right) f^{(k)}(x)$$
$$= \int_a^b r_x^n(t) f^{(n)}(t)dt. \quad (1.20)$$

Moreover, for all $k = 1, \ldots, n$,

$$\int_a^b r_x^k(t)dt = \frac{1}{k!}\int_a^b p(t)(t-x)^k dt. \quad (1.21)$$

Proof First of all note that the left-hand side of (1.20) becomes zero, if f is a polynomial of degree less than n; this easily follows from the expansion $f(t) = \sum_{k=0}^{n-1} \frac{f^{(k)}(x)}{k!}(t-x)^k$ for polynomials of degree less than n.

Next by induction on n we simultaneously prove that equalities (1.20) and (1.21) hold. For $n = 1$ equality (1.20) follows from Lemma 1.5. Since the left-hand side of (1.20) with $n = 2$ is zero for $f(t) = t$, using (1.20) for $n = 1$, we obtain

$$0 = \int_a^b p(t)t\,dt - \left(\int_a^b p(t)dt\right)x - \int_a^b p(t)(t-x)dt$$
$$= \int_a^b r_x^1(t)dt - \int_a^b p(t)(t-x)dt$$

and equality (1.21) for $k = 1$ follows.

Assume that equalities (1.20) and (1.21) hold for some $n = s \in \mathbb{N}$. Since the left-hand side of (1.20) with $n = s+2$ is zero for the function $f(t) = \frac{t^{s+1}}{(s+1)!}$, we obtain, using the inductive assumptions for $n = s$,

$$0 = \int_a^b p(t)\frac{t^{s+1}}{(s+1)!}dt - \sum_{k=0}^{s+1} \frac{1}{k!}\left(\int_a^b p(t)(t-x)^k dt\right)\frac{x^{s-k+1}}{(s-k+1)!}$$
$$= \int_a^b p(t)\frac{t^{s+1}}{(s+1)!}dt - \sum_{k=0}^{s-1} \frac{1}{k!}\left(\int_a^b p(t)(t-x)^k dt\right)\frac{x^{s-k+1}}{(s-k+1)!}$$
$$- \frac{1}{s!}\left(\int_a^b p(t)(t-x)^s dt\right)x - \frac{1}{(s+1)!}\int_a^b p(t)(t-x)^{s+1}dt$$
$$= \int_a^b r_x^s(t)\cdot t\,dt - \left(\int_a^b r_x^s(t)dt\right)x - \frac{1}{(s+1)!}\int_a^b p(t)(t-x)^{s+1}dt$$
$$= \int_a^b r_x^{s+1}(t)dt - \frac{1}{(s+1)!}\int_a^b p(t)(t-x)^{s+1}dt,$$

which proves (1.21) for $k = s + 1$. Finally,

$$\int_a^b p(t)f(t)dt - \sum_{k=0}^{s} \frac{1}{k!}\left(\int_a^b p(t)(t-x)^k dt\right) f^{(k)}(x)$$

$$= \int_a^b p(t)f(t)dt - \sum_{k=0}^{s-1} \frac{1}{k!}\left(\int_a^b p(t)(t-x)^k dt\right) f^{(k)}(x)$$

$$- \frac{1}{s!}\left(\int_a^b p(t)(t-x)^s dt\right) f^{(s)}(x)$$

$$= \int_a^b r_x^s(t) f^{(s)}(t)dt - \left(\int_a^b r_x^s(t)dt\right) f^{(s)}(x) = \int_a^b r_x^{s+1}(t) f^{(s+1)}(t)dt.$$

1.2.3.2 Ostrowski-Type Inequality for the Classes $W_q^n[a,b]$

For $n \in \mathbb{N}$ and $1 \leq q \leq \infty$ denote by $W_q^n[a,b]$ the class of continuous functions $f : [a,b] \to \mathbb{R}$ such that $f^{(n-1)}$ is absolutely continuous on $[a,b]$, and $\|f^{(n)}\|_{L_q[a,b]} \leq 1$.

Note that class $W_q^n[a,b]$ contains all polynomials of degree less than n. If $x \in [a,b]$ is fixed, p is an integrable on $[a,b]$ functions and

$$\sup_{f \in W_q^n[a,b]} \left|\int_a^b p(t)f(t)dt - \sum_{k=0}^{n-1} c_k f^{(k)}(x)\right| < \infty \tag{1.22}$$

for some numbers $c_0, \ldots, c_{n-1}$, then the value of the expression under the supremum must be zero for all polynomials of degree less than n. Setting f to be $t \mapsto 1, t \mapsto t - x, \ldots, t \mapsto (t-x)^{n-1}$, we get

$$c_k = \frac{1}{k!}\int_a^b p(t)(t-x)^k dt, k = 0, 1, \ldots, n-1.$$

This means that (1.22) can hold only for the above choice of the coefficients c_k, $k = 0, \ldots, n-1$. At the same time, for these values of the coefficients, the supremum is indeed finite and is found in the following theorem.

Theorem 1.3 *Let $n \in \mathbb{N}$, $1 \leq q \leq \infty$, $x \in [a,b]$ and an integrable on $[a,b]$ function p be given. Then*

$$\sup_{f \in W_q^n[a,b]} \left|\int_a^b p(t)f(t)dt - \sum_{k=0}^{n-1} \frac{1}{k!}\left(\int_a^b p(t)(t-x)^k dt\right) f^{(k)}(x)\right|$$

$$= \sup_{\|g\|_{L_q[a,b]} \leq 1} \left|\int_a^b r_x^n(t)g(t)dt\right| = \|r_x^n\|_{L_{q'}[a,b]}, \tag{1.23}$$

where $1/q + 1/q' = 1$ and r_x^n is defined in (1.19) with $w_k \equiv 1$, $k = 1, \ldots, n$.

Proof The right equality in (1.23) is true due to the Hölder inequality. To obtain the left one, it is sufficient to apply Lemma 1.4 with $S = T = [a,b]$, $X = Y = Z = M = \mathbb{R}$, $\mathcal{A} = \mathcal{B} = W_q^n[a,b]$, $\Lambda f = \int_a^b p(t)f(t)dt$,

$$If = \sum_{k=0}^{n-1} \frac{1}{k!} \left(\int_a^b p(t)(t-x)^k dt \right) f^{(k)}(x),$$

$\lambda f = f^{(n)}$ and $\varphi g = \left| \int_a^b r_x^n(t)g(t)dt \right|$. Inequality (1.15) becomes equality due to Lemma 1.7.

Related results with $p(t) \equiv 1$ can be found in [2, 45].

Note that inequality (1.23) can be stated for all $x \in [a,b]$ at once as follows. Assume that $S = T = [a,b]$, $X = Z = \mathbb{R}$, $\mathcal{A} = \mathcal{B} = W_q^n[a,b]$, $\lambda f = f^{(n)}$. Let $Y = M$ be the set of measurable essentially bounded on $[a,b]$ functions with pointwise addition and partial order;

$$h_Y(f,g) = s \mapsto |f(s) - g(s)|, s \in [a,b],$$

$\Lambda f = x \mapsto \int_a^b p(t)f(t)dt$ be the constant function,

$$If = x \mapsto \sum_{k=0}^{n-1} \frac{1}{k!} \left(\int_a^b p(t)(t-x)^k dt \right) f^{(k)}(x), \, x \in [a,b],$$

and $\varphi g = x \mapsto \left| \int_a^b r_x^n(t)g(t)dt \right|$. Applying Lemma 1.4 we obtain sharp inequality (1.23); the fact that $\sup \varphi(\lambda(\mathcal{A}))$ is well defined is contained in [76, Chapter IV, § 1].

1.2.3.3 Ostrowski-Type Inequality for the Classes $W^n H^\omega[a,b]$

Recall that a continuous non-decreasing subadditive function $\omega \colon [0,\infty) \to [0,\infty)$ that vanishes at zero is called a modulus of continuity. For $n \in \mathbb{N}$ denote by $W^n H^\omega[a,b]$ the class of continuous on $[a,b]$ functions f such that $f^{(n)} \in H^\omega[a,b]$, that is for all $h > 0$,

$$|f^{(n)}(x) - f^{(n)}(y)| \le \omega(h), \text{ whenever } |x-y| \le h.$$

Using the same arguments as in the proof of Theorem 1.3, we obtain

$$\sup_{f \in W^n H^\omega[a,b]} \left| \int_a^b p(t)f(t)dt - \sum_{k=0}^{n-1} \frac{1}{k!} \left(\int_a^b p(t)(t-x)^k dt \right) f^{(k)}(x) \right|$$

$$= \sup_{g \in H^\omega[a,b]} \left| \int_a^b r_x^n(t)g(t)dt \right|.$$

1.2 Ostrowski-Type Inequalities

Since the class $H^\omega[a, b]$ contains all constants, the right-hand side of the latter equality can be finite only in the case

$$\int_a^b r_x^n(t)dt = 0. \qquad (1.24)$$

Assume that p is non-negative on $[a, b]$ and $\int_a^b p(t)dt > 0$. Due to (1.21), condition (1.24) does not hold for any even n. On the other hand, for each odd n there exist $x \in [a, b]$ such that condition (1.24) holds (and such x is unique provided p is positive almost everywhere).

Theorem 1.4 *Let $n \in \mathbb{N}$ be odd, ω be a modulus of continuity, p be an integrable positive almost everywhere on $[a, b]$ function, and $x \in [a, b]$ be such that condition (1.24) holds. Then*

$$\sup_{f \in W^n H^\omega[a,b]} \left| \int_a^b p(t)f(t)dt - \sum_{k=0}^{n-1} \frac{1}{k!} \left(\int_a^b p(t)(t-x)^k dt \right) f^{(k)}(x) \right|$$
$$= \sup_{g \in H^\omega[a,b]} \left| \int_a^b r_x^n(t)g(t)dt \right| \leq \int_a^x |r_x^n(t)|\omega(\rho(t) - t)dt, \qquad (1.25)$$

where $\rho: [a, x] \to [x, b]$ is uniquely determined by the condition

$$\int_a^t r_x^n(s)ds = \int_a^{\rho(t)} r_x^n(s)ds.$$

If ω is concave, then the inequality in (1.25) becomes equality.

Proof The equality in (1.25) can be proved using the arguments from the proof of Theorem 1.3. Since n is odd, from the definition of the function r_x^n we obtain that r_x^n is non-positive on $[a, x]$ and non-negative on $[x, b]$. Taking into account equality (1.24), we may apply the Korneichuk–Stechkin lemma, see e.g. [55, § 7.1], which gives the inequality in (1.25). Moreover, it becomes equality in the case of a concave modulus of continuity ω.

1.2.3.4 Ostrowski-Type Inequality for Classes Defined by a General Differential Operator

Let operators D_k, $k = 1, \ldots, n$ be defined by (1.18). Denote by $\mathcal{D}_q^n$, $q \in [1, \infty]$, the space of continuous functions $f: [a, b] \to \mathbb{R}$ such that $\|D_n f\|_{L_q[a,b]} \leq 1$. Using Lemma 1.4 together with Lemma 1.6, we obtain the following result.

Theorem 1.5 *Let $n \in \mathbb{N}$, $1 \leq q \leq \infty$, $x \in [a, b]$ and an integrable on $[a, b]$ function p be given. Then*

$$\sup_{f \in \mathcal{D}_q^n[a,b]} \left| \int_a^b p(t)f(t)dt - \sum_{k=0}^{n-1} \left(\int_a^b r_x^k(t) w_{k+1}(t) dt \right) \frac{D_k f(x)}{w_{k+1}(x)} \right|$$

$$= \sup_{\|g\|_{L_q[a,b]} \leq 1} \left| \int_a^b r_x^n(t) g(t) dt \right| = \|r_x^n\|_{L_{q'}[a,b]},$$

where $1/q + 1/q' = 1$, r_x^n is defined in (1.19), and the operators D_k, $k = 1, \ldots, n$, are defined in (1.18).

1.2.4 Some Auxiliary Results

In this section we give some definitions and auxiliary results which will be used in Ostrowski-type and Kolmogorov-type inequalities later. These results are mainly from [29].

For $x, y \in \mathbb{R}^d$ denote by (x, y) the dot product of x and y. Let $K \subset \mathbb{R}^d$ be an open convex symmetric with respect to the origin θ bounded set with $\theta \in \text{int } K$. Denote by $\mathbb{K}$ the family of all such sets K. For $K \in \mathbb{K}$ and $x \in \mathbb{R}^d$ denote by $|x|_K$ the norm of x generated by the set K, i.e.,

$$|x|_K := \inf\{\lambda > 0 \colon x \in \lambda K\}.$$

If K is the unit ball in $\mathbb{R}_p^d$ i.e.,

$$K = \left\{ (x_1, \ldots, x_d) \in \mathbb{R}^d \colon \begin{cases} \sum_{k=1}^d |x_k|^p \leq 1, & p \in [1, \infty) \\ \max_{k=1,\ldots,d} |x_k| \leq 1, & p = \infty \end{cases} \right\}$$

then we write $|\cdot|_p$ instead of $|x|_K$.

A set $C \subset \mathbb{R}^d$ is called a *cone*, if $x \in C \implies \lambda x \in C$ for all $\lambda > 0$. By $\mathcal{C}$ we denote the set of all open convex cones in $\mathbb{R}^d$.

For $R > 0$, B_R denotes the Euclidean ball with center at the origin θ of $\mathbb{R}^d$. By S^{d-1} we denote the unit Euclidean sphere in $\mathbb{R}^d$ with the center θ. Let a set $C \in \mathcal{C}$ be given, and $\Omega \colon C \to \mathbb{R}$ be a non-negative homogeneous of degree 0 (i.e., $\Omega(\lambda x) = \Omega(x)$ for all $\lambda > 0$ and $x \in C$) integrable on $C \cap S^{d-1}$ with respect to the spherical measure function. This function Ω will be referred to as a *non-negative characteristic*.

For a set $K \in \mathbb{K}$, denote by K° its *polar set* (see e.g. [68, Section 14]),

$$K^\circ = \left\{ y \in \mathbb{R}^d \colon \sup_{x \in K}(x, y) \leq 1 \right\}.$$

It is well known that

$$|z|_{K^\circ} = \sup\{(x, z) \colon |x|_K \leq 1\}.$$

In particular for all $x, y \in \mathbb{R}^d$ (see e.g., [68, Sect. 15]), $(x, y) \leq |x|_K \cdot |y|_{K^\circ}$. From the definition of the set K° it also follows that $|y|_{K^\circ} = 1 \implies \sup_{x \in K}(x, y) = 1$.

1.2 Ostrowski-Type Inequalities

A $d-1$ dimensional hyperplane α is called a *supporting hyperplane* for a convex set K, if K is contained in one of the two closed half-spaces generated by α, and $\alpha \cap \partial K \neq \emptyset$. A supporting hyperplane α of a convex set K is called a *tangent hyperplane* at $x \in \partial K$, if $x \in \alpha$ and there is only one supporting hyperplane of the set K that contains x.

Lemma 1.8 *Let $K \in \mathbb{K}$, α be a supporting hyperplane of K, δ be the distance between α and the origin θ, and n be the unit (i.e., $|n|_2 = 1$) external normal of α. Then $|n|_{K^\circ} = \delta$. In particular, this equality holds if K is a polytope and α contains a face γ of K.*

Proof By definition, K lies on one side of the hyperplane α, hence

$$|n|_{K^\circ} = \sup\{(x, n) \colon |x|_K \leq 1\} \leq \delta.$$

On the other hand, if $y \in \alpha \cap \partial K$, and $\{y_m\} \subset K$ is a sequence of points converging to y, then $|y_m|_K \to 1$ and $(y_m, n) \to \delta$ as $m \to \infty$.

The following consequence of the coarea (see e.g. [44, Theorem 3.2.12]) formula will be needed. This result is essentially contained in [28].

Lemma 1.9 *Let $C \in \mathcal{C}$, $K \in \mathbb{K}$ be a polytope, γ be a face of $K \cap C$ that does not contain the origin θ and δ be the distance between θ and the plane that contains γ. For $(a, b) \subset (0, \infty)$ set $\Gamma := \bigcup_{\lambda \in (a,b)} \lambda \gamma$. For all integrable $g \colon \Gamma \to \mathbb{R}$*

$$\int_\Gamma g(x)dx = \delta \int_a^b \rho^{d-1} \int_\gamma g(\rho y) dy d\rho.$$

Proof Let n be the unit external normal of the face γ. Consider the function $u(x) = \frac{1}{\delta}(n, x)$. Then $|\nabla u(x)|_2 = \frac{1}{\delta}$ for all $x \in \mathbb{R}^d$, $\gamma \subset \{x \in \mathbb{R}^d \colon u(x) = 1\}$, and using the coarea formula we obtain

$$\int_\Gamma g(x)dx = \delta \int_\Gamma g(x)|\nabla u(x)|_2 dx = \delta \int_a^b \int_{\rho \cdot \gamma} g(y) dy d\rho$$

$$= \delta \int_a^b \rho^{d-1} \int_\gamma g(\rho z) dz d\rho.$$

Observe that the set γ is contained in a $d-1$-dimensional hyperplane and we can integrate over the $d-1$-dimensional Lebesgue measure (and write dz) instead of the $d-1$-dimensional Hausdorff measure H_{d-1}.

If $K \in \mathbb{K}$, then $|\cdot|_K$ is a finite convex homogeneous function on $\mathbb{R}^d$. According to [68, Theorem 25.5] the set $D \subset \mathbb{R}^d$ of points where $|\cdot|_K$ is differentiable is dense in $\mathbb{R}^d$ and its complement has Lebesgue measure zero. Due to homogeneity of $|\cdot|_K$, $\mathbb{R}^d \setminus D$ is a cone in $\mathbb{R}^d$, and hence the set $\partial K \setminus D$ has zero $d-1$-dimensional Hausdorff measure H_{d-1}. At each point $x \in \partial K \cap D$ the function $|\cdot|_K$ is differentiable, and hence K has a tangent hyperplane (with a normal $\nabla |x|_K$). Thus H_{d-1}-almost everywhere on ∂K we can define a function n that maps a point x to the external unit normal of the tangent hyperplane of K at x.

For a convex function f by $\partial f(x)$ we denote the *subgradient* of f at x i.e., the set of all vectors $x^* \in \mathbb{R}^d$ such that $f(z) \geq f(x) + (x^*, z - x)$ for all z from the domain of definition of f (see e.g., [68, Sect. 23]). If $K \in \mathbb{K}$ is a polytope, and x belongs to the interior of a face γ of K, then the subgradient $\partial |x|_K$ is a singleton that contains an orthogonal to γ vector; if $x \in \partial K$ is not in the interior of a face, then $\partial |x|_K$ contains vectors orthogonal to all supporting planes of K at x, in particular it contains vectors orthogonal to each of the adjacent to x faces of K.

Let Ω be a non-negative characteristic defined on a cone $C \in \mathcal{C}$. From the homogeneity of Ω it follows that for each $K \in \mathbb{K}$, Ω is integrable on the set $K \cap C$ with respect to $d-1$-dimensional Hausdorff measure H_{d-1}.

For $K \in \mathbb{K}$, $C \in \mathcal{C}$ and a non-negative characteristic Ω set

$$\Omega(K, C) := \int_{C \cap \partial K} |n(t)|_{K^\circ} \Omega(t) H_{n-1}(dt),$$

where n is the unit external normal of K. We need the following lemma.

Lemma 1.10 *Let $K \in \mathbb{K}$, $C \in \mathcal{C}$, and a non-negative characteristic Ω be given. Let $\{K_i\} \subset \mathbb{K}$, be a sequence of polytopes that converges to K in the Hausdorff metric. Then as $i \to \infty$*

$$\Omega(K_i, C) \to \Omega(K, C).$$

Proof The functions $|\cdot|_{K_i}$ converge to $|\cdot|_K$ uniformly on S^{d-1}, and due to positive homogeneity of the support functions, pointwisely on $\mathbb{R}^d$. For each $\xi \in S^{d-1}$, let $x_i(\xi)$, $i \in \mathbb{N}$, and $x(\xi)$ be the points where the ray $\{\lambda \xi : \lambda > 0\}$ intersects ∂K_i and ∂K respectively. Let $\varepsilon > 0$ be fixed. For each $\xi \in S^{d-1}$ such that $|\cdot|_K$ is differentiable at $x(\xi)$, applying [68, Theorem 24.5] to the sequence $\{x_i(\xi)\}$, we can find $N = N(\xi)$ such that for all $i > N$ one has

$$\partial |x_i(\xi)|_{K_i} \subset \partial |x(\xi)|_K + B_\varepsilon = \nabla |x(\xi)|_K + B_\varepsilon.$$

Hence if γ is a face of K_i that contains $x_i(\xi)$ and $n_\gamma \in \partial |x_i(\xi)|_{K_i}$ is its normal, then

$$|n_\gamma - \nabla |x(\xi)|_K|_2 \leq \varepsilon. \tag{1.26}$$

By [68, Theorem 24.7] the set $G := \bigcup_{x \in \partial K} \partial |x|_K$ is closed. Moreover, since the function $|\cdot|_K$ is positively homogeneous, it does not have extrema on ∂K, and hence $0 \notin G$. Thus G is separated from zero and due to (1.26) for almost all $\xi \in S^{d-1}$ one has

$$|n_i(x_i(\xi))|_{K_i^\circ} \to |n(x(\xi))|_{K^\circ} \text{ as } i \to \infty.$$

1.2 Ostrowski-Type Inequalities

Moreover, since $|n_i(x_i(\xi))|_{K_i^\circ}$ is the distance from the origin to a hyperplane that contains one of the faces of K_i, there exists $A > 0$ such that

$$|n_i(x_i(\xi))|_{K_i^\circ} \cdot |x_i(\xi)|_2^{d-1} < A$$

for all $\xi \in S^{d-1}$ and $i \in \mathbb{N}$. The functions $\xi \mapsto |n_i(x_i(\xi))|_{K_i^\circ}|x_i(\xi)|_2^{d-1}\Omega(\xi)$, $\xi \in S^{d-1}$, are majored by an integrable on $C \cap S^{d-1}$ function $\xi \mapsto A\Omega(\xi)$ and converge almost everywhere to the function $|n(x(\xi))|_{K^\circ}|x(\xi)|_2^{d-1}\Omega(\xi)$. Thus as $i \to \infty$,

$$\int_{C \cap \partial K_i} |n_i(s)|_{K_i^\circ} \Omega(s) H_{n-1}(ds) = \int_{C \cap S^{d-1}} |n_i(x_i(\xi))|_{K_i^\circ} |x_i(\xi)|_2^{d-1} \Omega(\xi) d\xi$$

$$\to \int_{C \cap S^{d-1}} |n(x(\xi))|_{K^\circ} |x(\xi)|_2^{d-1} \Omega(\xi) d\xi = \int_{C \cap \partial K} |n(t)|_{K^\circ} \Omega(t) H_{n-1}(dt),$$

as desired.

For $0 \leq a < b \leq +\infty$ and $p \in [1, \infty]$, denote by $\mathcal{L}_p(a, b)$ the space of all measurable functions $w \colon (a, b) \to \mathbb{R}_+$ with finite norm

$$\|w\|_{\mathcal{L}_p(a,b)} = \begin{cases} \left(\int_a^b t^{d-1} w^p(t) dt\right)^{1/p}, & p < \infty, \\ \operatorname{ess\,sup}_{t \in (a,b)} t^{d-1} w(t), & p = \infty. \end{cases} \quad (1.27)$$

For $(a, b) \subset (0, \infty)$ and $K \in \mathbb{K}$ we set

$$K(a, b) := \bigcup_{\lambda \in (a,b)} \lambda \cdot \partial K.$$

The following lemma is a key tool to prove the Ostrowski-type inequality for the class $H_K^\omega(K \cap C)$ defined in Sect. 1.2.5 (see Theorem 1.6 below), which in turn is a key step in the proof of a sharp Kolmogorov-type inequality formulated in Theorem 1.13.

Lemma 1.11 *Let $K \in \mathbb{K}$, $C \in \mathcal{C}$, $(a, b) \subset (0, \infty)$, $w \in \mathcal{L}_1(a, b)$ and Ω be a non-negative characteristics. Then*

$$\int_{C \cap K(a,b)} w(|t|_K) \Omega(t) dt = \Omega(K, C) \cdot \|w\|_{\mathcal{L}_1(a,b)}.$$

Proof Assume that K is a polytope first. Let γ, Γ, δ and n be as in Lemma 1.9. Using Lemmas 1.8 and 1.9 we obtain

$$\int_{\Gamma} w(|t|_K)\Omega(t)dt = \delta \int_a^b \rho^{d-1} \int_{\gamma} w(|\rho s|_K)\Omega(\rho s)ds d\rho$$

$$= \|w\|_{\mathcal{L}_1(a,b)} \int_{\gamma} |n|_{K^\circ}\Omega(s)ds.$$

Summing all such equalitites, we obtain the required equality.

Let $K \in \mathbb{K}$ be an arbitrary set now, and let $\{K_i\} \subset \mathbb{K}$ be a sequence of polytopes that converges to K in the Hausdorff metric.

For each $\xi \in S^{d-1}$ set $r(\xi) = \sup\{\lambda > 0 \colon \lambda \xi \in K\}$ and analogously set $r_i(\xi) = \sup\{\lambda > 0 \colon \lambda \xi \in K_i\}$. Switching to the spherical coordinates, and using homogeneity of Ω, we obtain

$$\int_{C \cap K(a,b)} w(|t|_K)\Omega(t)dt - \int_{C \cap K_i(a,b)} w(|t|_{K_i})\Omega(t)dt =$$

$$\int_{C \cap S^{d-1}} \int_{ar(\xi)}^{br(\xi)} \rho^{d-1} w\left(\frac{\rho}{r(\xi)}\right) \Omega(\xi) d\rho d\xi - \int_{C \cap S^{d-1}} \int_{ar_i(\xi)}^{br_i(\xi)} \rho^{d-1} w\left(\frac{\rho}{r_i(\xi)}\right) \Omega(\xi) d\rho d\xi$$

$$= \int_{C \cap S^{d-1}} \Omega(\xi) \int_a^b \left(r^d(\xi) - r_i^d(\xi)\right) \rho^{d-1} w(\rho) d\rho d\xi \to 0 \text{ as } i \to \infty,$$

since r_i uniformly on $C \cap S^{d-1}$ converges to r and hence the internal integral tends to 0 as $i \to \infty$ uniformly on $C \cap S^{d-1}$. Using Lemma 1.10 and already proved case of the lemma for polytopes, we obtain as $i \to \infty$

$$\int_{C \cap K(a,b)} w(|t|_K)\Omega(t)dt = \int_{C \cap K_i(a,b)} w(|t|_{K_i})\Omega(t)dt + o(1) =$$

$$= \Omega(K_i, C) \cdot \|w\|_{\mathcal{L}_1(a,b)} + o(1) = \Omega(K, C) \cdot \|w\|_{\mathcal{L}_1(a,b)} + o(1).$$

This finishes the proof of the lemma for arbitrary $K \in \mathbb{K}$.

The following lemma is a key tool to prove Theorem 1.7. Note that the techniques of its proof was gradually refined in several works. The case $d = 1$ is contained e.g., in [64, Theorem 4.1]. The case $K = B_1$ is contained in [31]; note that in this case, a switch to the polar coordinate system and back can be (and was) used instead of the arguments related to the coarea formula (see formula (1.31) below). In [22] the techniques was extended to the case $K = (-1, 1)^d$, which has the property that $\nabla |\cdot|_K$ is piecewise-constant and has constant Euclidean length H_{d-1}-almost everywhere on ∂K. In [22] this technique was extended to polytopal sets K, for which ∂K (up to a set of H_{d-1} zero measure) can be split into a finite number of subsets with constant Euclidean length of $\nabla |\cdot|_K$ (but the length may differ on different pieces). Finally, the following lemma was proved in [29].

Lemma 1.12 *Let $K \in \mathbb{K}$, $C \in \mathcal{C}$, $h > 0$, Ω be a non-negative characteristics and non-negative measurable functions $f \colon C \cap K(0, h) \to \mathbb{R}$, $w \colon (0, h) \to \mathbb{R}$ be such that $w \in$*

1.2 Ostrowski-Type Inequalities

$\mathcal{L}_1(u, h)$ for all $u \in (0, h)$, and the integral $\int_{C \cap K(0,h)} g_w(|t|_K) f(t) \Omega(t) dt$ exist, where

$$g_w : (0, h) \to \mathbb{R}, \ g_w(u) = \frac{1}{u^{d-1}} \int_u^h w(\rho) \rho^{d-1} d\rho. \tag{1.28}$$

Then

$$\int_0^1 \int_{C \cap K(0,h)} w(|s|_K) |s|_K f(\rho s) \Omega(s) ds d\rho = \int_{C \cap K(0,h)} g_w(|t|_K) f(t) \Omega(t) dt.$$

Proof Since $|\cdot|_K$ is a convex function on $\mathbb{R}^d$, by [68, Theorem 24.7] it is Lipschitz on the unit ball B_1. Since in addition $|\cdot|_K$ is positively homogeneous, $|\cdot|_K$ is Lipschitz on the whole space $\mathbb{R}^d$ (with the same Lipschitz constant). Thus by the Rademacher theorem, it is differentiable almost everywhere. Again by positively homogeneity of $|\cdot|_K$ we obtain that $\nabla |\lambda x|_K = \nabla |x|_K$ for all $\lambda > 0$.

Let $R > 0$ be such that $K \subset B_R$. For arbitrary $0 \neq x \in \mathbb{R}^d$ such that $\nabla |x|_K$ exists, set $v = \frac{x}{|x|_2}$. Then

$$|\nabla |x|_K|_2 \geq (\nabla |x|_K, v) = \lim_{s \to 0} \frac{|x + s \cdot v|_K - |x|_K}{s} \geq \frac{1}{R}. \tag{1.29}$$

On the other hand, according to [68, Theorem 24.7] applied to the function $|\cdot|_K$ and $S = \partial K$, we obtain that there exists $r > 0$ such that at each point $x \in \partial K$ where $|\cdot|_K$ is differentiable (and hence for each $x \in \mathbb{R}^d$, where $|\cdot|_K$ is differentiable), one has

$$|\nabla |x|_K|_2 \leq \frac{1}{r}. \tag{1.30}$$

Let $A \subset C$ be an open cone. From (1.29) and (1.30) it follows that the numbers R_A and r_A such that

$$\frac{1}{R_A} = \inf_x |\nabla |x|_K|_2, \ \frac{1}{r_A} = \sup_x |\nabla |x|_K|_2$$

are finite and positive, where the supremum and the infimum are taken over points $x \in A \cap \partial K$, where $|\cdot|_K$ is differentiable.

Applying the coarea formula, changing the order of integration, and then again the coarea formula, we obtain

$$\int_0^1 \int_{A \cap K(0,h)} w(|y|_K) |y|_K f(ty) \Omega(y) dy dt$$

$$\leq R_A \int_0^1 \int_{A \cap K(0,h)} w(|y|_K) |y|_K f(ty) \Omega(y) |\nabla |y|_K|_2 dy dt$$

$$= R_A \int_0^1 \int_0^h \int_{A \cap \rho \partial K} w(|x|_K) |x|_K f(tx) \Omega(x) H_{d-1}(dx) d\rho dt$$

$$= R_A \int_0^1 \int_0^h \int_{A \cap \partial K} \rho^{d-1} w(|\rho x|_K) |\rho x|_K f(t\rho x) \Omega(\rho x) H_{d-1}(dx) d\rho dt$$

$$= R_A \int_{A \cap \partial K} \Omega(x) \int_0^h \int_0^1 \rho^d w(\rho) f(t\rho x) dt d\rho H_{d-1}(dx)$$

$$= R_A \int_{A \cap \partial K} \Omega(x) \int_0^h \int_0^\rho \rho^{d-1} w(\rho) f(sx) ds d\rho H_{d-1}(dx)$$

$$= R_A \int_{A \cap \partial K} \Omega(x) \int_0^h f(sx) \int_s^h \rho^{d-1} w(\rho) d\rho ds H_{d-1}(dx)$$

$$= R_A \int_{A \cap \partial K} \Omega(x) \int_0^h s^{d-1} f(sx) g_w(s) ds H_{d-1}(dx)$$

$$= R_A \int_0^h \int_{A \cap \partial K} s^{d-1} f(sx) g_w(|sx|_K) \Omega(sx) H_{d-1}(dx) ds$$

$$= R_A \int_{A \cap K(0,h)} f(y) g_w(|y|_K) \Omega(y) |\nabla |y|_K|_2 dy$$

$$\leq \frac{R_A}{r_A} \int_{A \cap K(0,h)} f(y) g_w(|y|_K) \Omega(y) dy. \quad (1.31)$$

This estimate in particular implies that the integral on the left-hand side of (1.31) exists. Applying the same arguments, but using an estimate from below for $|\nabla |y|_K|_2$ in the first inequality, and from above for $|\nabla |y|_K|_2$ in the second inequality in (1.31), one can obtain the inequality

$$\int_0^1 \int_{A \cap K(0,h)} w(|y|_K) |y|_K f(ty) \Omega(y) dy dt$$

$$\geq \frac{r_A}{R_A} \int_{A \cap K(0,h)} f(y) g_w(|y|_K) \Omega(y) dy. \quad (1.32)$$

Next we fix an arbitrary $\varepsilon > 0$. Since $|\cdot|_K$ is differentiable almost everywhere in $\mathbb{R}^d$, we can find an open cone $A_\varepsilon \subset \mathbb{R}^d$ such that meas $A_\varepsilon \cap K < \varepsilon$ and $|\cdot|_K$ is differentiable on $\partial K \setminus A_\varepsilon$. Using (1.30) and (1.29) together with (1.31) we obtain that

$$\int_0^1 \int_{C \cap K(0,h) \cap A_\varepsilon} w(|y|_K) |y|_K f(ty) \Omega(y) dy dt = o(1), \varepsilon \to 0. \quad (1.33)$$

For the chosen ε and each $x \in \partial K$ by [68, Corollary 24.5.1] we find $\delta(x) > 0$ such that

$$\partial |z|_K \subset \partial |x|_K + B_\varepsilon \; \forall z : |z - x|_2 < \delta(x).$$

Thus [68, Theorem 25.1] implies that for all $x \in \partial K \setminus A_\varepsilon$

$$|\nabla |x|_K - \nabla |z|_K|_2 \leq \varepsilon \; \forall z \in (K \setminus A_\varepsilon) \cap (x + B_{\delta(x)}). \quad (1.34)$$

The family of open balls with center x and radius $\delta(x)$, $x \in \partial K \setminus A_\varepsilon$ is an open cover of the compact set $\partial K \setminus A_\varepsilon$, and hence it has a finite subcover. Let it be determined by points $x_1, \ldots, x_m$, $m \in \mathbb{N}$. Let $\overline{A}_i$ be the interior of the cone generated by the set $\partial K \cap (x_i + B_{\delta(x_i)}) \setminus A_\varepsilon$, $i = 1, \ldots, m$, and set $A_1 = \overline{A}_1$, $A_{i+1} = \overline{A}_{i+1} \setminus \cup_{j=1}^{i} A_j$, $i = 1, \ldots, m - 1$. From the construction it follows that the sets $C \cap K \cap A_i$, $i = 1, \ldots, m$ and $C \cap K \cap A_\varepsilon$ up to a set of zero measure form a partition of the set $C \cap K$, and from (1.29) and (1.34) it follows that

$$1 \leq \frac{R_{A_i}}{r_{A_i}} < 1 + \alpha(\varepsilon), i = 1, \ldots, m,$$

where $\alpha = o(1)$ as $\varepsilon \to 0$. Thus from (1.31), (1.32) applied to each of the sets $K \cap A_i \cap C$, $i = 1, \ldots, m$, and (1.33) we obtain that as $\varepsilon \to 0$,

$$\int_0^1 \int_{C \cap K(0,h)} w(|y|_K)|y|_K f(ty)\Omega(y) dy dt$$
$$= (1 + o(1)) \int_{C \cap K(0,h)} f(y) g_w(|y|_K) \Omega(y) dy + o(1),$$

which implies the required.

Lemma 1.13 *Let $K \in \mathbb{K}$ and $y \in \mathbb{R}^d$, $y \neq 0$ be such that $|\cdot|_K$ is differentiable at K. Then $(y, \nabla |y|_K) = |y|_K$ and*

$$|\nabla |y|_K|_{K^\circ} = 1. \tag{1.35}$$

Proof We prove (1.35) first. According to [68, Theorem 14.5], the function $|y|_K$ is equal to the support function of K°. Hence all subgradient vectors of $|\cdot|_K$ belong to ∂K° in virtue of [68, Corollary 23.5.3]. Thus $|\nabla |y|_K|_{K^\circ} = 1$, since $\nabla |y|_K$ is the unique subgradient of $|\cdot|_K$ at y (see [68, Theorem 25.1]).

The hyperplane orthogonal to $\nabla |y|_K$ that contains y is a supporting hyperplane of $|y|_K \cdot K$, and hence (the first equality holds due to (1.35))

$$1 = \sup_{x \in K}(x, \nabla |y|_K) = \frac{1}{|y|_K} \sup_{x \in |y|_K \cdot K}(x, \nabla |y|_K) = \frac{1}{|y|_K}(y, \nabla |y|_K).$$

Lemma 1.14 *Let $K \in \mathbb{K}$ and g be an integrable on $(0, 1)$ function. For the function*

$$f \colon K \to \mathbb{R}, \, f(y) = \int_0^{|y|_K} g(u) du, \, y \in K$$

almost everywhere (more precisely, at all points $y \in K$, where $\nabla |y|_K$ exists)

$$\nabla f(y) = g(|y|_K) \cdot \nabla |y|_K \text{ and } |\nabla f(y)|_{K^\circ} = |g(|y|_K)|. \tag{1.36}$$

Proof Let e_j be the j-th element of the usual basis in $\mathbb{R}^d$, $j = 1, \ldots, d$. Assume that $y \in K$ is such that $\nabla |y|_K$ exists. Consider a function $F(\eta) = \int_0^{|y+\eta e_j|_K} g(u) du$ of real variable η. Then

$$\frac{\partial f}{\partial x_j}(y) = F'(0) = g(|y|_K) \cdot \left(|y + \eta e_j|_K\right)'\big|_{\eta=0} = g(|y|_K)(\nabla |y|_K, e_j)$$

and the first equality in (1.36) follows. The second equality now follows from (1.35).

1.2.5 Classes of Functions with Given Majorant of Modulus of Continuity

For arbitrary $h > 0$, $K \in \mathbb{K}$ and $C \in \mathcal{C}$, we consider the following quantity

$$S(f; w, \Omega, h) = S(h) = \int_{hK \cap C} w(|t|_K)(f(t) - f(\theta))\Omega(t)dt,$$

where $f(\theta) := \lim_{x \to \theta} f(x)$, we assume the existence of this limit, and further conditions on the function f and weights w and Ω will be specified later. This quantity on the one hand is related to Ostrowski type inequalities, and on the other hand is intimately connected to fractional derivatives.

Let ω be a modulus of continuity i.e., $\omega : [0, \infty) \to [0, \infty)$ is a continuous non-decreasing semi-additive function such that $\omega(0) = 0$. For $K \in \mathbb{K}$ and $C \subset \mathbb{R}^d$ we consider the space $H_K^\omega(C)$ of functions $f : C \to \mathbb{R}$ such that the quantity

$$\|f\|_{H_K^\omega(C)} := \sup_{x, y \in C, x \neq y} \frac{|f(x) - f(y)|}{\omega(|x - y|_K)}$$

is finite. It is easy to see that all functions from $H_K^\omega(C)$ are continuous.

The following theorem is contained in [29]; the case when $C = \mathbb{R}^d$, and K is the unit Euclidean ball and $w(t) = t^{-(d+\alpha)}$, $\alpha \in (0, 1)$, was known earlier, see [24].

Theorem 1.6 *Assume that a set $K \in \mathbb{K}$, a cone $C \in \mathcal{C}$, a weight function w, a modulus of continuity ω, and a non-negative characteristic Ω are given. If $w \cdot \omega \in \mathcal{L}_1(0, h)$ for some $h > 0$, then*

$$\sup_{\|f\|_{H_K^\omega(hK \cap C)} \leq 1} S(f; w, \Omega, h) = \Omega(K, C) \cdot \|w \cdot \omega\|_{\mathcal{L}_1(0,h)}.$$

The supremum is attained on functions

$$f(t) = \pm \omega(|t|_K) + A, \, A \in \mathbb{R}. \tag{1.37}$$

Proof For each f such that $\|f\|_{H_K^\omega(hK \cap C)} \leq 1$, using Lemma 1.11 we obtain

1.2 Ostrowski-Type Inequalities

$$\left| \int_{hK \cap C} w(|t|_K)(f(t) - f(\theta))\Omega(t)dt \right| \leq \int_{hK \cap C} w(|t|_K)\omega(|t|_K)\Omega(t)dt$$

$$= \Omega(K, C) \cdot \|w \cdot \omega\|_{\mathcal{L}_1(0,h)}.$$

Moreover, the inequality becomes equality for the functions defined in (1.37).

It is now enough to show that for functions (1.37), $\|f\|_{H_K^\omega(hK \cap C)} \leq 1$ (in fact equality holds). It is sufficient to show this for the function $f(t) = \omega(|t|_K)$. For arbitrary $t, s \in hK \cap C$, using semi-additivity and monotonicity of ω we obtain

$$|f(t) - f(s)| = |\omega(|t|_K) - \omega(|s|_K)| \leq \omega(||t|_K - |s|_K|) \leq \omega(|t - s|_K),$$

as desired.

1.2.6 Classes of Functions with a Restriction on Their Gradient

For $p \in [1, \infty]$ and an open set $Q \subset \mathbb{R}^d$, we consider the Sobolev class $W^{1,p}(Q)$ that consists of all functions $f \in L_p(Q)$ such that all their partial derivatives (understood in the distributional sense) belong to $L_p(Q)$ (see e.g. [62, Section 6.7]). We also consider the classes $W^{1,p}_{\text{loc}}(Q)$ of functions that belong to $W^{1,p}(T)$ for each compact set $T \subset Q$.

We need the following lemma, which follows from the results in [62, Chapter 6.9].

Lemma 1.15 *Suppose $Q \subset \mathbb{R}^d$ is an open convex set, $f \in W^{1,1}_{\text{loc}}(Q)$ and $x, y \in Q$. Then*

$$f(y) - f(x) = \int_0^1 (y - x, \nabla f[(1-t)x + ty]) \, dt.$$

For a cone $C \in \mathcal{C}$, a non-negative characteristic $\Omega \colon C \to \mathbb{R}$, and an open set $Q \subset C$ that contains θ in its closure, by $L^\Omega_{\infty,p}(Q)$ we denote the class of continuous at θ functions $f \in L_\infty(Q)$ such that for all $i = 1, \ldots, d$, $\Omega^{\frac{1}{p}} \cdot f'_{x_i} \in L_p(Q)$. Due to equivalence of finite-dimensional norms, for each $K \in \mathbb{K}$ and $f \in L^\Omega_{\infty,p}(Q)$ one has $\Omega^{\frac{1}{p}} \cdot |\nabla f|_{K^\circ} \in L_p(Q)$. It is easy to see that $L^\Omega_{\infty,p}(Q) \subset W^{1,1}_{\text{loc}}(Q)$, and hence Lemma 1.15 holds for all functions $f \in L^\Omega_{\infty,p}(Q)$, provided Q is convex.

For $h > 0$ and $p \in [1, \infty]$ denote by $\mathcal{W}_p(0, h)$ the space of all non-negative functions $w \colon (0, h) \to \mathbb{R}$ such that $w \in \mathcal{L}_1(u, h)$ for all $u \in (0, h)$ and the function g_w defined in (1.28) belongs to $\mathcal{L}_p(0, h)$. If a non-zero weight w belongs to $\mathcal{W}_{p'}(0, h)$, then the integral $\int_0^h u^{(d-1)(1-p)} dt$ converges, which happens only for $p > d$.

Observe that for a convex bounded set Q, an embedding of the class $W^{1,p}(Q)$ into the space of bounded continuous on Q functions holds, see [1, Chapter 4]. Hence in the case $\Omega \equiv 1$, the class $L^\Omega_{\infty,p}(Q)$ consists of continuous functions.

Theorem 1.7 *Let $p \in (d, \infty]$, $C \in \mathcal{C}$, $K \in \mathbb{K}$, $\Omega \colon C \to \mathbb{R}$ be a non-negative characteristic, $h > 0$ and $f \in L_{\infty,p}^{\Omega}(hK \cap C)$. For each $w \in \mathcal{W}_{p'}(0, h)$ one has*

$$\left| \int_{hK \cap C} w(|y|_K)[f(y) - f(\theta)]\Omega(y)dy \right|$$
$$\leq (\Omega(K, C))^{\frac{1}{p'}} \|g_w\|_{\mathcal{L}_{p'}(0,h)} \|\Omega^{\frac{1}{p}} \cdot |\nabla f|_{K^\circ}\|_{L_p(hK \cap C)}. \quad (1.38)$$

The inequality is sharp. It becomes equality for the functions $a \cdot f + b$, where $a, b \in \mathbb{R}$ and

$$f(y) = \int_0^{|y|_K} g_w^{p'-1}(u)du, \ y \in hK \cap C. \quad (1.39)$$

Proof We begin with the case $p \in (d, \infty)$. Using Lemmas 1.15, 1.12, the Hölder inequality, and Lemma 1.11, we obtain

$$\left| \int_{hK \cap C} w(|y|_K)[f(y) - f(\theta)]\Omega(y)dy \right|$$
$$= \left| \int_{hK \cap C} w(|y|_K)\Omega(y) \int_0^1 (y, \nabla f(ty))dt dy \right| \leq$$
$$\int_{hK \cap C} \int_0^1 w(|y|_K)|y|_K |\nabla f(ty)|_{K^\circ}\Omega(y)dt dy = \int_{hK \cap C} |\nabla f(y)|_{K^\circ} g_w(|y|_K)\Omega(y)dy$$
$$\leq \|\Omega^{\frac{1}{p}} \cdot |\nabla f|_{K^\circ}\|_{L_p(hK \cap C)} \left(\int_{hK \cap C} g_w^{p'}(|y|_K)\Omega(y)dy \right)^{\frac{1}{p'}}$$
$$= (\Omega(K, C))^{\frac{1}{p'}} \|g_w\|_{\mathcal{L}_{p'}(0,h)} \|\Omega^{\frac{1}{p}} \cdot |\nabla f|_{K^\circ}\|_{L_p(hK \cap C)}. \quad (1.40)$$

By the definition, the function f defined in (1.39) is continuous. The fact that it belongs to the class $L_{\infty,p}^{\Omega}(hK \cap C)$, follows from equality

$$(p' - 1)p = p' \quad (1.41)$$

and Lemmas 1.11 and 1.14. Next we prove that both inequalities in (1.40) turn into equality for the function f defined in (1.39). Using Lemmas 1.14 and 1.13 we obtain

$$(y, \nabla f(ty)) = (y, g_w^{p'-1}(|ty|_K) \cdot \nabla |y|_K) = |\nabla f(ty)|_{K^\circ} \cdot |y|_K,$$

and the first inequality indeed becomes an equality. Using (1.41) together with Lemma 1.14 we obtain $|\nabla f(y)|_{K^\circ}^p = g_w^{p'}(|y|_K)$, and hence the second inequality in (1.40) also becomes an equality for f.

In the case $p = \infty$, $f(y) = |y|_K$, and the fact that chain (1.40) holds and all inequalities in it become equalities can be checked directly, under the agreement that $\Omega^0 \equiv 1$. Moreover, it is easy to see that $f \in L_{\infty,\infty}^{\Omega}(hK \cap C)$ in this case too.

1.2 Ostrowski-Type Inequalities

Observe that this theorem solves the extremal problem to find the quantity

$$\sup_{\|\nabla f|_{K^\circ}\|_{L_p(hK\cap C)}\leq 1} \left|\int_{hK\cap C} w(|y|_K)[f(y) - f(\theta)]\Omega(y)dy\right|,$$

which under an assumption of integrability of $w(|\cdot|_K)$ on $hK \cap C$ is the deviation between the operators $\Lambda f := \int_{hK\cap C} w(|y|_K)f(y)dy$ and $If := \Lambda(\chi_{hK\cap C}(\cdot)f(\theta))$, where χ is the characteristic function. Moreover, the proof of Theorem 1.7 follows the approach outlined in Lemma 1.4; the key step of the proof is a piece of inequality (1.40) that estimates from above the left-hand side of (1.40) by the quantity $\int_{hK\cap C} |\nabla f(y)|_{K^\circ} g_w(|y|_K)\Omega(y)dy$, which is a functional of $\lambda f := |\nabla f(y)|_{K^\circ}$ and becomes equality for a large enough set of functions $f \in L^\Omega_{\infty,p}(hK \cap C)$.

1.2.7 Classes of Functions with Bounded Norms of Their Laplacian

Let $\Delta = \frac{\partial^2}{\partial x_1^2} + \ldots + \frac{\partial^2}{\partial x_d^2}$ be the Laplace operator, where the derivatives are understood in the distributional sense. For an open set $Q \subset \mathbb{R}^d$ by $L^\Delta_{\infty,p}(Q)$ $1 \leq p \leq \infty$ we denote the set of functions $f \in L_\infty(Q)$ such that $\Delta f \in L_p(Q)$. The results of this section are contained in [30, 32].

1.2.7.1 Construction of Extremal Functions

We consider the function

$$G_h: (0, \infty) \to \mathbb{R}, \quad G_h(t) = \begin{cases} \frac{1}{\sigma_{d-1}(d-2)} \left(\frac{1}{t^{d-2}} - \frac{1}{h^{d-2}}\right)_+, & d \geq 3, \\ \frac{1}{2\pi}\left(\ln \frac{h}{t}\right)_+, & d = 2, \end{cases} \quad (1.42)$$

here $a_+ := \max\{a, 0\}$ and σ_{d-1} is the area of the unit sphere S^{d-1} in $\mathbb{R}^d$. The function $G_h(|\cdot|_2)$ it the Green function for the ball B_h from $\mathbb{R}^d$. For $\alpha > 0$ set

$$F_h(t) = \begin{cases} \int_t^h G_\rho(t) \frac{d\rho}{\rho^{\alpha+1}}, & t \in (0, h], \\ 0, & t > h. \end{cases} \quad (1.43)$$

If $p > d/2$, then $G_h(|y|_2) \in L_{p'}(\mathbb{R}^d)$, $p' = p/(p-1)$. If in addition, $0 < \alpha < 2 - d/p$, then $F_h(|y|_2) \in L_{p'}(\mathbb{R}^d)$. Since $(p'-1)p = p'$, we obtain that for all $c \in \mathbb{R}$, the function $y \mapsto |F_h(|y|_2) - c \cdot G_h(|y|_2)|^{p'-1}$ belongs to $L_p(\mathbb{R}^d)$, and hence is integrable on B_h.

For each $h > 0$ we set

$$\Phi_h(\rho) = F_h(\rho) - c_h \cdot G_h(\rho), \rho \in (0, h], \quad (1.44)$$

with the number $c = c_h$ chosen from the condition

$$\int_{B_h} \widetilde{\psi}_h(|y|_2) dy = 0, \tag{1.45}$$

where

$$\widetilde{\psi}_h(\rho) = \begin{cases} -|\Phi_h(\rho)|^{p'-1} \operatorname{sgn} \Phi_h(\rho), & \rho \in (0, h], \\ 0, & \rho > h. \end{cases}$$

We set

$$\widetilde{\varphi}_{h,2}(\rho) = \begin{cases} \frac{1}{2} \left(\int_0^\rho - \int_\rho^h \right) \int_t^h u^{d-1} \widetilde{\psi}_h(u) \, du \frac{dt}{t^{d-1}}, & \rho \in [0, h] \\ \frac{1}{2} \int_0^h \int_t^h u^{d-1} \widetilde{\psi}_h(u) \, du \frac{dt}{t^{d-1}}, & \rho > h, \end{cases}$$

let $\psi_h(y) = \widetilde{\psi}_h(|y|_2)$ and

$$\varphi_{h,2}(y) = \widetilde{\varphi}_{h,2}(|y|_2), \quad y \in \mathbb{R}^d. \tag{1.46}$$

We note that in the case $p = \infty$ the function $\varphi_{h,2}$ can be written explicitly and can be viewed as a multidimensional analogue of the Euler perfect spline of the second order, see [32].

Lemma 1.16 *If $p > d/2$ and $0 < \alpha < 2 - d/p$, then $\varphi_{h,2} \in L^{\Delta}_{\infty,p}(\mathbb{R}^d)$, it is a continuously differentiable function with piecewise-continuous second derivatives and*

$$\Delta \varphi_{h,2}(y) = -\psi_h(y). \tag{1.47}$$

Moreover, $\max_{y \in \mathbb{R}^d} \varphi_{h,2}(y) = -\min_{y \in \mathbb{R}^d} \varphi_{h,2}(y)$.

Proof First of all we prove that the function $\widetilde{\varphi}_{h,2}$ is well-defined on $[0, h]$ i.e., the integrals in its definition converge. Using (1.45), we obtain

$$\sigma_{d-1} \int_t^h u^{d-1} \widetilde{\psi}_h(u) du = \int_{S^{d-1}} d\xi \int_t^h u^{d-1} \widetilde{\psi}_h(|u\xi|_2) du$$
$$= \int_{B_h \setminus B_t} \widetilde{\psi}_h(|y|_2) dy = -\int_{B_t} \widetilde{\psi}_h(|y|_2) dy.$$

Hence

$$\sigma_{d-1} \left| \int_0^\rho \frac{dt}{t^{d-1}} \int_t^h u^{d-1} \widetilde{\psi}_h(u) du \right| = \left| \int_0^\rho \frac{dt}{t^{d-1}} \int_{B_t} \widetilde{\psi}_h(|y|_2) dy \right|$$
$$\leq \int_0^\rho \frac{1}{t^{d-1}} \left(\int_{B_t} |\widetilde{\psi}_h(|y|_2)|^p \, dy \right)^{\frac{1}{p}} \cdot (\text{meas } B_t)^{\frac{1}{p'}} dt$$
$$\leq (\text{meas } B_1)^{\frac{1}{p'}} \|\widetilde{\psi}_h(|\cdot|_2)\|_{L_p(B_h)} \cdot \int_0^\rho t^{\frac{d}{p'}-(d-1)} dt < \infty,$$

since $\frac{d}{p'} - (d-1) > -1 \iff p > \frac{d}{2}$.

1.2 Ostrowski-Type Inequalities

Taking into account that the Laplace transform for radial functions becomes

$$\Delta\varphi(\rho) = \varphi''(\rho) + \frac{d-1}{\rho}\varphi'(\rho),$$

it is now easy to verify equality (1.47).

Statement about smoothness of $\varphi_{h,2}$ and the last equality in the lemma follow from the definition of the function $\varphi_{h,2}$.

1.2.7.2 Ostrowski-Type Inequality

For $x \in \mathbb{R}^d$ and $h > 0$ we consider the following quantity

$$\widetilde{f}(x, h) = \frac{1}{\sigma_{d-1}} \int_{S^{d-1}} f(x + hy) dy, \qquad (1.48)$$

which is the average value of f over the sphere of radius h with center at θ. It is well known (see e.g. [38, Section 4.3]) that for a continuously differentiable function f that has piecewise-continuous second derivatives one has

$$f(x) = \widetilde{f}(x, h) + \int_{B_h} G_h(|y|_2)(-\Delta f(x \pm y)) \, dy. \qquad (1.49)$$

For any $f \in L_{\infty,p}^{\Delta}(\mathbb{R}^d)$ and arbitrary infinitely differentiable function φ with a compact support one has

$$\int_{\mathbb{R}^d} \varphi(x) f(x) \, dx = \int_{\mathbb{R}^d} f(x) \left[\widetilde{\varphi}(x, h) + \int_{B_h} G_h(|y|_2)(-\Delta\varphi(x \mp y)) \, dy \right] dx =$$

$$= \int_{\mathbb{R}^d} \varphi(x) \left[\widetilde{f}(x, h) + \int_{B_h} G_h(|y|_2)(-\Delta f(x \pm y)) \, dy \right] dx.$$

Hence for a function $f \in L_{\infty,p}^{\Delta}(\mathbb{R}^d)$ and almost all $x \in \mathbb{R}^d$ equality (1.49) holds.

For $h > 0$, $x \in \mathbb{R}^d$ and a function f such that $f \in L_{\infty,p}^{\Delta}(x + B_h)$ we set

$$S_h(f; x) = \int_{B_h} \frac{2f(x) - f(x+t) - f(x-t)}{|t|_2^{d+\alpha}} dt - 2c_h \sigma_{d-1}(f(x) - \widetilde{f}(x, h)).$$

Theorem 1.8 *Let $d/2 < p < \infty$ and $0 < \alpha < 2 - d/p$, or $p = \infty$ and $0 < \alpha < 2$. For arbitrary $h > 0$, function $f \in \bigcap_{y \in \mathbb{R}^d} L_{\infty,p}^{\Delta}(y + B_h)$ and almost all $x \in \mathbb{R}^d$*

$$|S_h(f; x)| \le 2\sigma_{d-1} \|\Phi_h(|\cdot|)\|_{L_{p'}(B_h)} \cdot \|\Delta f\|_{L_p(x+B_h)}$$

$$= 2\sigma_{d-1} h^{2-\alpha-\frac{d}{p}} \|\Phi_1(|\cdot|)\|_{L_{p'}(B_1)} \cdot \|\Delta f\|_{L_p(x+B_h)}, \qquad (1.50)$$

where the function Φ_h is defined in (1.44). The inequality is sharp. It becomes equality for the function $f = \varphi_{h,2}(\cdot + x)$.

Proof For simplicity of notations we assume that $x = \theta$ and equality (1.49) holds for $x = \theta$. Then

$$S_h(f;\theta) = \int_{B_h} \frac{2f(\theta) - f(t) - f(-t)}{|t|_2^{d+\alpha}} dt - 2c_h\sigma_{d-1}(f(\theta) - \widetilde{f}(\theta,h))$$

$$= \int_{S^{d-1}} d\xi \int_0^h s^{d-1} \frac{2f(\theta) - f(s\xi) - f(-s\xi)}{s^{d+\alpha}} ds - 2c_h\sigma_{d-1}(f(\theta) - \widetilde{f}(\theta,h))$$

$$= 2\int_{S^{d-1}} d\xi \int_0^h \frac{f(\theta) - f(s\xi)}{s^{1+\alpha}} ds - 2c_h\sigma_{d-1}(f(\theta) - \widetilde{f}(\theta,h))$$

$$= 2\sigma_{d-1} \int_0^h \frac{f(\theta) - \widetilde{f}(\theta,s)}{s^{1+\alpha}} ds - 2c_h\sigma_{d-1}(f(\theta) - \widetilde{f}(\theta,h))$$

$$= 2\sigma_{d-1} \left(\int_0^h \left(-\int_{B_s} G_s(|y|_2)\Delta f(y) dy \right) \frac{ds}{s^{1+\alpha}} + c_h \int_{B_h} G_h(|y|_2)\Delta f(y) dy \right)$$

$$= -2\sigma_{d-1} \int_{B_h} \Delta f(y) \Phi_h(|y|) dy,$$

since switching to the polar coordinates and changing the order of integrals, we obtain

$$\int_0^h \int_{B_s} G_s(|y|_2)\Delta f(y) dy \frac{ds}{s^{1+\alpha}} = \int_0^h \int_{S^{d-1}} \int_0^s \rho^{d-1} G_s(\rho)\Delta f(\rho\xi) d\rho d\xi \frac{ds}{s^{1+\alpha}}$$

$$= \int_{S^{d-1}} \int_0^h \left(\int_\rho^h G_s(\rho) \frac{ds}{s^{1+\alpha}} \right) \rho^{d-1} \Delta f(\rho\xi) d\rho d\xi$$

$$= \int_{S^{d-1}} \int_0^h F_h(\rho) \rho^{d-1} \Delta f(\rho\xi) d\rho d\xi = \int_{B_h} F_h(|y|)\Delta f(y) dy.$$

Applying the Hölder inequality we obtain the inequality in (1.50). Moreover, from (1.47) it follows that the inequality in (1.50) becomes equality for function (1.46), and hence is sharp.

It is easy to see that for the functions G_h and F_h defined in (1.42) and (1.43), one has

$$G_h(\rho) = h^{2-d}G_1(\rho/h) \text{ and } F_h(\rho) = h^{2-\alpha-d}F_1(\rho/h). \quad (1.51)$$

Moreover, from (1.51) it follows that for all $h > 0$ one has $c_h = h^{-\alpha}c_1$. Direct computations now show that the equality in (1.50) holds.

1.2.8 Classes of Random Processes

Let $\{\Omega, \mathcal{F}, P\}$ be a probability space. For a random variable η, defined on the probability space $\{\Omega, \mathcal{F}, P\}$, set $\|\eta\|_\infty := \operatorname{ess\,sup}_{w \in \Omega} |\eta(w)|$.

For $a > 0$ denote by $\mathcal{R}(a)$ the space of all random variables η on the space $\{\Omega, \mathcal{F}, P\}$ such that $\eta(w) \in [0, a]$ for all $w \in \Omega$.

In this section ω denotes a concave modulus of continuity and for brevity we denote by $H^\omega(a)$ the class of functions $x: [0, a] \to \mathbb{R}$ such that $|x(s) - x(t)| \leq \omega(|t - s|)$ for all $t, s \in [0, a]$.

For a fixed $\tau \in \mathcal{R}(a)$ denote by $\mathcal{H}_\tau^\omega(a)$ the set of all measurable random processes ξ_t, $t \in [0, a]$, defined on the probability space $\{\Omega, \mathcal{F}, P\}$, and such that for all $\eta \in \mathcal{R}(a)$

$$\mathbf{E}|\xi_\tau - \xi_\theta| \leq \omega(\|\tau - \theta\|_\infty). \tag{1.52}$$

Set $\mathcal{H}^\omega(a) := \bigcap_{\tau \in \mathcal{R}(a)} \mathcal{H}_\tau^\omega(a)$, so that $\mathcal{H}^\omega(a)$ is the class of measurable processes such that inequality (1.52) holds for all $\tau, \theta \in \mathcal{R}(a)$.

In the case, when $a = 1$, we write $H^\omega, \mathcal{H}^\omega, \mathcal{H}_\tau^\omega$ and $\mathcal{R}$ instead of $H^\omega(1), \mathcal{H}^\omega(1), \mathcal{H}_\tau^\omega(1)$ and $\mathcal{R}(1)$ respectively.

The following theorem gives an Ostrowski type inequality for random processes of the class $\mathcal{H}^\omega(a)$, see [58].

Theorem 1.9 *Let $a > 0$ and $\tau \in \mathcal{R}(a)$ be given. Set $t^* := \left\| \tau(\cdot) - \frac{a}{2} \right\|_\infty$. Then*

$$\sup_{\xi \in \mathcal{H}^\omega(a)} \mathbf{E} \left| \int_0^a \xi_t \, dt - a \cdot \xi_\tau \right| = \int_0^{\frac{a}{2} - t^*} \omega(s) \, ds + \int_0^{\frac{a}{2} + t^*} \omega(s) \, ds. \tag{1.53}$$

We prove the theorem in the case $a = 1$, the general case can be proved similarly. The proof of the theorem is given in the following paragraphs.

1.2.8.1 Some Remarks About the Proof of Theorem 1.9

It is enough to prove (1.53) for the case of a simple random variable τ. The general case can be obtained using approximation of τ by simple random variables.

Let $\Omega_1, \ldots, \Omega_n \in \mathcal{F}$ be pairwise disjoint sets with positive measures such that $P\left(\bigcup_{k=1}^n \Omega_k\right) = 1$, and $\tau(w) = \tau_k$ for $w \in \Omega_k$, $k = 1, \ldots, n$.

For a fixed $k \in \{1, \ldots, n\}$ set

$$\tau^*(w) := \begin{cases} \tau(w), & w \in \Omega \setminus \Omega_k, \\ 1 - \tau(w), & w \in \Omega_k. \end{cases}$$

Since together with arbitrary $\xi \in \mathcal{H}_\tau^\omega$ (or $\xi \in \mathcal{H}^\omega$), the process ξ_t^*, $t \in [0, 1]$,

$$\xi_t^*(w) := \begin{cases} \xi_t(w), & w \in \Omega \setminus \Omega_k, \\ \xi_{1-t}(w), & w \in \Omega_k, \end{cases}$$

belongs to $\mathcal{H}_{\tau^*}^\omega$ (to $\mathcal{H}^\omega$ respectively), and for almost all $w \in \Omega$

$$\left| \int_0^1 \xi_t \, dt - \xi_\tau \right| = \left| \int_0^1 \xi_t^* \, dt - \xi_{\tau^*}^* \right|,$$

one has

$$\sup_{\xi \in \mathcal{H}_\tau^\omega} \mathbf{E} \left| \int_0^1 \xi_t \, dt - \xi_\tau \right| = \sup_{\xi \in \mathcal{H}_{\tau^*}^\omega} \mathbf{E} \left| \int_0^1 \xi_t \, dt - \xi_{\tau^*} \right|$$

and

$$\sup_{\xi \in \mathcal{H}^\omega} \mathbf{E} \left| \int_0^1 \xi_t \, dt - \xi_\tau \right| = \sup_{\xi \in \mathcal{H}^\omega} \mathbf{E} \left| \int_0^1 \xi_t \, dt - \xi_{\tau^*} \right|.$$

Hence without loss of generality we may assume that

$$0 \leq \tau_k \leq \frac{1}{2}, k = 1, \ldots, n. \tag{1.54}$$

Moreover, we can also assume that

$$\tau_1 \leq \ldots \leq \tau_n. \tag{1.55}$$

Under the assumptions above, we have $t^* = \frac{1}{2} - \tau_1$ and the right hand side of (1.53) becomes

$$\int_0^{\tau_1} \omega(s) \, ds + \int_0^{1-\tau_1} \omega(s) \, ds. \tag{1.56}$$

1.2.8.2 Estimate From Above

Lemma 1.17 *Let the assumptions of Theorem 1.9 and* (1.54), (1.55) *hold. The inequality*

$$\mathbf{E} \left| \int_0^1 \xi_t \, dt - \xi_\tau \right| \leq \int_0^{\frac{1}{2} - t^*} \omega(s) \, ds + \int_0^{\frac{1}{2} + t^*} \omega(s) \, ds \tag{1.57}$$

holds for all $\xi \in \mathcal{H}_\tau^\omega$, in particular for all $\xi \in \mathcal{H}^\omega$.

Proof For all $\xi \in \mathcal{H}_\tau^\omega$ one has

$$\mathbf{E} \left| \int_0^1 \xi_t \, dt - \xi_\tau \right| = \sum_{k=1}^n \int_{\Omega_k} \left| \int_0^1 (\xi_t - \xi_{\tau_k}) \, dt \right| P(dw)$$

$$\leq \sum_{k=1}^n \int_{\Omega_k} \int_0^1 |\xi_t - \xi_{\tau_k}| \, dt \, P(dw) = \sum_{k=1}^n \int_0^1 \int_{\Omega_k} |\xi_t - \xi_{\tau_k}| \, P(dw) \, dt$$

1.2 Ostrowski-Type Inequalities

$$= \sum_{k=1}^{n} \left(\int_0^{\tau_k-\tau_1} \int_{\Omega_k} |\xi_t - \xi_{\tau_k}| P(dw)dt + \int_{\tau_k-\tau_1}^{\tau_k} \int_{\Omega_k} |\xi_t - \xi_{\tau_k}| P(dw)dt \right.$$
$$\left. + \int_{\tau_k}^{\tau_k+1-\tau_n} \int_{\Omega_k} |\xi_t - \xi_{\tau_k}| P(dw)dt + \int_{\tau_k+1-\tau_n}^{1} \int_{\Omega_k} |\xi_t - \xi_{\tau_k}| P(dw)dt \right)$$
$$= \sum_{k=1}^{n} \left(\int_0^{\tau_k-\tau_1} \int_{\Omega_k} |\xi_t - \xi_{\tau_k}| P(dw)dt + \int_{\tau_k+1-\tau_n}^{1} \int_{\Omega_k} |\xi_t - \xi_{\tau_k}| P(dw)dt \right)$$
$$+ \int_0^{\tau_1} \sum_{k=1}^{n} \int_{\Omega_k} |\xi_{\tau_k-s} - \xi_{\tau_k}| P(dw)ds + \int_0^{1-\tau_n} \sum_{k=1}^{n} \int_{\Omega_k} |\xi_{\tau_k+s} - \xi_{\tau_k}| P(dw)dt \quad (1.58)$$

Setting $\theta_s(w) := \tau(w) - s$, we obtain that $\|\theta_s - \tau\|_\infty = s$, $s \in [0, \tau_1]$, and hence

$$\int_0^{\tau_1} \sum_{k=1}^{n} \int_{\Omega_k} |\xi_{\tau_k-s} - \xi_{\tau_k}| P(dw)ds = \int_0^{\tau_1} \mathbf{E}|\xi_{\theta_s} - \xi_\tau|ds \leq \int_0^{\tau_1} \omega(s)ds.$$

Analogously

$$\int_0^{1-\tau_n} \sum_{k=1}^{n} \int_{\Omega_k} |\xi_{\tau_k+s} - \xi_{\tau_k}| P(dw)dt \leq \int_0^{1-\tau_n} \omega(s)ds.$$

Now set
$$\theta_s(w) := \begin{cases} \tau_k + s, & s \in [1-\tau_n, 1-\tau_k], \\ \tau_k + (1-\tau_1-\tau_k-s), & s \in (1-\tau_k, 1-\tau_1], \end{cases}$$

$w \in \Omega_k$, $k = 1, \ldots, n$. Since $\tau_1 + \tau_k \leq 1$, $k = 1, \ldots, n$, $|\theta_s - \tau| \leq s$ for almost all $w \in \Omega$ and $s \in [1-\tau_n, 1-\tau_1]$. Hence

$$\sum_{k=1}^{n} \left(\int_0^{\tau_k-\tau_1} \int_{\Omega_k} |\xi_t - \xi_{\tau_k}| P(dw)dt + \int_{\tau_k+1-\tau_n}^{1} \int_{\Omega_k} |\xi_t - \xi_{\tau_k}| P(dw)dt \right)$$
$$= \sum_{k=1}^{n} \left(\int_{1-\tau_k}^{1-\tau_1} \int_{\Omega_k} |\xi_{1-\tau_1-s} - \xi_{\tau_k}| P(dw)ds + \int_{1-\tau_n}^{1-\tau_k} \int_{\Omega_k} |\xi_{\tau_k+s} - \xi_{\tau_k}| P(dw)ds \right)$$
$$= \int_{1-\tau_n}^{1-\tau_1} \sum_{k=1}^{n} \int_{\Omega_k} |\xi_{\theta_s} - \xi_{\tau_k}| P(dw)ds = \int_{1-\tau_n}^{1-\tau_1} \mathbf{E}|\xi_{\theta_s} - \xi_\tau|ds \leq \int_{1-\tau_n}^{1-\tau_1} \omega(s)ds. \quad (1.59)$$

Finally, inequalities (1.58)–(1.59), together with observation (1.56), give inequality (1.57). The lemma is proved.

1.2.8.3 A Random Process Generated by a Function

The following lemma gives a way to generate random processes from the class $\mathcal{H}^\omega$, given a function $x \in H^\omega$.

Lemma 1.18 *Let $x \in \mathcal{H}^\omega$ and $F \in \mathcal{F}$, $P(F) > 0$, be given. Then the process $\xi_t = \xi_t(x, F)$, $t \in [0, 1]$,*

$$\xi_t(w) := \begin{cases} \frac{1}{P(F)} x(t), & w \in F, \\ 0, & w \in \Omega \setminus F \end{cases}$$

belongs to $\mathcal{H}^\omega$ and

$$\mathbf{E}\xi_t = x(t). \tag{1.60}$$

Proof In order to prove that $\xi_t \in \mathcal{H}^\omega$, it is enough to show that the inequality

$$\mathbf{E}\left|\xi_{\theta_1} - \xi_{\theta_2}\right| \leq \omega(\|\theta_1 - \theta_2\|_\infty)$$

holds for arbitrary two simple random variables θ_1 and θ_2.

Assume that the pairwise disjoint measurable sets $F_i \subset F$ with positive measures are such that $\theta_k(w) = \theta_i^k \in [0, 1]$, $w \in F_i$, $i = 1, \ldots, n$, $k = 1, 2$, and $P\left(\bigcup_{i=1}^n F_i\right) = P(F)$. Taking into account that ω is a non-decreasing concave function, one has

$$\mathbf{E}\left|\xi_{\theta_1} - \xi_{\theta_2}\right| = \sum_{i=1}^n \frac{P(F_i)}{P(F)} \left|x(\theta_i^1) - x(\theta_i^2)\right| \leq \sum_{i=1}^n \frac{P(F_i)}{P(F)} \omega\left(\left|\theta_i^1 - \theta_i^2\right|\right)$$

$$\leq \omega\left(\sum_{i=1}^n \frac{P(F_i)}{P(F)} \left|\theta_i^1 - \theta_i^2\right|\right) \leq \omega\left(\max_{i=1,\ldots,n} \left|\theta_i^1 - \theta_i^2\right|\right) \leq \omega(\|\theta_1 - \theta_2\|_\infty).$$

Equality (1.60) follows from the definition of the process.

1.2.8.4 Estimate From Below

Let the assumptions (1.54) and (1.55) hold. Consider the process $\xi_t^* \in \mathcal{H}^\omega$, built according to Lemma 1.18 with $F = \Omega_1$ and $x(\cdot) := \omega(|\cdot - \tau_1|) \in H^\omega$. Then

$$\mathbf{E}\left|\int_0^1 \xi_t^* dt - \xi_\tau^*\right| = \mathbf{E}\int_0^1 \xi_t^* dt = \int_0^1 \mathbf{E}\xi_t^* dt = \int_0^1 \omega(|t - \tau_1|) dt$$

$$= \int_0^{\tau_1} \omega(t) dt + \int_0^{1-\tau_1} \omega(t) dt,$$

which together with (1.56) gives the estimate

$$\sup_{\xi \in \mathcal{H}^\omega} \mathbf{E}\left|\int_0^1 \xi_t dt - \xi_\tau\right| \geq \int_0^{\frac{1}{2}-t^*} \omega(s) ds + \int_0^{\frac{1}{2}+t^*} \omega(s) ds.$$

1.2.9 Sets and Functions of Bounded Variation

1.2.9.1 Univariate Functions of Bounded Variation

In [39] the following Ostrowki-type inequality for univariate functions of bounded variation was proved.

Theorem 1.10 *For a function f with bounded on $[0, 1]$ variation and arbitrary $x \in [0, 1]$ the following inequality holds.*

$$\left| \int_0^1 f(t)dt - f(x) \right| \leq \max\{x, 1 - x\} \bigvee_0^1 f. \tag{1.61}$$

It is sharp in the sense that for each fixed x the number $\max\{x, 1 - x\}$ can not be decreased.

In order to give motivation for the definitions considered below and to outline the main idea of the proof of Theorems 1.11 and 1.12 below, we give a different then in [39] proof in the case of continuous functions of bounded variation.

Proof Let $f: [0, 1] \to \mathbb{R}$ be a continuous function of bounded variation. Denote by $n(f, y)$ the number of solutions of the equation

$$f(t) = y, t \in [0, 1], y \in \mathbb{R} \tag{1.62}$$

(we set $n(f, y) = 0$, if equation (1.62) has no solutions, and $n(f, y) = +\infty$, if the equation has infinite set of solutions). By the Banach indicatrix theorem, the function n is measurable and

$$\bigvee_0^1 f = \int_{\mathbb{R}} n(f, y) dy.$$

Both sides of inequality (1.61) do not change if f is substituted by $f + \beta$, $\beta \in \mathbb{R}$, thus we can assume that $f(x) = 0$. For $a \in \mathbb{R}$ set $\{f \geq a\} = \{t \in [0, 1]: f(t) \geq a\}$. It is easy to see that for all $a > 0$ (μ denotes the Lebesgue measure)

$$\mu\{f \geq a\} \leq \begin{cases} 0, & n(f, a) = 0, \\ \max\{x, 1 - x\}, & n(f, a) = 1, \\ 1, & n(f, a) \geq 2, \end{cases}$$

and hence $\mu\{f \geq a\} \leq \max\{x, 1 - x\} \cdot n(f, a)$ for all $a > 0$. Integrating this inequality we obtain

$$\int_0^1 f(t)dt \le \int_0^\infty \mu\{f \ge a\} da \le \int_0^\infty \max\{x, 1-x\} \cdot n(f, a) da$$

$$\le \max\{x, 1-x\} \bigvee_0^1 f,$$

as desired.

Observe that if f is an arbitrary function, then the family of level sets that contain an extremum is at most countable. Thus the value $n(f, y)$ can be defined as the number of connected components of the set of solutions of the equation $f(t) = y$ (instead of the number of solutions).

1.2.9.2 Multidimensional Sets and Multivariate Functions of Bounded Variation

Many ways to extend the notion of a function of bounded variation from the case of univariate to the case of multivariate functions are known, see e.g., [37] for a survey of different approaches for functions of two variables. We propose a new way for such a generalization. The approach is based on the Kronrod–Vitushkin variations (which, in turn, are based on the Banach indicatrix theorem). Unlike the Kronrod–Vitushkin variations, the proposed definition satisfies the following two properties: the variation of a function does not change after multiplication of its argument by a non-zero constant; the variation of a radial function is twice as big as the variation of the generating univariate function (see Lemma 1.19 for a precise statement).

For a set $F \subset \mathbb{R}^d$ we denote by $\overline{F}$ its closure; B^d denotes the unit Euclidean ball in $\mathbb{R}^d$. By $\mathbb{P}^{d-1}$ we denote the $d-1$ dimensional real projective space i.e., the set of all lines in $\mathbb{R}^d$ that contain θ. The measure of a set $A \subset \mathbb{P}^{d-1}$ is by definition equal to the spherical measure of the set $\bigcup_{l \in A} l \cap S^{d-1}$.

For each $r \in \mathbb{P}^{d-1}$ by $\Pi^{d-1}(r)$ we denote the hyperplane that contains θ and is orthogonal to the line r; $\Pi^{d-1}(r)$ is considered as a $d-1$-dimensional space with $d-1$-dimensional Lebesgue measure and Euclidean metric. For each $\beta \in \Pi^{d-1}(r)$ by $l(r, \beta)$ we denote the line that contains β and is parallel to r. By μ^k, $k \in \mathbb{N}$, we denote the k-dimensional Lebesgue measure in $\mathbb{R}^k$; by μ we denote the measure on the projective space $\mathbb{P}^{d-1}$.

Denote by $N(F)$ the number of connected components of the set $F \subset \mathbb{R}^d$; 0 for empty set, and $+\infty$ if the set of connected components is infinite. Variation of a compact set F in direction $r \in \mathbb{P}^{d-1}$ is by definition

$$v(F, r) := \operatorname*{ess\,sup}_{\beta \in \Pi^{d-1}(r)} N(F \cap l(r, \beta)).$$

For a compact set $F \subset \mathbb{R}^d$ and a number $p \in [1, \infty]$ set

1.2 Ostrowski-Type Inequalities

$$v_p(F) := \begin{cases} \left(\frac{1}{\mu \mathbb{P}^{d-1}} \int_{\mathbb{P}^{d-1}} v^p(F,r) dr\right)^{\frac{1}{p}}, & p \in [1, \infty), \\ \operatorname{ess\,sup}_{r \in \mathbb{P}^{d-1}} v(F,r), & p = \infty. \end{cases}$$

If $d = 1$ then for all $p \in [1, \infty]$ we set $v_p(F) = N(F)$.

Let a set $E \subset \mathbb{R}^d$ and a function $f : E \to \mathbb{R}$ be given. For $t \in \mathbb{R}$ the set

$$L(f; t) := \{x \in E : f(x) = t\}$$

is called a level set of the function f.

The variation of a continuous function is defined as follows. Let $E \subset \mathbb{R}^d$, $f : E \to \mathbb{R}$ be a continuous on a compact subset $F \subset E$ function, and $p \in [1, \infty]$. Set

$$v_p(f; F) := \int_{-\infty}^{\infty} v_p(F \cap L(f; t)) dt.$$

If $F = E$, then we write $v_p(f)$ instead of $v_p(f; E)$. In [56] it was shown that the functions under the integral signs in the definitions of the variations are measurable.

The following property of the variation will be needed, more properties were considered in [56].

Lemma 1.19 *Let $\varphi : [0, 1] \to \mathbb{R}$ be a continuous function and $d \in \mathbb{N}$. Let $f_\varphi : B^d \to \mathbb{R}$, $f_\varphi(x) = \varphi(|x|)$. Then for all $p \in [1, \infty]$*

$$v_p(f_\varphi; B^d) = 2 \cdot \bigvee_0^1 \varphi.$$

Proof In the case $d = 1$ the property follows from the Banach indicatrix theorem, so we can assume that $d \geq 2$. Let arbitrary $t \neq \varphi(0)$ be fixed. For arbitrary $r \in \mathbb{P}^{d-1}$ and $\beta \in \Pi^{d-1}(r)$ the number $N(L(f_\varphi; t) \cap l(r, \beta))$ can be obtained by the following procedure: consider the line $r = l(r, \theta)$ and mark points of the set $L(f_\varphi; t) \cap l(r, \theta)$; cut the interval $(-|\beta|, |\beta|)$ from the line and stick the points $-|\beta|$ and $|\beta|$ together; the number of components of marked points on the obtained "cut" line is equal to $N(L(f_\varphi; t) \cap l(r, \beta))$. This shows that for arbitrary β

$$N(L(f_\varphi; t) \cap l(r, \beta)) \leq N(L(f_\varphi; t) \cap l(r, \theta)). \tag{1.63}$$

From the choice of t it follows that

$$\theta \notin L(f_\varphi; t) \tag{1.64}$$

and hence there exists $\varepsilon > 0$ such that $B(\varepsilon) \cap L(f_\varphi; t) = \emptyset$. This implies that the set $L(f_\varphi; t) \cap l(r, \theta)$ does not contain points x with $|x| < \varepsilon$ and hence for all β such that $|\beta| < \varepsilon$ (1.63) becomes equality. This implies that $v(L(f_\varphi; t), r) = N(L(f_\varphi; t) \cap l(r, \theta))$.

From (1.64) it follows that $N(L(f_\varphi; t) \cap l(r, \theta)) = 2 \cdot N(L(\varphi; t))$. Equality $v_p(f_\varphi; B) = 2 \cdot \bigvee_0^1 \varphi$ follows from the Banach indicatrix theorem now.

1.2.9.3 Ostrowski-Type Inequalities

The following result is the main tool to prove Ostrowski type inequalities for functions and sets of bounded variation below.

Lemma 1.20 *Let $d \in \mathbb{N}$ and two sets $F, W \subset B^d$ be given. Assume that the following properties hold:*

1. *F is measurable and $\theta \notin \overline{F}$;*
2. *W is closed and $\theta \notin W$ and*
3. *If $x \in F$ and $y \in B^d \setminus F$, then $xy \cap W \neq \emptyset$.*

Then for all $p \in [1, \infty]$

$$\mu^d F \leq \frac{\mu^d B^d}{2} v_p(W). \tag{1.65}$$

The inequality is sharp in the sense that for arbitrary $\varepsilon > 0$ there exist sets F and W that satisfy the conditions above and such that

$$\mu^d F > \left(\frac{\mu^d B^d}{2} - \varepsilon\right) v_p(W).$$

If (1.65) becomes equality, then $\mu^d F = 0$.

We will prove Lemma 1.20 in the next subsections. Here we state two consequences of this theorem, which can be considered as Ostrowski type inequalities. We state them in the cases, when the domain of definition is the unit ball B^d, see [56]. Results for more general domains were obtained in [57].

Theorem 1.11 *Let $d \in \mathbb{N}$ and a continuous function $f: B^d \to \mathbb{R}$ be given. Then for all $p \in [1, \infty]$*

$$\left|\frac{1}{\mu^d B^d} \int_{B^d} f(x)dx - f(\theta)\right| \leq \frac{v_p(f)}{2}.$$

The inequality is sharp. It becomes equality only in the case when f is constant.

Proof From the definition of the variation it follows that $v_p(f + \beta) = v_p(f)$ for arbitrary $\beta \in \mathbb{R}$, hence we can assume that $f(\theta) = 0$ and it is sufficient to prove that

$$\int_{B^d} f(x)dx \leq \frac{\mu^d B^d}{2} v_p(f). \tag{1.66}$$

Set $\Gamma := \{(x, t) \in B^d \times [0, \infty) : f(x) \geq t\}$. Then

$$\int_{B^d} f(x)dx \leq \mu^{d+1}\Gamma = \int_{t\geq 0} \mu^d(\Gamma \cap \mathbb{R}_t^{d+1})dt, \quad (1.67)$$

where $\mathbb{R}_t^{d+1} = \{(x, t) : x \in \mathbb{R}^d\}$. For each $t > 0$ consider the sets $F := \Gamma \cap \mathbb{R}_t^{d+1}$ and $W := \Gamma(f) \cap \mathbb{R}_t^{d+1}$, where

$$\Gamma(f) := \{(x, t) \in \mathbb{R}^{d+1} : x \in E, f(x) = t\}.$$

Both F and W are closed sets that do not contain θ, since $f(\theta) = 0$. If $x \in F$ and $y \in B^d \setminus F$, then $f(x) \geq t$ and $f(y) < t$ and hence the segment xy contains a point z with $f(z) = t$, i. e. $xy \cap W \neq \emptyset$. This means that all conditions of Lemma 1.20 are satisfied and hence

$$\mu^d(\Gamma \cap \mathbb{R}_t^{d+1}) = \mu^d(F) \leq \frac{\mu^d B^d}{2} v_p(W) = \frac{\mu^d B^d}{2} v_p(L(f;t))$$

with equality possible only in the case when $\mu^d F = 0$. Taking into account (1.67) we obtain

$$\mu^{d+1}\Gamma \leq \frac{\mu^d B^d}{2} \int_{t\geq 0} v_p(L(f;t))dt \leq \frac{\mu^d B^d}{2} \int_{t\in\mathbb{R}} v_p(L(f;t))dt = \frac{\mu^d B^d}{2} v_p(f)$$

and inequality (1.66) is proved; moreover due to continuity of f we obtain that equality in (1.66) can hold only if $f \equiv 0$.

For all $\varepsilon > 0$ consider the function $\varphi_\varepsilon : [0, 1] \to \mathbb{R}$, $\varphi_\varepsilon(t) = 1$ for $t \geq \varepsilon$, $\varphi_\varepsilon(0) = 0$ and φ_ε is linear on $[0, \varepsilon]$. Due to Lemma 1.19 for the radial function $f_\varepsilon(x) : B^d \to \mathbb{R}$, $f_\varepsilon(x) = \varphi_\varepsilon(|x|)$, and arbitrary $p \in [1, \infty]$ $v_p(f_\varepsilon) = 2$; moreover $\int_{B^d} f_\varepsilon(x)dx \to \mu^d B^d$ as $\varepsilon \to 0$. This proves the sharpness of the stated inequality.

Theorem 1.12 *Let $d \in \mathbb{N}$ and a closed set $F \subset B^d$ be given. If $\theta \notin F$, then for all $p \in [1, \infty]$*

$$\mu^d F \leq \frac{\mu^d B^d}{2} v_p(F).$$

The inequality is sharp. If equality holds, then $\mu^d F = 0$.

Proof It is enough to apply Lemma 1.20 with $W = F$; all three conditions of Lemma 1.20 are satisfied.

For arbitrary $\varepsilon > 0$ consider a set $F_\varepsilon := \{x \in B^d : |x| \geq \varepsilon\}$. For all $p \in [1, \infty]$ $v_p(F_\varepsilon) = 2$; $\mu^d F_\varepsilon \to \mu^d B^d$ as $\varepsilon \to 0$. This proves that the stated inequality is sharp.

The proof of Theorem 1.11 again uses the scheme from Lemma 1.4. Indeed, let T be the unit ball B^d, $X = Y = M = \mathbb{R}$, $S = [0, \infty)$, $\mathcal{A}$ be the space of continuous functions $f : B^d \to \mathbb{R}$ with bounded variation $v_p(f) \leq 1$, $p \in [1, \infty]$, Z be the family of all closed

subsets of B^d. For each $s \in S$ set
$$\lambda f(s) = \{x \in B^d : s \leq |f(x) - f(\theta)|\} \in Z.$$

For the operators $\Lambda f = \int_{B^d} f(x)dx$, $If = \mu^d B^d \cdot f(\theta)$ and $\varphi(\lambda f) = \int_0^\infty \mu(\lambda f(s))\,ds$, one has
$$\left| \int_{B^d} f(x)dx - \mu B^d f(\theta) \right| \leq \int_{B^d} |f(x) - f(\theta)|dx = \varphi(\lambda f),$$
so inequality (1.15) holds for all $f \in \mathcal{A}$. It turns into equality for all functions
$$f \in \mathcal{B} := \left\{ f \in \mathcal{A} : f(x) \geq f(\theta) \text{ for all } x \in B^d \right\}.$$

Theorem 1.12 states that $\sup_{f \in \mathcal{A}} \varphi(\lambda f) = \frac{\mu B^d}{2}$ and there is an extremal sequence of functions from the set $\mathcal{B}$. Hence Lemma 1.4 implies the statement of Theorem 1.11.

1.2.9.4 Auxiliary Results

Lemma 1.21 *Let $d \in \mathbb{N}$, $d \geq 2$, $\varepsilon > 0$, $x \in \mathbb{R}^d$, $r \in \mathbb{P}^{d-1}$ and a measurable set $F \subset B^d(x, \varepsilon)$ be given. For arbitrary $A \in (0, 1)$ there exists $\alpha = \alpha(A) \in (0, 1)$ that does not depend on ε, x and r such that*
$$\mu^{d-1}\{\beta \in \Pi^{d-1}(r) : F \cap l(r, \beta) \neq \emptyset\} > A \cdot \mu^{d-1} B^{d-1}(\varepsilon)$$
whenever $\mu^d F > \alpha \cdot \mu^d B^d(\varepsilon)$.

Proof The fact that α does not depend on ε follows from the observation, that
$$\frac{\mu^d F}{\mu^d B^d(\varepsilon)} = \frac{\mu^d \left(\frac{1}{\varepsilon} F\right)}{\mu^d B^d}$$
and
$$\frac{\mu^{d-1}\{\beta \in \Pi^{d-1}(r) : F \cap l(r, \beta) \neq \emptyset\}}{\mu^{d-1} B^{d-1}(\varepsilon)}$$
$$= \frac{\mu^{d-1}\{\beta \in \Pi^{d-1}(r) : \frac{1}{\varepsilon} F \cap l(r, \beta) \neq \emptyset\}}{\mu^{d-1} B^{d-1}}. \quad (1.68)$$

The fact that α is independent of x and r is obvious. The existence of α follows from the equality
$$\mu^d F = \int_{\Pi^{d-1}(r) \cap B^d(y, \varepsilon)} \mu^1(l(r, \beta) \cap F) \mu^{d-1}(d\beta), \quad (1.69)$$
where $y \in \Pi^{d-1}(r)$ is such, that the line $l(r, y)$ contains x, and equality
$$\mu^{d-1}(\Pi^{d-1}(r) \cap B^d(y, \varepsilon)) = \mu^{d-1} B^{d-1}.$$

1.2 Ostrowski-Type Inequalities

Lemma 1.22 *Let $p \in [1, \infty)$, $A > 0$ and $B \in [0, A]$ be given. Then*

$$\frac{1}{A}\left(B + 2^p(A - B)\right) \geq \left(2 - \frac{B}{A}\right)^p.$$

Proof It is sufficient to prove that the function $\varphi(x) = 2^p + (1 - 2^p)x - (2 - x)^p$ is non-negative on $[0, 1]$. Since $\varphi(0) = \varphi(1) = 0$, the function φ' has at least one zero on $(0, 1)$. The function $\varphi'(x) = p(2 - x)^{p-1} + 1 - 2^p$ is decreasing on $[0, 1]$, hence has at most one zero on $(0, 1)$. This implies that $\varphi'(0) > 0$ and hence the function φ is increasing on $[0, x^*]$ and is decreasing on $[x^*, 1]$, where x^* is zero of φ' on $(0, 1)$; hence φ is non-negative on $[0, 1]$.

Denote by $\tilde{F}$ the set of all points $x \in F$ such that $\lim_{\delta \to +0} \frac{\mu^d(F \cap B^d(x,\delta))}{\mu^d B^d(\delta)} = 1$. Then $\tilde{F} \cap S^{d-1} = \emptyset$ and by Lebesgue density theorem

$$\mu^d \tilde{F} = \mu^d F. \tag{1.70}$$

Lemma 1.23 *Assume that conditions of Lemma 1.20 hold. If $r \in \mathbb{P}^{d-1}$ is such that $v(W, r) = 0$, then for arbitrary $\beta \in \Pi^{d-1}$ either $\tilde{F} \supset \operatorname{int} B^d \cap l(r, \beta)$, or $\tilde{F} \cap l(r, \beta) = \emptyset$.*

Proof Assume that for some $\beta \in \Pi^{d-1}(r)$ there exist $x \in \tilde{F} \cap l(r, \beta)$ and $y \in (\operatorname{int} B^d \cap l(r, \beta)) \setminus \tilde{F}$. From the definition of $\tilde{F}$ it follows that there exist $a > 0$ and a sequence $\rho_n \to 0$ as $n \to \infty$ such that $\mu^d(B^d(y, \rho_n) \setminus F) \geq a \cdot \mu^d B^d(\rho_n)$ for all $n \in \mathbb{N}$. From (1.69) (with F substituted by $B^d(y, \rho_n) \setminus F$) it follows that there exists $A > 0$ such that

$$\mu^{d-1} \Omega_1(\rho_n) > A \cdot \mu^{d-1} B^{d-1}(\rho_n) \tag{1.71}$$

for all $n \in \mathbb{N}$, where

$$\Omega_1(\rho_n) = \{\beta \in \Pi^{d-1}(r) : (B^d(y, \rho_n) \setminus F) \cap l(r, \beta) \neq \emptyset\}.$$

Since $x \in \tilde{F}$, there exists $\delta > 0$ such that for all $\rho < \delta$ $\mu^d(B^d(x, \rho) \cap F) \geq \alpha(1 - A) \cdot \mu^d B^d(\rho)$ (the number $\alpha(1 - A)$ is defined in Lemma 1.21). Lemma 1.21 implies that

$$\mu^{d-1} \Omega_2(\rho) > (1 - A) \cdot \mu^{d-1} B^{d-1}(\rho) \tag{1.72}$$

for all $\rho \leq \delta$, where

$$\Omega_2(\rho) = \{\beta \in \Pi^{d-1}(r) : B^d(x, \rho) \cap F \cap l(r, \beta) \neq \emptyset\}.$$

Choose n so big, that $\rho_n < \delta$. Then

$$\mu^{d-1} \Omega_1(\rho_n) + \mu^{d-1} \Omega_2(\rho_n) > \mu^{d-1} B^{d-1}(\rho_n) \tag{1.73}$$

due to (1.71) and (1.72). Moreover, since $x, y \in l(r, \beta)$, we receive that

$$\Omega_1(\rho_n), \Omega_2(\rho_n) \subset \Pi^{d-1}(r) \cap B^d(\beta, \rho_n) \tag{1.74}$$

and

$$\mu^{d-1}(\Pi^{d-1}(r) \cap B^d(\beta, \rho_n)) = \mu^{d-1}(B^{d-1}(\rho_n)). \tag{1.75}$$

Set $\Omega = \Omega_1(\rho_n) \cap \Omega_2(\rho_n)$. Then due to (1.73), (1.74) and (1.75) $\mu^{d-1}\Omega > 0$. But each line $l(r, \beta)$, $\beta \in \Omega$, contains a point from W due to Condition 3 of Lemma 1.20 and the definitions of the sets $\Omega_1(\rho_n)$ and $\Omega_2(\rho_n)$; this contradicts to assumption $v(W, r) = 0$ of the lemma.

Lemma 1.24 *Assume that conditions of Lemma 1.20 hold. Let $R \subset \mathbb{P}^{d-1}$ be such that $v(W, r) = 0$ for all $r \in R$. If R contains d lines that are not contained in any $d-1$-dimensional hyperplane, then $\mu^d(F) = 0$.*

Proof Due to (1.70) it is enough to prove that $\tilde{F} = \emptyset$. Let $r_1, \ldots, r_d$ be the lines from the statement of the lemma and let $\rho_1, \ldots, \rho_d$ be unit vectors parallel to these lines. Set $P := \left\{ \sum_{k=1}^d t_k \rho_k : t_k \in (-1, 1), k = 1, \ldots, d \right\}$, then P is an open in $\mathbb{R}^d$ set.

Consider arbitrary $x \in \text{int } B^d$. Choose $\varepsilon > 0$ such that $x + \varepsilon P \subset B^d$. Then for all points y from the segment θx, $P_y := y + \varepsilon P \subset B^d$. $\bigcup_{y \in \theta x} P_y$ is an open cover of a compact set θx, hence it contains a finite subcover $P_1, P_2, \ldots, P_m$, $m \in \mathbb{N}$. From Lemma 1.23 it follows that for each $s = 1, \ldots, m$ either $P_s \subset \tilde{F}$, or

$$P_s \cap \tilde{F} = \emptyset. \tag{1.76}$$

Since $\theta \notin \tilde{F}$ we obtain that (1.76) holds for each $s = 1, \ldots, m$ and hence $x \notin \tilde{F}$.

1.2.9.5 Proof of Lemma 1.20

Proof If $v(W, r) \geq 2$ for almost all $r \in \mathbb{P}^{d-1}$, then $v_p(W) \geq 2$ and inequality (1.65) holds. It is strict because Condition 1 of Lemma 1.20 holds. If there is a set $R \subset \mathbb{P}^{d-1}$ of positive measure such that $v(W, r) = 0$ for all $r \in R$, then $\mu^d F = 0$ due to Lemma 1.24 and inequality (1.65) holds.

Assume there exists $R \subset \mathbb{P}^{d-1}$, $\mu R > 0$, such that $v(W, r) = 1$ for all $r \in R$ and $v(W, r) \geq 2$ for almost all $r \in \mathbb{P}^{d-1} \setminus R$. Then

$$v_p(W) \geq 2 - \frac{\mu R}{\mu S^{d-1}}. \tag{1.77}$$

Really, if $p = \infty$, then $v_\infty(W) \geq 2$ in the case $\mu R < \mu \mathbb{P}^{d-1}$ and $v_\infty(W) = 1$ in the case $\mu R = \mu \mathbb{P}^{d-1} = \mu S^{d-1}$. In both cases (1.77) holds. In the case $p \in [1, \infty)$

1.3 Inequalities for Derivatives

$$v_p(W) \geq \left(\frac{1}{\mu S^{d-1}}\left(\mu R + 2^p \cdot (\mu S^{d-1} - \mu R)\right)\right)^{\frac{1}{p}} \geq 2 - \frac{\mu R}{\mu S^{d-1}}$$

due to Lemma 1.22.

Conditions 1 and 2 of the lemma imply that there exists $\varepsilon > 0$ such that $B^d(\varepsilon) \cap W = \emptyset$ and $B^d(\varepsilon) \cap F = \emptyset$. Set $\Lambda := \bigcup_{r \in R}(r \cap B^d)$. Below we prove that

$$\mu^d(\Lambda \cap F) < \frac{\mu^d \Lambda}{2}. \tag{1.78}$$

In order to prove (1.78) it is enough to show that

$$\mu^d(\Lambda \cap \tilde{F}) < \frac{\mu^d \Lambda}{2} \tag{1.79}$$

due to (1.70). Consider arbitrary $r \in R$. Then all points of the intersection $r \cap \tilde{F}$ are from one side of $r \cap B^d(\varepsilon)$. This fact can be proved using arguments similar to the proof of Lemma 1.23. Denote by χ the characteristic function of the set $\Lambda \cap \tilde{F}$. Then $\chi(x) = 0$ for all $|x| < \varepsilon$ and $\chi(x) + \chi(-x) \leq 1$ for all $x \in \Lambda$. This implies (1.79).

Finally, having (1.78), we can write

$$\mu^d F \leq \mu^d(F \cap \Lambda) + \mu^d(B^d \setminus \Lambda) < \mu^d B^d - \frac{1}{2}\mu^d \Lambda$$
$$= \mu^d B^d - \frac{1}{2} \cdot \frac{\mu^d B^d}{\mu S^{d-1}}\mu R = \frac{\mu^d B^d}{2}\left(2 - \frac{\mu R}{\mu S^{d-1}}\right).$$

The latter together with (1.77) proves (1.65).

The same example as in Theorem 1.12 shows that inequality (1.65) is sharp.

1.3 Inequalities for Derivatives

1.3.1 Hypersingular Integral Operator

Inequalities for derivatives is an important topic in approximation theory. It has a rich more than century-long history and is still actively researched. We refer to [27] for many results related to the Kolmogorov-type inequalities for univariate functions and the derivatives of integer orders. Kolmogorov-type inequalities for multivariate functions for derivatives of both integer and fractional orders were also heavily studied. A survey on Kolmogorov-type inequalities for fractional derivatives for functions of one and many variables is contained in [64, Chapter 2]. In this section we focus on Kolmogorov-type and Nagy-type inequalities that use Ostrowski-type inequalities in their proofs.

In this section we follow the notations from Sects. 1.2.4, 1.2.5 and 1.2.6.

For a set $K \in \mathbb{K}$, cone $C \in \mathcal{C}$ and a non-negative characteristic Ω we consider the following integral operator

$$D_K^{w,\Omega} f(x) := \int_C w(|t|_K)(f(x+t) - f(x))\Omega(t)dt, \, x \in C, \quad (1.80)$$

where $w \colon (0, \infty) \to \mathbb{R}$ is some non-negative weight; precise definitions for the classes of function where $D_K^{w,\Omega}$ is defined and conditions on the functions w and Ω will be given later. Note that in the case when K is the Euclidean unit ball in $\mathbb{R}^d$, $\Omega \equiv 1$, $C = \mathbb{R}^d$ and $w(t) = t^{-(d+\alpha)}$, $0 < \alpha < 1$, we obtain the Riesz derivative D^α of order α, see [69, §26.7].

In [23] the univariate case $d = 1$ with $C = (0, \infty)$, $w(t) = t^{-(1+\alpha)}$, $\alpha \in (0, 1)$, $\Omega \equiv 1$, $K = (-1, 1)$ was considered. In this case operator $D_K^{w,\Omega}$ becomes the Marchaud (see e.g. [69, §5.4]) fractional derivative, and a sharp Kolmogorov-type inequality that estimates $\|D_K^{w,\Omega} f\|_{L_\infty(\mathbb{R})}$ via $\|f\|_{L_\infty(\mathbb{R})}$ and $\|f\|_{H^\omega(\mathbb{R})}$ was obtained. In the same situation, a sharp inequality that estimates $\|D_K^{w,\Omega} f\|_{L_\infty(\mathbb{R}_+)}$ via $\|f\|_{L_\infty(\mathbb{R}_+)}$ and $\|f'\|_{L_s(\mathbb{R}_+)}$, $1 < s < \infty$, $0 < \alpha < 1 - \frac{1}{s}$, is contained in [64, §2.4].

In the case when $C = \mathbb{R}^d$, $K = B_1$, and $w(t) = t^{-(d+\alpha)}$, a sharp inequality that estimates $\|D_K^{w,\Omega} f\|_{L_\infty(\mathbb{R}^d)}$ via $\|f\|_{L_\infty(\mathbb{R}^d)}$ and $\|f\|_{H^\omega(\mathbb{R}^d)}$, where the modulus of continuity ω is such that $\int_0^1 \frac{\omega(t)}{t^{1+\alpha}} dt$ converges, was obtained in [24].

In the case when $C = \mathbb{R}^d$, $K = B_1$, and $w(t) = t^{-(d+\alpha)}$, a sharp inequality that estimates $\|D_K^{w,\Omega} f\|_{L_\infty(\mathbb{R}^d)}$ via $\|f\|_{L_\infty(\mathbb{R}^d)}$ and $\||\Omega^{\frac{1}{s}} \cdot \nabla f|_2\|_{L_s(\mathbb{R}^d)}$, $s > d$, $0 < \alpha < 1 - \frac{d}{s}$ is contained in [67]. The corresponding result for $\Omega \equiv 1$ was obtained earlier in [31].

Finally, the case when $\Omega \equiv 1$ and the cone C is generated by a finite set of points was considered in [28]. Below we follow [29], where the results from [28] were further refined in several directions: first of all, Ω was allowed to be non-unit, and the cone C—to be an arbitrary convex cone; moreover, the additional conditions imposed in [28] on the weight w for the case, when K is not a polytope (by a *polytope* we mean the convex hull of a finite number of points) was removed. The results of [24] were also generalized by allowing C to be an arbitrary convex cone, K to be a unit ball of an arbitrary norm in $\mathbb{R}^d$, and weight w to belong to a rather large set of functions.

Below on the classes $H_K^\omega(C) \cap L_\infty(C)$ and $L_{\infty,p}^\Omega(C)$ (see Sects. 1.2.5 and 1.2.6) together with hypersingular operator (1.80) we also consider the operator

$$D_{K,h}^{w,\Omega} f(x) := \int_{C \setminus hK} w(|t|_K)(f(x+t) - f(x))\Omega(t)dt, \, h > 0. \quad (1.81)$$

Lemma 1.25 *For any function* $w \in \mathcal{L}_1(h, \infty)$ *the operator* $D_{K,h}^{w,\Omega} \colon L_\infty(C) \to L_\infty(C)$ *is bounded and*

$$\|D_{K,h}^{w,\Omega}\| = 2 \cdot \Omega(K, C) \cdot \|w\|_{\mathcal{L}_1(h,\infty)}.$$

The supremum in the definition of the norm of operator $D_{K,h}^{w,\Omega}$ *is attained on the function*

1.3 Inequalities for Derivatives

$$f(x) = \begin{cases} \omega(|x|_K) - \frac{1}{2}\omega(h), & |x|_K \leq h, \\ \frac{1}{2}\omega(h), & |x|_K > h, \end{cases} \quad (1.82)$$

which is continuous and belongs to $H_K^\omega(C) \cap L_\infty(C)$, and on

$$f(x) = \begin{cases} \int_0^{|x|_K} g_w^{p'-1}(u)du - \frac{1}{2}\int_0^h g_w^{p'-1}(u)du, & |x|_K \leq h, \\ \frac{1}{2}\int_0^h g_w^{p'-1}(u)du, & |x|_K > h, \end{cases} \quad (1.83)$$

provided additionally $w \in \mathcal{W}_{p'}(0,h)$, $p \in (d, \infty]$. Under this additional assumption on w, function (1.83) is continuous and belongs to $L_{\infty,p}^\Omega(C)$.

Proof Using Lemma 1.11 for each $f \neq 0$ and $x \in C$, we obtain

$$\frac{1}{2\|f\|_{L_\infty(C)}} \cdot \left| \int_{C \setminus hK} w(|t|_K)(f(x+t) - f(x))\Omega(t)dt \right| \leq \int_{C \setminus hK} w(|t|_K)\Omega(t)dt$$
$$= \|w\|_{\mathcal{L}_1(h,\infty)} \cdot \Omega(K,C).$$

This implies

$$\|D_{K,h}^{w,\Omega}\| \leq 2 \cdot \|w\|_{\mathcal{L}_1(h,\infty)} \cdot \Omega(K,C).$$

The fact that function (1.82) belongs to $H_K^\omega(C)$ can be proved similarly to the proof that function (1.37) belongs to $H_K^\omega(hK \cap C)$. For arbitrary $K \in \mathbb{K}$ and the function f defined in (1.82), the function $D_{K,h}^{w,\Omega} f$ is continuous at θ, since for arbitrary $x \in C$, due to Lemma 1.11

$$\left| D_{K,h}^{w,\Omega} f(x) - D_{K,h}^{w,\Omega} f(\theta) \right|$$
$$= \left| \int_{C \setminus hK} w(|t|_K)(f(x+t) - f(x) - f(t) + f(\theta))\Omega(t)dt \right|$$
$$\leq \int_{C \setminus hK} w(|t|_K)(|f(x+t) - f(t)| + |f(x) - f(\theta)|)\Omega(t)dt$$
$$\leq 2\omega(|x|_K) \left| \int_{C \setminus hK} w(|t|_K)\Omega(t)dt \right| \to 0 \text{ as } x \to \theta.$$

For the function f defined in (1.82), we have $\|f\|_{L_\infty(C)} = \frac{1}{2}\omega(h)$ and

$$\|D_{K,h}^{w,\Omega}\| \geq \frac{2}{\omega(h)} D_{K,h}^{w,\Omega} f(\theta) = \frac{2}{\omega(h)} \int_{C \setminus hK} w(|t|_K)(f(t) - f(\theta))\Omega(t)dt$$
$$= 2 \int_{C \setminus hK} w(|t|_K)\Omega(t)dt = 2\|w\|_{\mathcal{L}_1(h,\infty)} \cdot \Omega(K,C),$$

which proves the statement of the lemma for function (1.82).

Observe that function (1.83) is continuous on C and constant outside a compact set. Thus it is uniformly continuous, and hence the same arguments as for function (1.82) imply continuity of $D_{K,h}^{w,\Omega} f$ at θ for function (1.83). Hence

$$\|D_{K,h}^{w,\Omega} f\|_{L_\infty(C)} \geq D_{K,h}^{w,\Omega} f(\theta) = \int_{C \setminus hK} w(|t|_K)(f(t) - f(\theta))\Omega(t)dt$$

$$= 2\|f\|_{L_\infty(C)} \int_{C \setminus hK} w(|t|_K)\Omega(t)dt = 2\|f\|_{L_\infty(C)} \|w\|_{\mathcal{L}_1(h,\infty)} \cdot \Omega(K,C),$$

which implies the desired.

1.3.2 Kolmogorov-Type Inequalities for Classes $H_K^\omega(C)$

The following theorem is contained in [29]; in the partial case when $C = \mathbb{R}^d$, $w(t) = t^{-(d+\alpha)}$, $\alpha \in (0, 1)$, and K is the unit Euclidean ball it is contained in [24].

Theorem 1.13 *Let a set $K \in \mathbb{K}$, a cone $C \in \mathcal{C}$, and a homogeneous characteristic Ω be given. Then for a modulus of continuity ω and a weight w such that for some $h > 0$, $w \cdot \omega \in \mathcal{L}_1(0, h)$ and $w \in \mathcal{L}_1(h, \infty)$ the inequality*

$$\|D_K^{w,\Omega} f\|_{L_\infty(C)} \leq \Omega(K,C) \cdot \Big(2\|f\|_{L_\infty(C)}\|w\|_{\mathcal{L}_1(h,\infty)} + \|f\|_{H_K^\omega(C)}\|w \cdot \omega\|_{\mathcal{L}_1(0,h)}\Big) \quad (1.84)$$

holds for each function $f \in H_K^\omega(C) \cap L_\infty(C)$ and is sharp. The inequality becomes equality on function (1.82).

Proof For arbitrary $x \in C$ using Theorem 1.6 we obtain the following estimate

$$\left|(D_K^{w,\Omega} f - D_{K,h}^{w,\Omega})(x)\right| \leq \int_{hK \cap C} w(|t|_K)|f(x+t) - f(x)|\Omega(t)dt$$

$$\leq \|f\|_{H_K^\omega(x+hK \cap C)} \cdot \Omega(K,C) \cdot \|w\|_{\mathcal{L}_1(0,h)} \leq \|f\|_{H_K^\omega(C)} \cdot \Omega(K,C) \cdot \|w\|_{\mathcal{L}_1(0,h)}.$$

Using the triangle inequality

$$\|D_K^{w,\Omega} f\|_{L_\infty(C)} \leq \|(D_K^{w,\Omega} - D_{K,h}^{w,\Omega})f\|_{L_\infty(C)} + \|D_{K,h}^{w,\Omega} f\|_{L_\infty(C)},$$

and applying the obtained estimate together with Lemma 1.25, we obtain inequality (1.84).

For all $x \in C$ and $t \in hK \cap C$, on the one hand

$$|f(x+t) - f(x) - f(t) + f(\theta)| \leq |f(x+t) - f(x)| + |f(t) - f(\theta)| \leq 2\omega(|t|_K),$$

1.3 Inequalities for Derivatives

and on the other hand

$$|f(x+t) - f(x) - f(t) + f(\theta)| \leq |f(x+t) - f(t)| + |f(x) - f(\theta)| \leq 2\omega(|x|_K).$$

For a fixed $x \in C$ set $\omega_x(\delta) := 2\min\{\omega(\delta), \omega(|x|_K)\}$, $\delta \geq 0$. Using Lemma 1.11 we obtain as $x \to \theta$

$$\left|(D_K^\Omega - D_{K,h}^{w,\Omega})f(x) - (D_K^\Omega - D_{K,h}^{w,\Omega})f(\theta)\right|$$
$$\leq \int_{hK \cap C} w(|t|_K) |f(x+t) - f(x) - f(t) + f(\theta)| \Omega(t) dt$$
$$\leq \int_{hK \cap C} w(|t|_K) \omega_x(|t|_K) \Omega(t) dt = \Omega(K, C) \cdot \|w \cdot \omega_x\|_{\mathcal{L}_1(0,h)} \to 0.$$

Thus $(D_K^\Omega - D_{K,h}^{w,\Omega})f$ is continuous at θ. In the proof of Lemma 1.25 it was shown that $D_{K,h}^{w,\Omega} f$ is continuous at θ. Thus $D_K^\Omega f$ is continuous at θ.

The function f defined in (1.82) is extremal in Lemma 1.25 and its restriction to $hK \cap C$ is extremal in Theorem 1.6, hence

$$\|D_K^{w,\Omega} f\|_{L_\infty(C)} \geq |D_K^{w,\Omega} f(\theta)| = D_K^{w,\Omega} f(\theta) = (D_K^{w,\Omega} - D_{K,h}^{w,\Omega})f(\theta) + D_{K,h}^{w,\Omega} f(\theta)$$
$$= \Omega(K, C) \cdot \left(2\|f\|_{L_\infty(C)} \|w\|_{\mathcal{L}_1(h,\infty)} + \|f\|_{H_K^\omega(C)} \|w \cdot \omega\|_{\mathcal{L}_1(0,h)}\right),$$

which finishes the proof of the theorem.

1.3.3 Kolmogorov-Type Inequalities for Classes $L_{\infty,p}^\Omega(C)$

This theorem is contained in [29]. Partial cases of the following theorem are contained in [22, 31, 67].

Theorem 1.14 *Let $p \in (d, \infty]$, $C \in \mathcal{C}$, $K \in \mathbb{K}$, $\Omega \colon C \to \mathbb{R}$ be a non-negative characteristic, and $f \in L_{\infty,p}^\Omega(C)$. For each $h > 0$ and $w \in \mathcal{W}_{p'}(0, h) \cap \mathcal{L}_1(h, \infty)$ one has*

$$\|D_K^{w,\Omega} f\|_{L_\infty(C)} \leq 2 \cdot \Omega(K, C) \cdot \|w\|_{\mathcal{L}_1(h,\infty)} \|f\|_{L_\infty(C)}$$
$$+ (\Omega(K, C))^{\frac{1}{p'}} \|gw\|_{\mathcal{L}_{p'}(0,h)} \|\Omega^{\frac{1}{p}} \cdot |\nabla f|_{K^\circ}\|_{L_p(C)}. \quad (1.85)$$

The inequality is sharp. It becomes equality on the functions $a \cdot f$, where $a \in \mathbb{R}$ and f is defined in (1.83).

Proof The proof of inequality (1.85) and the fact that for function (1.83), $D_K^{w,\Omega} f(\theta)$ is equal to the right-hand side of (1.85), can be done using the same arguments as in the proof of Theorem 1.13, because function (1.83) is extremal in Lemma 1.25 and its restriction to $hK \cap C$ is extremal in Theorem 1.7. In the proof of Lemma 1.25 it was shown that for function (1.83), $D_{K,h}^{w,\Omega} f$ is continuous at θ, thus to finish the proof of the theorem, it is sufficient to prove that $(D_K^{w,\Omega} - D_{K,h}^{w,\Omega})f$ is also continuous at θ. For a fixed $x \in \mathbb{R}^d$, applying Theorem 1.6 we obtain

$$|(D_K^{w,\Omega} - D_{K,h}^{w,\Omega})f(x) - (D_K^{w,\Omega} - D_{K,h}^{w,\Omega})f(\theta)|$$
$$= \left| \int_{hK \cap C} w(|t|_K)([f(x+t) - f(t)] - [f(x) - f(\theta)])\Omega(t) dt \right|$$
$$\leq (\Omega(K,C))^{\frac{1}{p'}} \|g_w\|_{L_{p'}(0,h)} \|\Omega^{\frac{1}{p}}(\cdot) \cdot |\nabla f(x+\cdot) - \nabla f(\cdot)|_{K^\circ}\|_{L_p(hK \cap C)}. \quad (1.86)$$

Due to Lemma 1.11, $\Omega \in L_1(hK \cap C)$; due to (1.41), Lemmas 1.11 and 1.14, $|\nabla f|_{K^\circ} \in L_p(hK \cap C)$ and $\Omega^{\frac{1}{p}} |\nabla f|_{K^\circ} \in L_p(hK \cap C)$. Moreover,

$$\int_{hK \cap C} \int_{hK \cap C} \Omega(y) |\nabla f(x+y)|_{K^\circ}^p dy dx$$
$$= \int_{hK \cap C} \Omega(y) \int_{hK \cap C} |\nabla f(x+y)|_{K^\circ}^p dx dy$$
$$= \int_{hK \cap C} \Omega(y) \| |\nabla f|_{K^\circ} \|_{L_p(y+hK \cap C)}^p dy \leq \|\Omega\|_{L_1(hK \cap C)} \| |\nabla f|_{K^\circ} \|_{L_p(hK \cap C)}^p,$$

where the last inequality is true, since ∇f is zero outside $hK \cap C$. This implies that $\Omega^{\frac{1}{p}}(\cdot) |\nabla f(\cdot + x)|_{K^\circ} \in L_p(hK \cap C)$ for almost all $x \in hK \cap C$, and hence the right-hand side of (1.86) is finite for almost all x.

For arbitrary $\varepsilon > 0$ and any compact set $T \subset (hK \cap C) \setminus B_\varepsilon$ such that $|\cdot|_K$ is differentiable on T, the function ∇f is uniformly continuous on T due to Lemma 1.14, continuity of the function g_w on sets separated from 0 and continuity of the gradient $\nabla |\cdot|_K$ (see [68, Theorem 25.5]). Thus the quantity $|\nabla f(x+\cdot) - \nabla f(\cdot)|_{K^\circ}$ can be made arbitrarily small on the set $T_\varepsilon := \{y \in T : y + B_\varepsilon \subset T\}$ by choosing x close enough to θ. Moreover, by absolute continuity of the integral the quantity $\|\Omega^{\frac{1}{p}}(\cdot) \cdot |\nabla f(x+\cdot) - \nabla f(\cdot)|_{K^\circ}\|_{L_p(hK \cap C \setminus T_\varepsilon)}$ can be made arbitrary small by appropriate choices of T and ε. This implies that the right-hand side of (1.86) tends to 0 as $x \to \theta$, which implies continuity of $(D_K^{w,\Omega} - D_{K,h}^{w,\Omega})f$ at θ.

1.3.4 Kolmogorov-Type Inequalities for Classes $L_{\infty,p}^{\Delta}(\mathbb{R}^d)$

For $0 < \alpha < 2$, we consider the following integral operator

$$(D^\alpha f)(x) := \int_{\mathbb{R}^d} \frac{2f(x) - f(x-t) - f(x+t)}{|t|_2^{d+\alpha}} dt. \quad (1.87)$$

1.3 Inequalities for Derivatives

Up to a normalization factor, D^α is the Riesz derivative of the order α, see e.g., [69, Sect. 25]. Note that the Riesz derivative can be considered as a functional power $(-\Delta)^{\alpha/2}$ of the Laplace operator.

1.3.4.1 Bounded Operator D_h^α

Together with operator (1.87) we consider the operator $D_h^\alpha f(x) \colon L_\infty(\mathbb{R}^d) \to L_\infty(\mathbb{R}^d)$,

$$D_h^\alpha f(x) = \int_{\mathbb{R}^d \setminus B_h} \frac{2f(x) - f(x-t) - f(x+t)}{|t|_2^{d+\alpha}} dt + 2c_h \sigma_{d-1}(f(x) - \widetilde{f}(x,h)), \tag{1.88}$$

where c_h is chosen from condition (1.45), and $\widetilde{f}(x, h)$ is defined in (1.48).

The next lemma shows that this operator is bounded and computes its norm.

Lemma 1.26 *Let $h > 0$, $p > d/2$ and $0 < \alpha < 2 - d/p$, or $p = \infty$ and $0 < \alpha < 2$. Then*

$$\|D_h^\alpha\| = 4\sigma_{d-1} h^{-\alpha} \left(\frac{1}{\alpha} + c_1 \right).$$

The supremum in the definition of the norm of the operator D_h^α is attained on the function $\varphi_{h,2}$.

Proof For any function $f \in L_\infty(\mathbb{R}^d)$ and $h > 0$

$$\|D_h^\alpha f\|_{L_\infty(\mathbb{R}^d)} = \left\| \int_{\mathbb{R}^d \setminus B_h} \frac{2f(\cdot) - f(\cdot + t) - f(\cdot - t)}{|t|^{d+\alpha}} dt \right.$$
$$\left. + 2c_h \sigma_{d-1}\left(f(\cdot) - \widetilde{f}(\cdot, h)\right) \right\|_{L_\infty(\mathbb{R}^d)} \le$$
$$\le 4\|f\|_{L_\infty(\mathbb{R}^d)} \left\{ \int_{\mathbb{R}^d \setminus B_h} \frac{dt}{|t|^{d+\alpha}} + c_h \sigma_{d-1} \right\} = 4\|f\|_{L_\infty(\mathbb{R}^d)} \sigma_{d-1} \left(\frac{h^{-\alpha}}{\alpha} + c_h \right)$$
$$= 4\|f\|_{L_\infty(\mathbb{R}^d)} \sigma_{d-1} h^{-\alpha} \left(\frac{1}{\alpha} + c_1 \right).$$

This implies

$$\|D_h^\alpha\| \le 4\sigma_{d-1} h^{-\alpha} \left(\frac{1}{\alpha} + c_1 \right). \tag{1.89}$$

Recall that the function $\varphi_{h,2}$ defined in (1.46) is continuous and hence $D_h^\alpha \varphi_{h,2}$ is also continuous. Thus for each $h > 0$ one has

$$\|D_h^\alpha \varphi_{h,2}\|_{L_\infty(\mathbb{R}^d)} \ge |D_h^\alpha \varphi_{h,2}(0)|$$
$$= \left| \int_{\mathbb{R}^d \setminus B_h} \frac{2\varphi_{h,2}(0) - 2\varphi_{h,2}(y)}{|y|^{d+\alpha}} dy + 2c_h \sigma_{d-1}(\varphi_{h,2}(0) - \widetilde{\varphi}_{h,2}(0)) \right| =$$

$$= 4\|\varphi_{h,2}\|_{L_\infty(\mathbb{R}^d)} \left(\int_{\mathbb{R}^d \setminus B_h} \frac{dy}{|y|^{d+\alpha}} + c_h \sigma_{d-1} \right)$$

$$= 4\|\varphi_{h,2}\|_{L_\infty(\mathbb{R}^d)} \sigma_{d-1} h^{-\alpha} \left(\frac{1}{\alpha} + c_1 \right),$$

hence,

$$\|D_h^\alpha\| \geq 4\sigma_{d-1} h^{-\alpha} \left(\frac{1}{\alpha} + c_1 \right). \tag{1.90}$$

Now the statement of the lemma follows from inequalities (1.89) and (1.90).

1.3.4.2 Kolmogorov-Type Inequalities

Theorem 1.15 *Let $d/2 < p < \infty$, $0 < \alpha < 2 - d/p$ or $p = \infty$ and $0 < \alpha < 2$. Then for an arbitrary function $f \in L^\Delta_{\infty,p}(\mathbb{R}^d)$ and each $h > 0$ one has*

$$\|D^\alpha f\|_{L_\infty(\mathbb{R}^d)} \leq 2\sigma_{d-1} h^{2-\alpha-\frac{d}{p}} \|\Phi_1(|\cdot|)\|_{L_{p'}(B_1)} \cdot \|\Delta f\|_{L_p(\mathbb{R}^d)}$$

$$+ 4\sigma_{d-1} h^{-\alpha} \left(\frac{1}{\alpha} + c_1 \right) \|f\|_{L_\infty(\mathbb{R}^d)}. \tag{1.91}$$

Inequality (1.91) becomes equality for the function $\varphi_{h,2}$.

Proof Using Lemma 1.26 and Theorem 1.8 and the triangle inequality

$$\|D^\alpha f\|_{L_\infty(\mathbb{R}^d)} \leq \|D^\alpha f - D_h^\alpha f\|_{L_\infty(\mathbb{R}^d)} + \|D_h^\alpha f\|_{L_\infty(\mathbb{R}^d)},$$

we obtain inequality (1.91) for all $h > 0$ and $f \in L^\Delta_{\infty,p}(\mathbb{R}^d)$. It becomes equality for the function $\varphi_{h,2}$, since it is extremal both in Lemma 1.26 and Theorem 1.8, and is constant outside B_h, which implies the equality

$$\|\Delta \varphi_{h,2}\|_{L_p(\mathbb{R}^d)} = \|\Delta \varphi_{h,2}\|_{L_p(B_h)}.$$

Using standard arguments (e.g., by minimizing the right-hand side of (1.91) with respect to h) and homogeneity of the functions involved in the obtained Kolmogorov-type inequality, one can obtain a Kolmogorov-type inequality in the multiplicative form. The constant may be written in terms of α, p, d, c_1 and $\|\Phi_1(|\cdot|)\|_{L_{p'}(B_1)}$, or in terms of the function $\varphi_{1,2}$. We formulate the theorem using the latter approach.

Theorem 1.16 *Let $d/2 < p < \infty$ and $0 < \alpha < 2 - d/p$ or $p = \infty$ and $0 < \alpha < 2$. Then for all functions $f \in L^\Delta_{\infty,p}(\mathbb{R}^d)$ one has*

$$\|D^\alpha f\|_{L_\infty(\mathbb{R}^d)} \leq \frac{\|D^\alpha \varphi_{1,2}\|_{L_\infty(\mathbb{R}^d)}}{\|\varphi_{1,2}\|_{L_\infty(\mathbb{R}^d)}^{\frac{2-\alpha-d/p}{2-d/p}}} \|f\|_{L_\infty(\mathbb{R}^d)}^{\frac{2-\alpha-d/p}{2-d/p}} \|\Delta f\|_{L_p(\mathbb{R}^d)}^{\frac{\alpha}{2-d/p}}.$$

1.3 Inequalities for Derivatives

The inequality is sharp and becomes equality for the functions $f(t) = a\varphi_{h,2}(t)$, $h > 0$, $a \in \mathbb{R}$.

1.3.5 Nagy-Type Inequalities

In 1941 for all admissible parameters p, q, s, Nagy [74] proved sharp inequalities of the following type

$$\|x\|_{L_q(\mathbb{R})} \leq K \cdot \|x\|_{L_p(\mathbb{R})}^{\alpha} \cdot \|x'\|_{L_s(\mathbb{R})}^{1-\alpha}.$$

In the periodic case inequalities of Nagy type are contained in [25, 26, 53]. The results of Sects. 1.3.5, 1.3.6, and 1.3.7 are contained in [17–19].

1.3.5.1 Notations

A tripe (X, ρ, μ), where (X, ρ) is a metric space with a Borel measure μ will be called a metric space with measure. We assume that X is a commutative monoid (i.e., an associative and commutative binary operation $+$ is defined on X, and there exists an element $\theta \in X$ such that $x + \theta = \theta + x = x$ for all $x \in X$) such that for each measurable set $Q \subset X$ and each $x \in X$ one has

$$\mu(x + Q) = \mu(Q).$$

Suppose that for all $x, y \in X$,

$$\rho(x + y, x) \leq \rho(y, \theta).$$

Everywhere below $B_h = B_h(\theta)$ is an open ball of radius $h > 0$ with center θ; we suppose that $0 < \mu(B_h) < \infty$ and $B_h \neq \{\theta\}$ for all $h > 0$.

For $p \in [1, \infty)$ by $L_p(X)$ we denote the space of measurable functions $f \colon Q \to \mathbb{R}$ such that the function $|f|^p$ is integrable on X with the corresponding norm; $L_\infty(X)$ denotes the space of measurable essentially bounded on X functions. By $L_{\mathrm{loc}}(X)$ we denote the space of all functions $f \colon X \to \mathbb{R}$ that are integrable on each open ball of X. In the space $L_{\mathrm{loc}}(X)$ we introduce a family of seminorms

$$\lfloor f \rceil_h = \sup_{x \in X} \left| \int_{x + B_h} f(u) \mu(du) \right|, \quad h > 0,$$

and a seminorm

$$\lfloor f \rceil = \sup_{h > 0} \lfloor f \rceil_h.$$

By $L_{\lfloor \cdot \rceil_h}(X)$ ($L_{\lfloor \cdot \rceil}(X)$) we denote the family of functions $f \in L_{\mathrm{loc}}(X)$ with a finite seminorm $\lfloor \cdot \rceil_h$ (resp. $\lfloor \cdot \rceil$). It is clear that the space $L_1(X)$ is contained in each of these sets.

By $C_b(X)$ we denote the space of continuous functions $f \colon X \to \mathbb{R}$ that have a finite norm

$$\|f\|_{C_b(X)} = \sup_{x \in X} |f(x)|;$$

by $\mathcal{B}(X)$ we denote the space of bounded functions $f : X \to \mathbb{R}$ with a norm

$$\|f\|_{\mathcal{B}(X)} = \sup_{x \in X} |f(x)|. \tag{1.92}$$

In this section we assume that the measure μ is such that each continuous function belongs to $L_{\text{loc}}(X)$.

Let ω be a modulus of continuity. Recall that by $H^\omega(X)$ we denote the space of functions $f : X \to \mathbb{R}$ such that

$$\|f\|_{H^\omega(X)} := \sup_{x,y \in X, x \neq y} \frac{|f(x) - f(y)|}{\omega(\rho(x,y))} < \infty.$$

We also consider the case, when X is some open convex cone C in the space $\mathbb{R}^d$, $d \geq 1$, the metric ρ is determined by a set K i.e., $\rho(x,y) = |x - y|_K$, and $\mu = m$ is the Lebesgue measure in the space $\mathbb{R}^d$ (see Sect. 1.2.4 for the definitions of the norms $|\cdot|_K$ and $|\cdot|_{K^\circ}$). In this case we write (C, K, m) instead of (X, ρ, μ), $B_h = hK \cap C$, and the introduced seminorms $\lfloor \cdot \rceil_h$ and $\lfloor \cdot \rceil$ become

$$\lfloor f \rceil_h = \sup_{x \in C} \left| \int_{hK \cap C} f(x+u) du \right|, h > 0, \text{ and } \lfloor f \rceil = \sup_{h > 0} \lfloor f \rceil_h.$$

1.3.5.2 Averaging Operator

For each $h > 0$ we define an operator $S_h : L_{\lfloor \cdot \rceil_h}(X) \to \mathcal{B}(X)$ by the following rule:

$$S_h f(x) = \frac{1}{\mu B_h} \int_{B_h} f(x+u) \mu(du). \tag{1.93}$$

It is clear that this operator is bounded and

$$\|S_h\|_{L_{\lfloor \cdot \rceil_h}(X) \to \mathcal{B}(X)} = \frac{1}{\mu(B_h)}. \tag{1.94}$$

In the case, when $(X, \rho, \mu) = (C, K, m)$, due to absolute continuity of the Lebesgue integral, the image $S_h f$ of each locally integrable function f, is continuous (and hence measurable) bounded function. Thus in this case we can count that the operator S_h acts into the space $L_\infty(C)$.

For $h > 0$ define a function $g_h : (0, h) \to \mathbb{R}$,

$$g_h(u) = \frac{1}{d \cdot \mu(K \cap C)} \left(\frac{1}{u^{d-1}} - \frac{u}{h^d} \right).$$

Using Lemma 1.11 we obtain that for all $p \in (d, \infty]$

1.3 Inequalities for Derivatives

$$\|g_h(|\cdot|_K)\|_{L_{p'}(hK\cap C)} = (d\cdot\mu(K\cap C))^{\frac{1}{p'}}\|g_h\|_{\mathcal{L}_{p'}(0,h)},$$

where, as in (1.27) for $1 \leq p < \infty$,

$$\|w\|_{\mathcal{L}_p(0,h)} = \left(\int_0^h t^{d-1}w^p(t)dt\right)^{1/p}.$$

Thus the next lemma is a partial case of Theorem 1.7.

Lemma 1.27 *Let $p \in (d, \infty]$, $h > 0$ and $f \in W^{1,p}(hK \cap C)$. Then*

$$|f(\theta) - S_h f(\theta)| \leq \|g_h(|\cdot|_K)\|_{L_{p'}(hK\cap C)} \||\nabla f|_{K^\circ}\|_{L_p(hK\cap C)}.$$

The inequality is sharp. It becomes equality for the functions $\alpha \cdot f + \beta$, where $\alpha, \beta \in \mathbb{R}$ and

$$f(y) = \int_0^{|y|_K} g_h^{p'-1}(u)du, \ y \in hK \cap C.$$

Note that the quantity $\|g_h(|\cdot|_K)\|_{L_{p'}(hK\cap C)}$ can be expressed via the Euler beta function B.

Lemma 1.28 *For $p \in (d, \infty]$ one has*

$$\|g_h(|\cdot|_K)\|_{L_{p'}(hK\cap C)} = (d\cdot\mu(K\cap C))^{1/p'}\|g_h\|_{\mathcal{L}_{p'}(0,h)} = A\mu^{-\frac{1}{p}}(K\cap C)h^{1-\frac{d}{p}},$$

where

$$A = A(d, p) = d^{-1}B^{\frac{1}{p'}}\left(1 - \frac{(d-1)p'}{d}, p'+1\right). \quad (1.95)$$

In particular, for $p = \infty$,

$$\|g_h(|\cdot|_K)\|_{L_1(hK\cap C)} = \frac{d\cdot h}{d+1}.$$

Proof The first equality follows from 1.11. The second one can be obtained via direct calculations. The last equality follows from (1.95) and properties of the function B.

1.3.5.3 Nagy-Type Inequalities in the Spaces $W^{1,p}(C)$

Theorem 1.17 *If $h > 0$, $p \in (d, \infty]$ and $f \in W^{1,p}(C) \cap L_{\rfloor\cdot\lceil_h}(C)$, then $f \in L_\infty(C)$ and the following inequality holds:*

$$\|f\|_{L_\infty(C)} \leq \|f - S_h f\|_{L_\infty(C)} + \|S_h\|_{L_{\rfloor\cdot\lceil_h}(C)\to L_\infty(C)}\cdot\rfloor f\lceil_h$$

$$\leq \|g_h(|\cdot|_K)\|_{L_{p'}(hK\cap C)} \||\nabla f|_{K^\circ}\|_{L_p(C)} + \mu^{-1}(K\cap C)h^{-d}\rfloor f\lceil_h. \quad (1.96)$$

Inequality (1.96) is sharp. It becomes equality on the functions

$$\alpha \cdot (f_{e,h} + \beta), \text{ where } \alpha > 0, \beta \geq -\frac{1}{2} f_{e,h}(0), \tag{1.97}$$

$$f_{e,h}(y) = \begin{cases} \int_{|y|_K}^h g_h^{p'-1}(u)du, & y \in hK \cap C, \\ 0, & y \in C \setminus hK. \end{cases} \tag{1.98}$$

For each function $f \in W^{1,p}(C) \cap L_{\rfloor,\lceil}(C)$ the following multiplicative inequality holds

$$\|f\|_{L_\infty(C)} \leq a(d,p)\mu^{-\frac{\alpha}{d}}(K \cap C)\lfloor f \rceil^{1-\alpha} \|\nabla f|_{K^\circ}\|_{L_p(C)}^\alpha, \tag{1.99}$$

where

$$\alpha = \frac{pd}{p+(p-1)d}, \quad a(d,p) = \left(\frac{(p-d)A(d,p)}{pd}\right)^\alpha \left(\frac{pd}{p-d}+1\right), \tag{1.100}$$

and $A(d,p)$ is defined in (1.95). Inequality (1.99) is sharp, it becomes equality on each of the functions $f_{e,h}$, $h > 0$.

Taking into account Lemma 1.28, inequality (1.96) can be rewritten in the form

$$\|f\|_{L_\infty(C)} \leq A(d,p)\mu^{-\frac{1}{p}}(K \cap C)h^{1-\frac{d}{p}} \|\nabla f|_{K^\circ}\|_{L_p(C)} + \mu^{-1}(K \cap C)h^{-d}\lfloor f \rceil_h,$$

where $A(d,p)$ is defined in (1.95). From the statement about sharpness of inequality (1.99) it follows that it can also be rewritten as follows:

$$\|f\|_{L_\infty(C)} \leq \frac{\|f_{e,1}\|_{L_\infty(C)}}{\lfloor f_{e,1} \rceil^{1-\alpha} \|\nabla f_{e,1}|_{K^\circ}\|_{L_p(C)}^\alpha} \lfloor f \rceil^{1-\alpha} \|\nabla f|_{K^\circ}\|_{L_p(C)}^\alpha.$$

Proof For each $x \in C$ one has

$$|f(x)| \leq \left| f(x) - \frac{1}{\mu(hK \cap C)} \int_{hK \cap C} f(x+y)dy \right|$$

$$+ \left| \frac{1}{\mu(hK \cap C)} \int_{hK \cap C} f(x+y)dy \right|. \tag{1.101}$$

Using Lemma 1.27 for the function $F(y) = f(x+y)$, $y \in hK \cap C, x \in C$, and Lemma 1.28, one obtains

$$\left| f(x) - \frac{1}{\mu(hK \cap C)} \int_{hK \cap C} f(x+y)dy \right|$$

$$= \left| \frac{1}{\mu(hK \cap C)} \int_{hK \cap C} (F(\theta) - F(y))dy \right|$$

$$\leq (d \cdot \mu(K \cap C))^{\frac{1}{p'}} \|g_h\|_{\mathcal{L}_{p'}(0,h)} \|\nabla F|_{K^\circ}\|_{L_p(hK \cap C)}$$

1.3 Inequalities for Derivatives

$$= A\mu^{-\frac{1}{p}}(K \cap C)h^{1-\frac{d}{p}} \||\nabla f|_{K^\circ}\|_{L_p(x+hK\cap C)}$$

$$\leq A\mu^{-\frac{1}{p}}(K \cap C)h^{1-\frac{d}{p}} \||\nabla f|_{K^\circ}\|_{L_p(C)}. \quad (1.102)$$

Hence

$$\|f - S_h f\|_{L_\infty(C)} \leq A\mu^{-\frac{1}{p}}(K \cap C)h^{1-\frac{d}{p}} \||\nabla f|_{K^\circ}\|_{L_p(C)}. \quad (1.103)$$

It is easy to see that

$$\|S_h\|_{L_{\rfloor \cdot \lceil_h}(C) \to L_\infty(C)} \leq \mu^{-1}(K \cap C)h^{-d}. \quad (1.104)$$

From inequalities (1.103) and (1.104) together with Lemma 1.28, inequality (1.96) follows. Next we prove its sharpness. Since the function g_h is non-negative,

$$|x|_K \leq |y|_K \implies f_{e,h}(x) \geq f_{e,h}(y). \quad (1.105)$$

Moreover, the function $f_{e,h}$ is continuous. Hence $\|f_{e,h}\|_{L_\infty(C)} = f(\theta)$ and, moreover, inequality (1.101) becomes equality for $x = \theta$ and the function $f_{e,h}$. Moreover, due to Lemma 1.27, the first inequality in (1.102) for $x = \theta$ becomes equality for the function $f_{e,h}$. Notice that the function $f_{e,h}$ vanishes outside the set $hK \cap C$, and hence

$$\||\nabla f_{e,h}|_{K^\circ}\|_{L_p(hK\cap C)} = \||\nabla f_{e,h}|_{K^\circ}\|_{L_p(C)},$$

thus the second inequality in (1.102) also becomes equality. Finally, taking into account (1.105), we obtain $\rfloor f_{e,h} \lceil_h = \int_{hK\cap C} f_{e,h}(y) dy$, which finishes the proof of sharpness of inequality (1.96).

From (1.96) it follows that for all $f \in W^{1,p}(C) \cap L_{\rfloor \cdot \lceil}(C)$ and $h > 0$ one has

$$\|f\|_{L_\infty(C)} \leq A(d,p)\mu^{-\frac{1}{p}}(K \cap C)h^{1-\frac{d}{p}} \||\nabla f|_{K^\circ}\|_{L_p(C)} + \mu^{-1}(K \cap C)h^{-d} \rfloor f\lceil. \quad (1.106)$$

Moreover, for each $h > 0$, $\rfloor f_{e,h} \lceil_h = \rfloor f_{e,h} \lceil$, and hence inequality (1.106) is sharp and becomes equality for the function $f_{e,h}$. If β satisfies the inequalities from (1.97), then after the substitution of $f_{e,h}$ by $f_{e,h} + \beta$ both side of inequality (1.106) change by β, hence remain equal. Finally, the equality remain true if the extremal function is multiplied by a positive constant. Thus inequality (1.106) becomes equality on each of the functions (1.97).

Minimizing the right-hand side of (1.106) with respect to h i.e., choosing

$$h = \left(\frac{p d \mu^{-\frac{1}{p'}}(K \cap C) \rfloor f \lceil}{(p-d)A(d,p) \||\nabla f|_{K^\circ}\|_{L_p(C)}} \right)^{\frac{p'}{p'+d}},$$

one obtains inequality (1.99).

Next we prove sharpness of inequality (1.99). Set $q = \frac{1}{\alpha}$ and $q' = \frac{1}{1-\alpha}$. For each $h > 0$ consider the numbers

$$u = \left(qA(d,p)\mu^{-\frac{1}{p}}(K \cap C)h^{1-\frac{d}{p}} \||\nabla f|_{K^\circ}\|_{L_p(C)} \right)^{\frac{1}{q}}$$

and
$$v = \left(q'\mu^{-1}(K \cap C)h^{-d}\rfloor f\lceil\right)^{\frac{1}{q'}}.$$

Using inequality (1.99), Young's inequality and straightforward computations, we obtain

$$\|f\|_{L_\infty(C)} \leq a(d,p)\mu^{-\frac{\alpha}{d}}(K \cap C)\rfloor f\lceil^{1-\alpha}\||\nabla f|_{K^\circ}\|_{L_p(C)}^{\alpha} = uv \leq \frac{u^q}{q} + \frac{v^{q'}}{q'}$$

$$= A(d,p)\mu^{-\frac{1}{p}}(K \cap C)h^{1-\frac{d}{p}}\||\nabla f|_{K^\circ}\|_{L_p(C)} + \mu^{-1}(K \cap C)h^{-d}\rfloor f\lceil.$$

Since inequality (1.106) becomes equality for the function $f_{e,h}$, we obtain that inequality (1.99) also becomes equality on the function $f_{e,h}$, and hence is sharp.

Corollary 1.3 *If $h > 0$, $p > d$ and $f \in W^{1,p}(C) \cap L_1(C)$, then $f \in L_\infty(C)$ and the following inequalities hold:*

$$\|f\|_{L_\infty(C)} \leq \|g_h(|\cdot|_K)\|_{L_{p'}(hK\cap C)} \||\nabla f|_{K^\circ}\|_{L_p(C)} + \mu^{-1}(K \cap C)h^{-d}\|f\|_{L_1(C)},$$

$$\|f\|_{L_\infty(C)} \leq a(d,p)\mu^{-\frac{\alpha}{d}}(K \cap C)\|f\|_{L_1(C)}^{1-\alpha}\||\nabla f|_{K^\circ}\|_{L_p(C)}^{\alpha},$$

where α and $a(d,p)$ are defined in (1.100). Both inequalities are sharp. The additive one becomes equality on the function $f_{e,h}$, the multiplicative one—on each of the functions $f_{e,h}$, $h > 0$.

Proof Since $L_1(C) \subset L_{\rfloor\cdot\lceil}(C) \subset L_{\rfloor\cdot\lceil_h}(C)$, $\rfloor f\lceil_h \leq \rfloor f\lceil \leq \|f\|_{L_1(C)}$ for each of the functions $f \in L_1(C)$, and $\|f_{e,h}\|_{L_1(C)} = \rfloor f_{e,h}\lceil_h = \rfloor f_{e,h}\lceil$ for each $h > 0$, the statement of the corollary follows from Theorem 1.17.

1.3.6 Kolmogorov-Type Inequalities for Charges

By $\mathfrak{N}(X)$ we denote the family of charges ν defined on the family of all μ–measurable subsets of X and that are absolutely continuous with respect to the measure μ, see e.g., [34, Chap. 5]. By the Radon–Nikodym theorem, for a charge $\nu \in \mathfrak{N}(X)$ there exists an integrable function $f: X \to \mathbb{R}$ such that for arbitrary measurable set $Q \subset X$

$$\nu(Q) = \int_Q f(x)\mu(dx). \tag{1.107}$$

This function f is called the Radon–Nikodym derivative of the charge ν with respect to the measure μ and will be denoted by $D_\mu \nu$. The family $\mathfrak{N}(X)$ is a linear space with respect to the standard addition and multiplication by a real number. Define a family of seminorms $\{\rfloor \cdot \lceil_h, \ h > 0\}$ as follows:

$$\rfloor \nu \lceil_h = \|\nu(\cdot + B_h)\|_{\mathcal{B}(X)},$$

1.3 Inequalities for Derivatives

where the norm $\|\cdot\|_{\mathcal{B}(X)}$ was defined in (1.92), and let $\rceil\nu\lfloor := \sup_{h>0} \rceil\nu\lfloor_h$. It is clear that if a charge ν and a function f are related via (1.107), then

$$\rceil\nu\lfloor_h = \rfloor f\lceil_h, h > 0, \text{ and } \rceil\nu\lfloor = \rfloor f\lceil.$$

For $h > 0$ by $\mathfrak{N}_{\rceil\cdot\lfloor_h}(X)$ ($\mathfrak{N}_{\rceil\cdot\lfloor}(X)$) we denote the set of charges $\nu \in \mathfrak{N}(X)$ with a finite seminorm $\rceil\cdot\lfloor_h$ (resp. $\rceil\cdot\lfloor$).

In the case, when $(X, \rho, \mu) = (C, K, m)$, we obtain

$$\rceil\nu\lfloor_h = \sup_{x \in C} |\nu(x + hK)|,$$

and, moreover, the function $x \mapsto \nu(x + hK)$ is continuous.

Theorem 1.18 *If $h > 0$ and the charge $\nu \in \mathfrak{N}_{\rceil\cdot\lfloor_h}(C)$ is such that $D_\mu \nu \in W^{1,p}(C)$, then*

$$\|D_\mu \nu\|_{L_\infty(C)} \leq \|D_\mu \nu - \overline{S}_h \nu\|_{L_\infty(C)} + \|\overline{S}_h\|_{\mathfrak{N}_{\rceil\cdot\lfloor_h}(C) \to L_\infty(C)} \rceil\nu\lfloor_h$$
$$\leq A\mu^{-\frac{1}{p}}(K \cap C) h^{1-\frac{d}{p}} \||\nabla D_\mu \nu|_{K^\circ}\|_{L_p(C)} + \mu^{-1}(K \cap C) h^{-d} \rceil\nu\lfloor_h, \quad (1.108)$$

where $A(d, p)$ is defined in (1.95), and

$$\overline{S}_h \nu(x) = \frac{\nu(x + hK \cap C)}{h^d \mu(K \cap C)}.$$

Inequality (1.108) is sharp. It becomes equality on the charge $\nu_{e,h}$ that is determined by the equality $D_\mu \nu_{e,h} = f_{e,h}$, where the function $f_{e,h}$ is defined in (1.98). If the charge $\nu \in \mathfrak{N}_{\rceil\cdot\lfloor}(C)$ is such that $D_\mu \nu \in W^{1,p}(C)$, then the following multiplicative inequality holds

$$\|D_\mu \nu\|_{L_\infty(C)} \leq a(d, p) \mu^{-\frac{\alpha}{d}}(K \cap C) \rceil\nu\lfloor^{1-\alpha} \||\nabla D_\mu \nu|_{K^\circ}\|^\alpha_{L_p(C)}, \quad (1.109)$$

where α and $a(d, p)$ are defined in (1.100). Inequality (1.109) is sharp. It becomes equality for each charge ν such that $D_\mu \nu = f_{e,h}$, $h > 0$.

Proof It is enough to apply Theorem 1.17 to the function $f = D_\mu \nu$ and to take into account that $\overline{S}_h \nu = S_h f$.

1.3.7 Inequalities for a Mixed Derivative of a Function

1.3.7.1 Assumptions and Notations

In this section $(X, \rho, \mu) = (C, K, m)$, $C = \mathbb{R}^d_{m,+} := \mathbb{R}^m_+ \times \mathbb{R}^{d-m}, 0 \leq m \leq d$, $K = (-1, 1)^d$ so that $|x|_K = |x|_\infty = \max_{i=1,\ldots,d} |x_i|$. Then $B_h = hK \cap C = (0, h)^m \times (-h, h)^{d-m}$ and $\mu(hK \cap C) = 2^{d-m} h^d$.

Let $\mathbf{I} = (1, \ldots, 1) \in \mathbb{R}^d$. For a locally integrable function $f: X \to \mathbb{R}$ set

$$\partial_{\mathbf{I}} f = \frac{\partial^d f}{\partial x_1 \ldots \partial x_d}, \qquad (1.110)$$

where the derivatives are understood in the distributional sense.

Let $\{e_i\}_{i=1}^d$ be the standard basis in $\mathbb{R}^d$ and for $h > 0$

$$\Delta_{i,h}^+ f(x) := f(x + he_i) - f(x), \text{ and } \Delta_{i,h} f(x) := f(x + he_i) - f(x - he_i).$$

According to the Fubini theorem for almost all $x \in \mathbb{R}_{m,+}^d$ one has

$$\int_{x+hK \cap C} \partial_{\mathbf{I}} f(u) du = (\Delta_{1,h}^+ \circ \ldots \circ \Delta_{m,h}^+ \circ \Delta_{m+1,h} \circ \ldots \circ \Delta_{d,h}) f(x). \qquad (1.111)$$

Define an operator $\mathfrak{S}_{h,m}: L_\infty(C) \to L_\infty(C)$, setting

$$\mathfrak{S}_{h,m} f(x) = \frac{1}{2^{d-m} h^d} \left(\Delta_{1,h}^+ \circ \ldots \circ \Delta_{m,h}^+ \circ \Delta_{m+1,h} \circ \ldots \circ \Delta_{d,h} \right) f(x).$$

It is easy to see that

$$\|\mathfrak{S}_{h,m}\|_{L_\infty(C) \to L_\infty(C)} = 2^m h^{-d}, \qquad (1.112)$$

and for arbitrary function $f \in L_\infty(\mathbb{R}_{m,+}^d)$ such that $\partial_{\mathbf{I}} f \in W^{1,p}(\mathbb{R}_{m,+}^d)$, one has

$$\mathfrak{S}_{h,m} f(x) = \frac{1}{\mu(hK \cap C)} \int_{x+hK \cap C} \partial_{\mathbf{I}} f(u) du$$

$$= \frac{1}{\mu(hK \cap C)} \int_{hK \cap C} \partial_{\mathbf{I}} f(x+u) du = S_h \partial_{\mathbf{I}} f(x), \qquad (1.113)$$

where the operator S_h is defined in (1.93).

1.3.7.2 Classes $W^{1,p}(\mathbb{R}_{m,+}^d)$

Theorem 1.19 *For $h > 0$ and a function $f \in L_\infty(\mathbb{R}_{m,+}^d)$ such that $\partial_{\mathbf{I}} f \in W^{1,p}(\mathbb{R}_{m,+}^d)$, one has*

$$\|\partial_{\mathbf{I}} f\|_{L_\infty(\mathbb{R}_{m,+}^d)} \le \|\partial_{\mathbf{I}} f - \mathfrak{S}_{h,m} f\|_{L_\infty(\mathbb{R}_{m,+}^d)} + \|\mathfrak{S}_{h,m}\| \|f\|_{L_\infty(\mathbb{R}_{m,+}^d)}$$

$$\le A(d,p) h^{1-\frac{d}{p}} 2^{\frac{m-d}{p}} \|\nabla \partial_{\mathbf{I}} f |_{K^\circ} \|_{L_p(\mathbb{R}_{m,+}^d)} + 2^m h^{-d} \|f\|_{L_\infty(\mathbb{R}_{m,+}^d)}, \qquad (1.114)$$

where $A(d,p)$ is defined in (1.95). The inequality (1.114) can be rewritten in the multiplicative form:

$$\|\partial_{\mathbf{I}} f\|_{L_\infty(\mathbb{R}_{m,+}^d)} \le a(d,p) 2^{\alpha\left(\frac{m}{d} - \frac{d}{p}\right)} \|f\|_{L_\infty(\mathbb{R}_{m,+}^d)}^{1-\alpha} \|\nabla \partial_{\mathbf{I}} f|_{K^\circ}\|_{L_p(\mathbb{R}_{m,+}^d)}^\alpha, \qquad (1.115)$$

1.3 Inequalities for Derivatives

where α and $a(d, p)$ are defined in (1.100). For $m = 0$ and $m = 1$ inequalities (1.114) and (1.115) are sharp.

Proof Taking into account $\mu(K \cap C) = 2^{d-m}$, (1.103), (1.112) and (1.113), one obtains

$$\|\partial_I f\|_{L_\infty(C)} \leq \|\partial_I f - \mathfrak{S}_{h,m} f\|_{L_\infty(C)} + \|\mathfrak{S}_{h,m}\|_{L_\infty(C) \to L_\infty(C)} \|f\|_{L_\infty(C)}$$

$$= \|\partial_I f - S_h \partial_I f\|_{L_\infty(C)} + 2^m h^{-d} \|f\|_{L_\infty(C)}$$

$$\leq A(d, p) \cdot \mu^{-\frac{1}{p}}(K \cap C) h^{1-\frac{d}{p}} \||\nabla \partial_I f|_{K^\circ}\|_{L_p(C)} + 2^m h^{-d} \|f\|_{L_\infty(C)}$$

$$= A(d, p) \cdot 2^{\frac{m-d}{p}} h^{1-\frac{d}{p}} \||\nabla \partial_I f|_{K^\circ}\|_{L_p(C)} + 2^m h^{-d} \|f\|_{L_\infty(C)}$$

and inequality (1.114) is proved. Using inequality (1.99) for the function $\partial_I f$ as well as equality (1.111), one has

$$\|\partial_I f\|_{L_\infty(C)} \leq a(d, p) \mu^{-\frac{\alpha}{d}}(K \cap C) \lfloor \partial_I f \rceil^{1-\alpha} \||\nabla \partial_I f|_{K^\circ}\|_{L_p(C)}^\alpha$$

$$\leq a(d, p) \mu^{-\frac{\alpha}{d}}(K \cap C) \left(2^d \|f\|_{L_\infty(C)}\right)^{1-\alpha} \||\nabla \partial_I f|_{K^\circ}\|_{L_p(C)}^\alpha$$

$$= a(d, p) 2^{\frac{(m-d)\alpha}{d} + d(1-\alpha)} \|f\|_{L_\infty(C)}^{1-\alpha} \||\nabla \partial_I f|_{K^\circ}\|_{L_p(C)}^\alpha$$

$$= a(d, p) 2^{\alpha \left(\frac{m}{d} - \frac{d}{p}\right)} \|f\|_{L_\infty(C)}^{1-\alpha} \||\nabla \partial_I f|_{K^\circ}\|_{L_p(C)}^\alpha,$$

and inequality (1.115) is proved.

Next we prove sharpness of inequalities (1.114) and (1.115) for $m = 0$. Consider the function

$$F_{e,h}(x) = \int_0^{x_1} \ldots \int_0^{x_d} f_{e,h}(u) du,$$

where the function $f_{e,h}$ is defined in (1.98). For this function one has $\partial_I F_{e,h} = f_{e,h}$, $|\nabla \partial_I F_{e,h}(\cdot)|_{K^\circ} = |\nabla f_{e,h}(\cdot)|_{K^\circ}$ and due to symmetry considerations

$$2^d \|F_{e,h}\|_{L_\infty(C)} = 2^d \int_{(0,h)^d} f_{e,h}(u) du = \int_{(-h,h)^d} f_{e,h}(u) du = \lfloor f_{e,h} \rceil_h = \lfloor f_{e,h} \rceil. \quad (1.116)$$

Since inequalities (1.96) and (1.99) become equalities on the function $f_{e,h}$, then taking into account (1.116), we obtain sharpness of inequalities (1.114) and (1.115) for $m = 0$.

Finally, we prove sharpness of inequalities (1.114) and (1.115) in the case $m = 1$. In this case $hK \cap C = (0, h) \times (-h, h)^{d-1}$ and $\mu(K \cap C) = 2^{d-1}$. There exists a number $a \in (0, h)$ such that

$$\int_{(0,a) \times (-h,h)^{d-1}} f_{e,h}(u) du = \int_{(a,h) \times (-h,h)^{d-1}} f_{e,h}(u) du. \quad (1.117)$$

Consider the function

$$G_{e,h}(x) = \int_a^{x_1} \int_0^{x_2} \cdots \int_0^{x_d} f_{e,h}(u)du.$$

For it one has $\partial_{\mathbf{I}} G_{e,h} = f_{e,h}$, $|\nabla \partial_{\mathbf{I}} G_{e,h}(\cdot)|_{K^\circ} = |\nabla f_{e,h}(\cdot)|_{K^\circ}$.

The hyperplanes $x_1 = a$ and $x_j = 0$, $j = 2, \ldots, d$, split the set $(0, h) \times (-h, h)^{d-1}$ into 2^d boxes $\Pi_1, \ldots, \Pi_{2^d}$; moreover, due to the symmetry of the plot of the function $f_{e,h}$ with respect to the coordinate hyperplanes and due to equality (1.117), we obtain

$$\int_{\Pi_i} f_{e,h}(u)du = \frac{1}{2^d} \int_{(0,h) \times (-h,h)^{d-1}} f_{e,h}(u)du, \, i = 1, \ldots, 2^d,$$

hence

$$\|G_{e,h}\|_{L_\infty(C)} = \frac{1}{2^d} \int_{(0,h) \times (-h,h)^{d-1}} f_{e,h}(u)du = \frac{1}{2^d} \lfloor f_{e,h} \rceil_h = \frac{1}{2^d} \lfloor f_{e,h} \rceil.$$

Due to this equality and the fact that inequalities (1.96) and (1.99) become equalities for the function $f_{e,h}$, we obtain sharpness of inequalities (1.114) and (1.115) for $m = 1$.

1.3.8 Approximation of Unbounded Operators by Bounded Ones and Related Problems

1.3.8.1 General Facts

Let X and Y be linear spaces equipped with a seminorm $\|\cdot\|_X$ and a norm $\|\cdot\|_Y$ respectively. By $\mathcal{L}(X, Y)$ we denote the space of all linear bounded operators $S \colon X \to Y$. Let $A \colon X \to Y$ be an operator (not necessarily linear) with a domain $D_A \subset X$. Let also $\mathfrak{M} \subset D_A$ be some class of elements. The problem to find the modulus of continuity of the operator A on the class $\mathfrak{M}$ i.e., the function

$$\Omega(\delta) = \Omega(A, \mathfrak{M}; \delta) := \sup\{\|Ax\|_Y : x \in \mathfrak{M}, \|x\|_X \le \delta\}, \, \delta \ge 0, \quad (1.118)$$

is an abstract version of the problem about the Landau-Kolmogorov type inequalities.

In Introduction we formulated a series of related problems. Here we formulate some general results on the connection of these problems. The Stechkin problem is intimately connected to Landau-Kolmogorov type inequalities. The following theorem describes this connection, see [71].

Theorem 1.20 *For any $x \in \mathfrak{M} \subset X$ and arbitrary $S \in \mathcal{L}(X, Y)$ the following inequality holds*

$$\|Ax\| \le \|Ax - Sx\| + \|S\|\|x\| \le U(A, S; \mathfrak{M}) + \|S\| \cdot \|x\|, \quad (1.119)$$

and, therefore for all $x \in \mathfrak{M}$ and $N > 0$,

$$\|Ax\| \le E_N(A, \mathfrak{M}) + N\|x\|.$$

1.3 Inequalities for Derivatives

If in addition there exist $\overline{S} \in \mathcal{L}(X, Y)$ *and* $\overline{x} \in \mathfrak{M}$ *such that both inequalities in* (1.119) *turn into equalities, then*

$$\Omega(\|\overline{x}\|) = \|A\overline{x}\| \text{ and } E_{\|\overline{S}\|}(A, \mathfrak{M}) = U(A, \overline{S}; \mathfrak{M}) = \|A\overline{x}\| - \|\overline{S}\| \|\overline{x}\|.$$

Thus the element $\overline{x}$ *is extremal for problem* (1.118), *and the operator* $\overline{S}$ *is optimal for problem* (1.3).

Remark 1.1 In Stechkin's article [71] it is assumed that X and Y are Banach spaces. However, as it is easy to see, completeness, and even presence of a norm in X is not necessary. It is sufficient to have a seminorm in X. Completeness of Y is also not necessary.

Let $\mathfrak{M} \subset D(A)$ and $S \in \mathcal{L}(X, Y)$. For $\delta \geq 0$ the value

$$U_\delta(A, S, \mathfrak{M}) := \sup\{\|Ax - S\eta\|_Y : x \in \mathfrak{M}, \eta \in X, \|x - \eta\|_X \leq \delta\},$$

if called the deviation of the operator $S \in \mathcal{L}(X, Y)$ from the operator A on the class $\mathfrak{M}$, where elements are known with error δ. The problem of optimal recovery of the operator A by linear operators on such class $\mathfrak{M}$ consists of finding the quantity

$$\mathcal{E}_\delta(A, \mathfrak{M}) := \inf_{S \in \mathcal{L}(X,Y)} U_\delta(A, S, \mathfrak{M})$$

and an operator S, on which the infimum on the right-hand side is attained, if it exists.

The connection of this problem with the Landau-Kolmogorov-type inequalities and the Stechkin problem is given by the following theorem, see e.g. [5, Theorem 2.1].

Theorem 1.21 *If* $\mathfrak{M}$ *is a convex centrally-symmetric set and* A *is a homogeneous operator, then for all* $\delta > 0$

$$\Omega(A, \mathfrak{M}; \delta) \leq \mathcal{E}_\delta(A, \mathfrak{M}) \leq \inf_{N>0} \{E_N(A, \mathfrak{M}) + N\delta\}.$$

If in addition there exist an element $\overline{x} \in \mathfrak{M}$ *and an operator* $\overline{S}$ *such that both inequalities in* (1.119) *become equalities, then*

$$\mathcal{E}_{\|\overline{x}\|}(A, \mathfrak{M}) = \Omega(A, \mathfrak{M}; \|\overline{x}\|) = \|A\overline{x}\|.$$

1.3.8.2 Extremal Problems for Hypersingular Integral Operators

In Sect. 1.3.1 we studied a hypersingular integral operator $D_K^{w,\Omega}$ defined in (1.80) and a truncated operator $D_{K,h}^{w,\Omega}$ defined in (1.81).

Theorem 1.22 *Let a set* $K \in \mathbb{K}$, *a cone* $C \in \mathcal{C}$, *and a homogeneous characteristic* Ω *be given. Assume also that a modulus of continuity* ω *and a weight* w *are such that for all* $h > 0$, $w \cdot \omega \in \mathcal{L}_1(0, h)$ *and* $w \in \mathcal{L}_1(h, \infty)$. *If*

$$N \in \begin{cases} (0, \infty), & D_K^{w,\Omega} \text{ is unbounded,} \\ \left(0, \|D_K^{w,\Omega}\|\right), & \text{otherwise,} \end{cases}$$

and h_N *is such that* $\|w\|_{\mathcal{L}_1(h_N, \infty)} = \frac{N}{2\Omega(K,C)}$, *then*

$$E_N(D_K^{w,\Omega}, \mathfrak{M}) = \Omega(K, C) \cdot \|w \cdot \omega\|_{\mathcal{L}_1(0, h_N)}, \tag{1.120}$$

where $\mathfrak{M} = \{f \in H_K^\omega(C) \cap L_\infty(C) \colon \|f\|_{H_K^\omega(C)} \leq 1\}$. *The extremal operator is* $D_{K, h_N}^{w,\Omega}$.

Proof From the choice of h_N and Lemma 1.25, we have $\|D_{K, h_N}^{w,\Omega}\| = N$. Moreover, from the proof of Theorem 1.13 is follows that the condition of Theorem 1.20 are satisfied, and hence $E_N(D_K^{w,\Omega}, \mathfrak{M}) = U(D_K^{w,\Omega}, D_{K,h_N}^{w,\Omega}; \mathfrak{M})$. Now equality (1.120) follows from Theorem 1.6.

Analogously using Theorems 1.7 and 1.14 we obtain the following result.

Theorem 1.23 *Let* $p \in (d, \infty]$, $C \in \mathcal{C}$, $K \in \mathbb{K}$, $\Omega \colon C \to \mathbb{R}$ *be a non-negative characteristic, and a weight w be such that for each $h > 0$ and $w \in \mathcal{W}_{p'}(0, h) \cap \mathcal{L}_1(h, \infty)$. For N and h_N as in Theorem 1.22,*

$$E_N(D_K^{w,\Omega}, \mathfrak{N}) = (\Omega(K, C))^{\frac{1}{p'}} \|g_w\|_{\mathcal{L}_{p'}(0, h_N)},$$

where $\mathfrak{N} = \{f \in L_{\infty, p}^\Omega(C) \colon \|\Omega^{\frac{1}{p}} \cdot |\nabla f|_{K^\circ}\|_{L_p(C)} \leq 1\}$. *The extremal operator is* $D_{K, h_N}^{w,\Omega}$.

1.3.8.3 Extremal Problems on Classes of Charges

Theorem 1.24 *For $p \in (d, \infty]$ set $X = W^{1,p}(C) \cap L_{].[}(C)$, let $\mathfrak{M}$ be the family of functions $f \in X$ such that $\||\nabla f|_{K^\circ}\|_{L_p(C)} \leq 1$, and $J \colon X \to L_\infty(C)$ be the embedding operator. For all $\delta > 0$ the following equalities hold:*

$$\mathcal{E}_\delta(J, \mathfrak{M}) = \Omega(J, \mathfrak{M}; \delta) = \frac{a(d, p)}{\mu^{\frac{\alpha}{d}}(K \cap C)} \delta^{1-\alpha}, \tag{1.121}$$

where α and $a(d, p)$ are defined in (1.100). For all $N > 0$,

$$E_N(D_\mu, \mathfrak{M}) = A(d, p) \mu^{-\frac{1}{d}}(K \cap C) N^{\frac{1}{p} - \frac{1}{d}}, \tag{1.122}$$

where the quantity $A(d, p)$ is defined in (1.95).

1.3 Inequalities for Derivatives

Proof First of all note that if inequality (1.99) turns into equality for some function f, then it becomes equality for the function $c \cdot f$, $c \in \mathbb{R}$, and hence from sharpness of (1.99) if follows that the second equality in (1.121) holds.

From sharpness of (1.38) and Theorem 1.20 it follows that for all $h > 0$, $E_{\|S_h\|}(J, \mathfrak{M}) = \|g_h(|\cdot|_K)\|_{L_{p'}(hK \cap C)}$. Using equality (1.94) and Lemma 1.28 we obtain (1.122). Finally, the first equality in (1.121) follows from Theorem 1.21.

The following two theorems can be obtained from Theorems 1.18 and 1.19 using the same arguments.

Theorem 1.25 *Denote by $\mathfrak{M}$ the family of charges $v \in \mathfrak{N}_{\rceil \cdot \lfloor}(C)$ such that*

$$\||\nabla D_\mu v|_{K^\circ}\|_{L_p(C)} \le 1.$$

For all $\delta > 0$ the following equalities hold:

$$\mathcal{E}_\delta(D_\mu, \mathfrak{M}) = \Omega(D_\mu, \mathfrak{M}; \delta) = \frac{a(d, p)}{\mu^{\frac{\alpha}{d}}(K \cap C)} \delta^{1-\alpha},$$

where α and $a(d, p)$ are defined in (1.100). For all $N > 0$,

$$E_N(D_\mu, \mathfrak{M}) = A(d, p)\mu^{-\frac{1}{d}}(K \cap C) N^{\frac{1}{p} - \frac{1}{d}},$$

where $A(d, p)$ is defined in (1.95).

Recall that the differential operator $\partial_\mathbf{I} f$ was defined in (1.110).

Theorem 1.26 *Let $p > d$ and $m = 0$ or $m = 1$. Denote by $\mathfrak{M}$ the family of functions $f \in L_\infty(\mathbb{R}^d_{m,+})$ such that $\||\nabla \partial_\mathbf{I} f|_1\|_{L_p(\mathbb{R}^d_{m,+})} \le 1$. For all $\delta > 0$ the following equalities hold.*

$$\mathcal{E}_\delta(\partial_\mathbf{I}, \mathfrak{M}) = \Omega(\partial_\mathbf{I}, \mathfrak{M}; \delta) = a(d, p) 2^{\alpha\left(\frac{m}{d} - \frac{d}{p}\right)} \delta^{1-\alpha},$$

where α and $a(d, p)$ are defined in (1.100). For all $N > 0$

$$E_N(\partial_\mathbf{I}, \mathfrak{M}) = A(d, p) \cdot 2^{\frac{m}{d} - \frac{d}{p}} N^{\frac{1}{p} - \frac{1}{d}}.$$

where $A(d, p)$ is defined in (1.95).

1.4 Optimization of Cubature Formulae

1.4.1 Optimization of Cubature Formulae on Multivariate Sobolev Classes

The results of this section in the case $K = (-1, 1)^d$ were obtained in [22]. In the case of arbitrary $K \in \mathbb{K}$ they are new.

1.4.1.1 Statement of the Problem and Extremal Functions

Let Q be a bounded open set, $p > d$ and $K \in \mathbb{K}$ be given. We consider the problem of optimal recovery for the integral $\int_Q f(x)dx$ using the values of the function

$$f \in W_p^K(Q) := \left\{ f \in W^{1,p}(Q) \colon \| |\nabla f|_{K^\circ} \|_{L_p(Q)} \leq 1 \right\}$$

at $n \in \mathbb{N}$ points $x_1, \ldots, x_n \in Q$. Any function $\Phi \colon \mathbb{R}^n \to \mathbb{R}$ is called a method of recovery. For given points $x_1, \ldots, x_n \in Q$, the error of recovery of the integral using information $f(x_1), \ldots, f(x_n)$ by the method Φ is defined by the following equality

$$e(W_p^K(Q), \Phi, x_1, \ldots, x_n) := \sup_{f \in W_p^K(Q)} \left| \int_Q f(x)dx - \Phi(f(x_1), \ldots, f(x_n)) \right|.$$

The problem of the optimal recovery of the integral is to find the best error of recovery

$$E_n(W_p^K(Q)) := \inf_{x_1, \ldots, x_n \in Q} \inf_{\Phi \colon \mathbb{R}^n \to \mathbb{R}} e(W_p^K(Q), \Phi, x_1, \ldots, x_n), \quad (1.123)$$

the best method of recovery, and the best position of the information points $x_1, \ldots, x_n$ (i.e., such method $\Phi \colon \mathbb{R}^n \to \mathbb{R}$ and points $x_1, \ldots, x_n \in Q$, for which the infima in (1.123) are attained, if they exist).

Note, that it is sufficient to consider only linear methods of recovery in (1.123). The existence of an optimal linear method of recovery is well known in many situations. See for example [70]. We will not prove it here.

Let $x_1, \ldots, x_n \in Q$ and $h > 0$. The function f, defined by (1.39), is well defined on the boundary ∂K and is constant there; hence we can continuously extend the function f to all of $\mathbb{R}^d$ by setting $f(y)$ equal to the value of f on the boundary of K for all $y \notin K$. For all $y \in \mathbb{R}^d$, we set

$$f_h(x_1, \ldots, x_n; y) := \min_{k=1,\ldots,n} f(y - x_k).$$

It is easy to see that $f_h(x_1, \ldots, x_n; y) \in W^{1,p}(Q)$ for all $p \in (d, \infty]$.

1.4.1.2 Simple Domains

Let $K \in \mathbb{K}$ and $n \in \mathbb{N}$. A set $Q \subset \mathbb{R}^d$ is called *n-simple*, if there exist points $\bar{x}_1, \ldots, \bar{x}_n \in Q$ such that for

$$h := \left(\frac{\operatorname{meas} Q}{n \cdot \operatorname{meas} K} \right)^{\frac{1}{d}} \quad (1.124)$$

the sets

$$C_k := \{ x \in \mathbb{R}^d : |x - \bar{x}_k|_K < h \}, \ k = 1, \ldots, n,$$

are pairwise disjoint and $\operatorname{meas}\left[Q \setminus \bigcup_{k=1}^n C_k \right] = 0$. Such set Q will also be called $(\bar{x}_1, \ldots, \bar{x}_n)$-*simple*.

Lemma 1.29 *Let $n \in \mathbb{N}$, points $\bar{x}_1, \ldots, \bar{x}_n \in \mathbb{R}^d$ and a $(\bar{x}_1, \ldots, \bar{x}_n)$-simple domain Q be given. If h is defined by* (1.124), *then for arbitrary points $x_1, \ldots, x_n \in Q$*

$$\int_Q f_h(x_1, \ldots, x_n; y) dy \geq \int_Q f_h(\bar{x}_1, \ldots, \bar{x}_n; y) dy \geq 0 \quad (1.125)$$

and for all $p \in (d, \infty]$

$$\||\nabla f_h(x_1, \ldots, x_n)|_{K^\circ}\|_{L_p(Q)} \leq \||\nabla f_h(\bar{x}_1, \ldots, \bar{x}_n)|_{K^\circ}\|_{L_p(Q)}. \quad (1.126)$$

Proof First, we prove inequality (1.125). For all $\lambda \geq 0$ and arbitrary $x_1, \ldots, x_n \in Q$, we consider the set

$$S(x_1, \ldots, x_n; \lambda) := \{ y \in Q : f_h(x_1, \ldots, x_n; y) \leq \lambda \}.$$

Definition of the function $f_h(x_1, \ldots, x_n; y)$ implies that the set $S(x_1, \ldots, x_n; \lambda)$ is the intersection of Q with the union of n sets $C_k(x_k, \lambda) := x_k + \mu(\lambda) K$, where μ is a non-decreasing function of λ. Moreover, if $\lambda_0 := \inf\{ \lambda \geq 0 : \operatorname{meas} S(\bar{x}_1, \ldots, \bar{x}_n; \lambda) = \operatorname{meas} Q \}$, then the sets $C_k(\bar{x}_k, \lambda)$ that define the set $S(\bar{x}_1, \ldots, \bar{x}_n; \lambda)$ are pairwise disjoint for all $\lambda < \lambda_0$ and $S(\bar{x}_1, \ldots, \bar{x}_n; \lambda) = Q$ for $\lambda \geq \lambda_0$. This implies that for all $\lambda \geq 0$,

$$\operatorname{meas} S(\bar{x}_1, \ldots, \bar{x}_n; \lambda) \geq \operatorname{meas} S(x_1, \ldots, x_n; \lambda)$$

and, hence,

$$\operatorname{meas} \{ y \in Q : f_h(x_1, \ldots, x_n; y) > \lambda \} \geq \operatorname{meas} \{ y \in Q : f_h(\bar{x}_1, \ldots, \bar{x}_n; y) > \lambda \}.$$

The latter inequality implies the first inequality in (1.125) (see [72, §1.1]). The second inequality in (1.125) follows from the definition of the functions $f_h(x_1, \ldots, x_n)$.

Next, we prove inequality (1.126). For $k = 1, \ldots, n$, we set

$$A_k := \{ x \in Q : |x - x_k|_K < |x - x_s|_K, \ \forall s \neq k \}.$$

From the definition of the function $f_h(x_1, \ldots, x_n)$, it follows that

$$f_h(x_1, \ldots, x_n; x) = f_e(x - x_k)$$

on $B_k := \{x \in A_k : |x - x_k|_K < h\}$, $k = 1, \ldots, n$, and

$$|\nabla f_h(x_1, \ldots, x_n; x)|_{K^\circ} = 0 \tag{1.127}$$

almost everywhere on the set $Q \setminus \left(\bigcup_{k=1}^n B_k\right)$. For all $k = 1, \ldots, n$,

$$\||\nabla f_h(x_1, \ldots, x_n; \cdot)|_{K^\circ}\|_{L_p(B_k)} = \||\nabla f_e(\cdot - x_k)|_{K^\circ}\|_{L_p(B_k)}$$
$$\leq \||\nabla f_e(\cdot)|_{K^\circ}\|_{L_p(hK)} = \||\nabla f_e(\cdot - \bar{x}_k)|_{K^\circ}\|_{L_p(\bar{x}_k + hK)}$$
$$= \||\nabla f_h(\bar{x}_1, \ldots, \bar{x}_n; \cdot)|_{K^\circ}\|_{L_p(\bar{x}_k + hK)}.$$

The latter together with (1.127) implies inequality (1.126). The lemma is proved.

1.4.1.3 Optimal Recovery of the Integral

Applying Theorem 1.7 to $\Omega \equiv 1$, $C = \mathbb{R}^d$, $w \equiv 1$ we can find a constant $c(d, p)$ such that inequality

$$\int_{hK} |f(x) - f(\theta)| dx \leq c(d, p) \cdot h^{1 + \frac{d}{p'}} \||\nabla f|_{K^\circ}\|_{L_p(hK)}$$

is sharp on $W_p^K(hK)$ (cf. with Lemma 1.28). The following theorem generalizes [22, Theorem 4], where the case $K = (-1, 1)^d$ was considered.

Theorem 1.27 *Let $K \in \mathbb{K}$, $n \in \mathbb{N}$, and an n-simple domain $Q \subset \mathbb{R}^d$ be given. For $p \in (d; \infty]$*

$$E_n\left(W_p^K(Q)\right) = \frac{c(d, p)}{n^{\frac{1}{d}}} \left[\frac{\text{meas } Q}{\text{meas } K}\right]^{\frac{1}{d} + \frac{1}{p'}}.$$

If Q is $(\bar{x}_1, \ldots, \bar{x}_n)$-simple, then the optimal information set is $\{\bar{x}_k\}_{k=1}^n$, and the best recovery method is

$$\Phi^*(f(\bar{x}_1), \ldots, f(\bar{x}_n)) = \frac{\text{meas } Q}{n} \sum_{k=1}^n f(\bar{x}_k).$$

Proof Let h be defined in (1.124). For arbitrary $f \in W_p^K(Q)$,

$$\left|\int_Q f(x) dx - \frac{\text{meas } Q}{n} \sum_{j=1}^n f(\bar{x}_j)\right| \leq \sum_{j=1}^n \int_{hK + \bar{x}_j} |f(x) - f(\bar{x}_j)| dx$$
$$\leq c(d, p) \cdot h^{1 + \frac{d}{p'}} \sum_{j=1}^n \||\nabla f|_{K^\circ}\|_{L_p(hK + \bar{x}_j)}$$

1.4 Optimization of Cubature Formulae

$$\leq c(d,p) \cdot h^{1+\frac{d}{p'}} n^{\frac{1}{p'}} \| |\nabla f|_{K^\circ}\|_{L_p(Q)} \leq \frac{c(d,p)}{n^{\frac{1}{d}}} \left[\frac{\operatorname{meas} Q}{\operatorname{meas} K}\right]^{\frac{1}{d}+\frac{1}{p'}}.$$

For arbitrary $x_1, \ldots, x_n$, using Lemma 1.29,

$$\sup_{f \in W_p^K(Q)} \left| \int_Q f(x)dx - \sum_{k=1}^n c_k f(x_k) \right| \geq \sup_{\substack{f \in W_p^K(Q), \\ f(x_k)=0, k=1,\ldots,n}} \left| \int_Q f(x)dx \right|$$

$$\geq \frac{1}{\| |\nabla f_h(x_1,\ldots,x_n)|_{K^\circ}\|_{L_p(Q)}} \int_Q f_h(x_1,\ldots,x_n;x)dx$$

$$\geq \frac{1}{\| |\nabla f_h(\bar{x}_1,\ldots,\bar{x}_n)|_{K^\circ}\|_{L_p(Q)}} \int_Q f_h(\bar{x}_1,\ldots,\bar{x}_n;x)dx = \frac{c(d,p)}{n^{\frac{1}{d}}} \left[\frac{\operatorname{meas} Q}{\operatorname{meas} K}\right]^{\frac{1}{d}+\frac{1}{p'}}.$$

Finally, taking into account the existence of the optimal linear method of recovery and arbitrariness of $x_1, \ldots, x_n \in Q$ and $c_1, \ldots, c_n \in \mathbb{R}$, we obtain the estimate for $E_n\left(W_p^K(Q)\right)$ from below, which completes the proof of the theorem.

Theorem 1.27 gives an optimal cubature formula for simple domains. It can be used to obtain asymptotically optimal cubature formulae for more complicated domains, see e.g. [22, 28, 60]. Asymptotically optimal cubature formulae for various classes of multivariate functions can be found in [8–10, 35, 36].

1.4.2 Optimization of Cubature Formulae on Classes of Random Processes

We consider the problem of optimal recovery of the integral

$$\operatorname{Int} \xi_t := \int_0^1 \xi_t dt$$

on the class of random processes $\mathcal{H}^\omega$, given the information operator

$$J(\xi_t) = (\xi_{\tau_1}, \ldots, \xi_{\tau_n}),$$

where $n \in \mathbb{N}$, $\tau_k = \tau + t_k$, $\tau \in \mathcal{R}$, and $\mathbf{t} = (t_1, t_2, \ldots, t_n)$, where the numbers $0 = t_1 < \ldots < t_n$ are such that $\tau + t_n \leq 1$ almost everywhere. The error of recovery is measured in the space $\mathcal{R}$ with metric $(\zeta, \eta) \mapsto \mathbf{E}|\zeta - \eta|$. Everywhere below we write $E(\tau, \mathbf{t})$ instead of $\mathcal{E}(\operatorname{Int}, \mathcal{H}^\omega, J, \mathcal{R})$ (see (1.4) in Introduction). For $t \geq 0$ set

$$I(t) := \int_0^t \omega(s)ds.$$

The following theorem is contained in [58].

Theorem 1.28 Let $n \in \mathbb{N}$, $\tau \in \mathcal{R}$ and the numbers $0 = t_1 < \ldots < t_n$ be such that $\tau + t_n \leq 1$ almost everywhere. Set $\tau_k := \tau + t_k$, $k = 1, \ldots, n$, and $t^* := \left\| \tau - \frac{1-t_n}{2} \right\|_\infty$. Then

$$E(\tau, \mathbf{t}) = 2\sum_{k=1}^{n-1} I\left(\frac{t_{k+1} - t_k}{2}\right) + I\left(\frac{1-t_n}{2} - t^*\right) + I\left(\frac{1-t_n}{2} + t^*\right). \quad (1.128)$$

The optimal recovery method is $U = \sum_{k=1}^{n} c_k^* \xi_{\tau_k}$, where $c_1^* = \tau + \frac{t_2 - t_1}{2}$, $c_k^* = \frac{t_{k+1} - t_{k-1}}{2}$, $k = 2, \ldots, n-1$ and $c_n^* = 1 - \tau - \frac{t_n + t_{n-1}}{2}$.

In [42], the problem of integral optimal recovery was considered in the case when τ is a constant on an analogue of the class $\mathcal{H}^\omega$. The proof of this theorem will follow from the results of subsequent paragraphs.

1.4.2.1 Auxiliary Result

Lemma 1.30 Let $a > 0$, $\tau \in \mathcal{R}(a)$ and $b > 0$ be such that $\tau + b \leq a$ almost everywhere. For a process $\xi \in \mathcal{H}^\omega(a)$ set

$$\zeta_t(w) := \begin{cases} \xi_t(w) - \xi_\tau(w), & 0 \leq t \leq \tau(w), \\ \xi_{t+b}(w) - \xi_{\tau+b}(w), & \tau(w) < t \leq a - b. \end{cases}$$

Then $\zeta \in \mathcal{H}_\tau^\omega(a - b)$ and $\zeta_\tau \equiv 0$.

Proof Equality $\zeta_\tau \equiv 0$ follows from the definition of the process ζ. For a random variable $\theta \in \mathcal{R}(a - b)$ set

$$\tilde{\theta}(w) = \begin{cases} \theta(w), & \theta(w) \leq \tau(w) \\ \theta(w) + b, & \theta(w) > \tau(w) \end{cases} \text{ and } \tilde{\tau}(w) = \begin{cases} \tau(w), & \theta(w) \leq \tau(w) \\ \tau(w) + b, & \theta(w) > \tau(w). \end{cases}$$

Then

$$\mathbf{E}|\zeta_\theta - \zeta_\tau| = \mathbf{E}|\zeta_\theta| = \mathbf{E}|\xi_{\tilde{\theta}} - \xi_{\tilde{\tau}}| \leq \omega(\|\tilde{\theta} - \tilde{\tau}\|_\infty) = \omega(\|\theta - \tau\|_\infty).$$

Hence $\zeta \in \mathcal{H}_\tau^\omega(a - b)$ and the lemma is proved.

1.4.2.2 Estimate from Above

In this paragraph we prove that

$$E(\tau, \mathbf{t}) \leq 2\sum_{k=1}^{n-1} I\left(\frac{t_{k+1} - t_k}{2}\right) + I\left(\frac{1-t_n}{2} - t^*\right) + I\left(\frac{1-t_n}{2} + t^*\right). \quad (1.129)$$

1.4 Optimization of Cubature Formulae

Set $\alpha_0 := 0$, $\alpha_k := \tau + \frac{t_k + t_{k+1}}{2}$, $k = 1, \ldots, n-1$, and $\alpha_n = 1$. Then $c_k^* = \alpha_k - \alpha_{k-1}$, $k = 1, \ldots, n$. Hence

$$E(\tau, \mathbf{t}) \leq \sup_{\xi \in \mathcal{H}^\omega} \mathbf{E} \left| \int_0^1 \xi_t \, dt - \sum_{k=1}^n c_k^* \xi_{\tau_k} \right| = \sup_{\xi \in \mathcal{H}^\omega} \mathbf{E} \left| \sum_{k=1}^n \int_{\alpha_{k-1}}^{\alpha_k} (\xi_t - \xi_{\tau_k}) \, dt \right|$$

$$\leq \sup_{\xi \in \mathcal{H}^\omega} \mathbf{E} \left| \int_0^{\alpha_1} (\xi_t - \xi_{\tau_1}) \, dt + \int_{\alpha_{n-1}}^1 (\xi_t - \xi_{\tau_n}) \, dt \right|$$

$$+ \sup_{\xi \in \mathcal{H}^\omega} \sum_{k=2}^{n-1} \mathbf{E} \int_{\alpha_{k-1}}^{\alpha_k} \left| \xi_t - \xi_{\tau_k} \right| dt. \quad (1.130)$$

Let $\xi \in \mathcal{H}^\omega$ and $k \in \{2, \ldots, n-1\}$. Then

$$\mathbf{E} \int_{\alpha_{k-1}}^{\alpha_k} \left| \xi_t - \xi_{\tau_k} \right| dt = \mathbf{E} \int_{\tau + \frac{t_{k-1} + t_k}{2}}^{\tau + \frac{t_k + t_{k+1}}{2}} \left| \xi_t - \xi_{\tau_k} \right| dt$$

$$= \mathbf{E} \int_{\tau_k + \frac{t_{k-1} - t_k}{2}}^{\tau_k + \frac{t_{k+1} - t_k}{2}} \left| \xi_t - \xi_{\tau_k} \right| dt = \mathbf{E} \int_{\frac{t_{k-1} - t_k}{2}}^{\frac{t_{k+1} - t_k}{2}} \left| \xi_{\tau_k + t} - \xi_{\tau_k} \right| dt$$

$$= \int_{\frac{t_{k-1} - t_k}{2}}^{\frac{t_{k+1} - t_k}{2}} \mathbf{E} \left| \xi_{\tau_k + t} - \xi_{\tau_k} \right| dt \leq \int_{\frac{t_{k-1} - t_k}{2}}^{\frac{t_{k+1} - t_k}{2}} \omega(|t|) \, dt$$

$$= \int_0^{\frac{t_k - t_{k-1}}{2}} \omega(t) \, dt + \int_0^{\frac{t_{k+1} - t_k}{2}} \omega(t) \, dt. \quad (1.131)$$

For arbitrary $\xi \in \mathcal{H}^\omega$,

$$\mathbf{E} \left| \int_0^{\alpha_1} (\xi_t - \xi_{\tau_1}) \, dt + \int_{\alpha_{n-1}}^1 (\xi_t - \xi_{\tau_n}) \, dt \right| \leq \mathbf{E} \left| \int_\tau^{\tau + \frac{t_2}{2}} (\xi_t - \xi_{\tau_1}) \, dt \right|$$

$$+ \mathbf{E} \left| \int_0^\tau (\xi_t - \xi_{\tau_1}) \, dt + \int_{\tau + t_n}^1 (\xi_t - \xi_{\tau_n}) \, dt \right| + \mathbf{E} \left| \int_{\tau + \frac{t_n + t_{n-1}}{2}}^{\tau + t_n} (\xi_t - \xi_{\tau_n}) \, dt \right|$$

$$\leq \mathbf{E} \left| \int_0^\tau (\xi_t - \xi_{\tau_1}) \, dt + \int_{\tau + t_n}^1 (\xi_t - \xi_{\tau_n}) \, dt \right| + \int_0^{\frac{t_2 - t_1}{2}} \mathbf{E} \left| \xi_{\tau_1 + t} - \xi_{\tau_1} \right| dt$$

$$+ \int_0^{\frac{t_n - t_{n-1}}{2}} \mathbf{E} \left| \xi_{\tau_n - t} - \xi_{\tau_n} \right| dt \leq \mathbf{E} \left| \int_0^\tau (\xi_t - \xi_{\tau_1}) \, dt + \int_{\tau + t_n}^1 (\xi_t - \xi_{\tau_n}) \, dt \right|$$

$$+ \int_0^{\frac{t_2 - t_1}{2}} \omega(t) \, dt + \int_0^{\frac{t_n - t_{n-1}}{2}} \omega(t) \, dt. \quad (1.132)$$

Due to Lemmas 1.30 and 1.17,

$$\mathbf{E}\left|\int_0^\tau (\xi_t - \xi_{\tau_1})\,dt + \int_{\tau+t_n}^1 (\xi_t - \xi_{\tau_n})\,dt\right| \leq \sup_{\substack{\zeta \in \mathcal{H}_\tau^\omega(1-t_n),\\ \zeta_\tau \equiv 0}} \mathbf{E}\left|\int_0^{1-t_n} \zeta_t\,dt\right|$$

$$\leq \int_0^{\frac{1-t_n}{2}-t^*} \omega(s)\,ds + \int_0^{\frac{1-t_n}{2}+t^*} \omega(s)\,ds$$

The latter inequality, together with inequalities (1.130), (1.131) and (1.132) give the estimate from above (1.129).

1.4.2.3 Estimate from Below

Below we prove that

$$E(\tau, \mathbf{t}) \geq 2\sum_{k=1}^{n-1} I\left(\frac{t_{k+1} - t_k}{2}\right) + I\left(\frac{1 - t_n}{2} - t^*\right) + I\left(\frac{1 - t_n}{2} + t^*\right). \tag{1.133}$$

It is enough to prove this inequality for the case of simple random variable τ such that assumptions (1.54) and (1.55) hold.

For each $\varphi\colon \mathcal{R}^n \to \mathcal{R}$, taking into account that the class $\mathcal{H}^\omega$ is centrally symmetric, one has

$$\sup_{\xi \in \mathcal{H}^\omega} \mathbf{E}\left|\int_0^1 \xi_t\,dt - \varphi(\xi_{\tau_1}, \ldots, \xi_{\tau_n})\right| \geq \sup_{\substack{\xi \in \mathcal{H}^\omega, \xi_{\tau_k} \equiv 0,\\ k=1,\ldots,n}} \mathbf{E}\left|\int_0^1 \xi_t\,dt - \varphi(0)\right|$$

$$= \sup_{\substack{\xi \in \mathcal{H}^\omega, \xi_{\tau_k} \equiv 0,\\ k=1,\ldots,n}} \max\left(\mathbf{E}\left|\int_0^1 \xi_t\,dt - \varphi(0)\right|, \mathbf{E}\left|\int_0^1 (-\xi_t)\,dt - \varphi(0)\right|\right)$$

$$\geq \frac{1}{2} \sup_{\substack{\xi \in \mathcal{H}^\omega, \xi_{\tau_k} \equiv 0,\\ k=1,\ldots,n}} \left(\mathbf{E}\left|\int_0^1 \xi_t\,dt - \varphi(0)\right| + \mathbf{E}\left|\int_0^1 \xi_t\,dt + \varphi(0)\right|\right)$$

$$\geq \sup_{\substack{\xi \in \mathcal{H}^\omega, \xi_{\tau_k} \equiv 0,\\ k=1,\ldots,n}} \mathbf{E}\left|\int_0^1 \xi_t\,dt\right| = \sup_{\substack{\xi \in \mathcal{H}^\omega, \xi_{\tau_k} \equiv 0,\\ k=1,\ldots,n}} \mathbf{E}\int_0^1 \xi_t\,dt,$$

hence

$$E(\tau, \mathbf{t}) \geq \sup_{\substack{\xi \in \mathcal{H}^\omega, \xi_{\tau_k} \equiv 0,\\ k=1,\ldots,n}} \int_0^1 \mathbf{E}\xi_t\,dt. \tag{1.134}$$

Set $s_0 := 0$, $s_k := \tau_1 + \frac{t_k + t_{k+1}}{2}$, $k = 1, \ldots, n-1$ and $s_n := 1$. Using Lemma 1.18, define a random process $\xi_t^* := \xi_t(\Omega_1, x)$, where

$$x(t) := \omega(|t - (\tau_1 + t_k)|), \, t \in [s_{k-1}, s_k), \, k = 1, \ldots, n. \tag{1.135}$$

From the equivalent definition

1.4 Optimization of Cubature Formulae

$$x(t) = \min_{k=1,\ldots,n} \omega(|t - (\tau_1 + t_k)|), t \in [0, 1],$$

if follows that $x(t) \in H^\omega$, hence $\xi_t^* \in \mathcal{H}^\omega$. Moreover, since $x(\tau_1 + t_k) = 0, k = 1, \ldots, n$, one has $\xi_{\tau_k}^* \equiv 0, k = 1, \ldots, n$, and hence, due to (1.134),

$$E(\tau, \mathbf{t}) \geq \int_0^1 \mathbf{E}\xi_t^* dt = \int_0^1 x(t) dt.$$

Evaluating the right hand side of the latter inequality, using representation (1.135) and the fact that $t^* = \frac{1-t_n}{2} - \tau_1$, we obtain the right hand side of (1.133).

1.4.2.4 Measurement Times Optimization

We consider the problem of optimization of the information set $\{\tau_1, \ldots, \tau_n\}$, in order to minimize the error of recovery. We consider the random process ξ_t as some physical quantity and the random variables ξ_{τ_k} to be the measurements of this quantity at (possibly random) times $\tau_k, k = 1, \ldots, n$.

It appears that if the error of recovery is measured by the error for the "worst" function, then the possibility to choose time for measurements randomly does not give benefits compared to the case, when the measurements are done at some fixed, non-random times. More precisely, the following statement holds.

Corollary 1.4 *Under the assumptions of Theorem 1.28,*

$$\inf_{\tau_1, \ldots, \tau_n} E(\tau, \mathbf{t}) = 2nI\left(\frac{1}{2n}\right).$$

The optimal measurement times are given by $\tau_k = \frac{2k-1}{2n}, k = 1, \ldots, n$.

Proof Recall that $I(t) = \int_0^t \omega(s) ds$. Since ω is non-decreasing, $I(\cdot)$ is a convex function. Then for arbitrary $\alpha_1, \ldots \alpha_{2n} > 0$ one has

$$\sum_{s=1}^{2n} I(\alpha_s) \geq 2nI\left(\frac{1}{2n}\sum_{s=1}^{2n} \alpha_s\right)$$

and the statement of the corollary follows from (1.128).

Let now the measurements be done by such a device that the first measurement is triggered by some random event (which occurs at the random time τ_1) and each of the rest $n-1$ measurements are done at time $\tau_k = \tau_1 + t_k$ i.e., in t_k time units after the first measurement, $k = 2, \ldots, n$. The following statement optimizes the choice of the numbers $t_2, \ldots, t_n$, given the information about τ_1.

Theorem 1.29 *Let the assumptions of Theorem 1.28 hold and*

$$m := \operatorname*{ess\,inf}_{w \in \Omega} \tau(w), \quad M := \operatorname*{ess\,sup}_{w \in \Omega} \tau(w).$$

If

$$(2n-1)m + M \geq 1, \tag{1.136}$$

then

$$\inf_{t_2,\ldots,t_n} E(\tau, \mathbf{t}) = (2n-1)I\left(\frac{1-M}{2n-1}\right) + I(M)$$

and the infimum is attained for $t_k = \frac{2(k-1)(1-M)}{2n-1}$, $k = 2, \ldots, n$.

If

$$(2n-1)M + m \leq 1, \tag{1.137}$$

then

$$\inf_{t_2,\ldots,t_n} E(\tau, \mathbf{t}) = (2n-1)I\left(\frac{1-m}{2n-1}\right) + I(m)$$

and the infimum is attained for $t_k = \frac{2(k-1)(1-m)}{2n-1}$, $k = 2, \ldots, n$. *Otherwise,*

$$\inf_{t_2,\ldots,t_n} E(\tau, \mathbf{t}) = (2n-2)I\left(\frac{1-m-M}{2n-2}\right) + I(m) + I(M)$$

and the infimum is attained for $t_k = \frac{(k-1)(1-m-M)}{n-1}$, $k = 2, \ldots, n$.

The proof of this statement will be given in subsequent paragraphs.

1.4.2.5 Auxiliary Results

Recall that a vector $a \in \mathbb{R}^d$ majorizes a vector $b \in \mathbb{R}^d$ (denoted by $a \succ b$), iff $\sum_{i=1}^{k} a_{[i]} \geq \sum_{i=1}^{k} b_{[i]}$, $k = 1, \ldots, d-1$, and $\sum_{i=1}^{d} a_i = \sum_{i=1}^{d} b_i$, where $a_{[i]}$ and $b_{[i]}$ denote the i-th biggest coordinates of the vectors a and b respectively.

Karamata's inequality [49] states that for every convex function f and vectors $x, y \in \mathbb{R}^d$ such that $x \succ y$, one has

$$\sum_{k=1}^{d} f(x_k) \geq \sum_{k=1}^{d} f(y_k).$$

We need the following lemma.

Lemma 1.31 *Let* $x, y \in \mathbb{R}^d$ *be such that* $x \succ y$ *and* $a \in \mathbb{R}$. *Then*

$$(x_1, \ldots, x_d, a) \succ (y_1, \ldots, y_d, a). \tag{1.138}$$

1.4 Optimization of Cubature Formulae

Proof It is well known, see for example [6, Theorem 2.1] that $x \succ y$ if and only if there exists a double stochastic matrix A such that $y = Ax$. Then the matrix $B = \begin{pmatrix} A & 0 \\ 0 & 1 \end{pmatrix}$ is also double stochastic and $\begin{pmatrix} y \\ a \end{pmatrix} = B \begin{pmatrix} x \\ a \end{pmatrix}$, hence (1.138) holds. The lemma is proved.

A vector $s = (s_1, \ldots, s_n) \in \mathbb{R}^n$ with $s_k \geq 0$, $k = 1, \ldots, n-1$, $s_n \geq M$, $\sum_{k=1}^n s_k = 1$ will be called admissible.

For an admissible vector $s = (s_1, \ldots, s_n)$, set

$$s^M := \left(\frac{s_1}{2}, \frac{s_1}{2}, \frac{s_2}{2}, \frac{s_2}{2}, \ldots, \frac{s_{n-1}}{2}, \frac{s_{n-1}}{2}, s_n - M, M \right)$$

and

$$s^m := \left(\frac{s_1}{2}, \frac{s_1}{2}, \frac{s_2}{2}, \frac{s_2}{2}, \ldots, \frac{s_{n-1}}{2}, \frac{s_{n-1}}{2}, s_n - m, m \right).$$

The following lemmas will be used during the proof of Theorem 1.29.

Lemma 1.32 *Let inequality (1.136) hold and $s \in \mathbb{R}^n$ be admissible. Set*

$$L := \Big(\underbrace{\frac{1-M}{2n-1}, \ldots, \frac{1-M}{2n-1}}_{2n-1}, M \Big).$$

Then $s^M \succ L$. Moreover, if

$$s_n \geq M + m, \tag{1.139}$$

then $s^m \succ L$.

Proof Note that for arbitrary $(\alpha_1, \ldots, \alpha_d) \in \mathbb{R}^d$, $\alpha := \frac{1}{d} \sum_{k=1}^d \alpha_k$, one has $(\alpha_1, \ldots, \alpha_d) \succ (\alpha, \alpha, \ldots, \alpha) \in \mathbb{R}^d$. Hence, due to Lemma 1.31, for arbitrary admissible vector s, $s^M \succ L$.

If (1.139) holds, then $(s_n - m, m) \succ (s_n - M, M)$, hence, due to Lemma 1.31, $s^m \succ s^M \succ L$. The lemma is proved.

Lemma 1.33 *Let inequality (1.137) hold and $s \in \mathbb{R}^n$ be admissible. Set*

$$L := \Big(\underbrace{\frac{1-m}{2n-1}, \ldots, \frac{1-m}{2n-1}}_{2n-1}, m \Big).$$

Then $s^m \succ L$. If
$$s_n \leq M + m, \tag{1.140}$$
then $s^M \succ L$.

The proof is similar to the proof of Lemma 1.32.

Lemma 1.34 *Let neither of inequalities* (1.136) *and* (1.137) *hold and* $s \in \mathbb{R}^n$ *be admissible. Set*
$$L := \Big(\underbrace{\frac{1-m-M}{2n-2}, \ldots, \frac{1-m-M}{2n-2}}_{2n-2}, m, M \Big).$$

If inequality (1.139) *holds, then* $s^m \succ L$. *If inequality* (1.140) *holds, then* $s^M \succ L$.

Proof From the conditions of the lemma it follows that $m < \frac{1-M-m}{2n-2} < M$. Let inequality (1.139) hold. Then
$$\Big(\frac{s_1}{2}, \frac{s_1}{2}, \frac{s_2}{2}, \frac{s_2}{2}, \ldots, \frac{s_{n-1}}{2}, \frac{s_{n-1}}{2}, s_n - m \Big) \succ \Big(\underbrace{\frac{1-s_n}{2n-2}, \ldots, \frac{1-s_n}{2n-2}}_{2n-1}, s_n - m \Big)$$
$$\succ \Big(\underbrace{\frac{1-m-M}{2n-2}, \ldots, \frac{1-m-M}{2n-2}}_{2n-2}, M \Big),$$

where the first majorization follows from Lemma 1.31 and the second one follows from the inequalities $s_n - m \geq M > \frac{1-M-m}{2n-2}$. The inequality $s^m \succ L$ now follows from Lemma 1.31.

The second statement of the lemma follows from similar arguments, using the inequalities $s_n - M \leq m < \frac{1-M-m}{2n-2}$. The lemma is proved.

1.4.2.6 Proof of Theorem 1.29

Proof The vector with coordinates $s_k := t_{k+1} - t_k, k = 1, \ldots, n-1, s_n := 1 - t_n$ is admissible. Obviously, the numbers $t_k, k = 1, \ldots, n$, are uniquely determined by an admissible vector $s \in \mathbb{R}^n$.

Note that if (1.139) holds, then $\|\tau - \frac{s_n}{2}\|_\infty = \frac{s_n}{2} - m$ and hence, due to Theorem 1.28,
$$E(\tau, \mathbf{t}) = 2 \sum_{k=1}^{n-1} I\Big(\frac{s_k}{2}\Big) + I(m) + I(s_n - m).$$

In the case, when (1.140), $\left\|\tau - \frac{s_n}{2}\right\|_\infty = M - \frac{s_n}{2}$ and hence, due to Theorem 1.28,

$$E(\tau, \mathbf{t}) = 2\sum_{k=1}^{n-1} I\left(\frac{s_k}{2}\right) + I(M) + I(s_n - M).$$

The estimates from below for the value of $E(\tau, \mathbf{t})$ follow from Lemmas 1.32, 1.33 and 1.34 and Karamata's inequality. Thus it is sufficient to show that the estimates from below are attained.

Let inequality (1.136) hold. Then $\frac{1-M}{2n-1} \leq m$ and hence for the set $s_k^* = \frac{2(1-M)}{2n-1}$, $k = 1, \ldots, n-1$ and $s_n^* = M + \frac{1-M}{2n-1}$, inequality (1.140) holds, thus

$$E(\tau, \mathbf{t}^*) = (2n-1)I\left(\frac{1-M}{2n-1}\right) + I(M),$$

where the vector $\mathbf{t}^*$ is determined by the numbers s_k^*, $k = 1, \ldots, n$.

Let inequality (1.137) hold. Then $\frac{1-m}{2n-1} \geq M$ and hence for the set $s_k^* = \frac{2(1-m)}{2n-1}$, $k = 1, \ldots, n-1$ and $s_n^* = m + \frac{1-m}{2n-1}$, inequality (1.139) holds, thus

$$E(\tau, \mathbf{t}^*) = (2n-1)I\left(\frac{1-m}{2n-1}\right) + I(m).$$

Finally, let neither of inequalities (1.136) and (1.137) hold. Then for the set $s_k^* = \frac{(1-m-M)}{n-1}$, $k = 1, \ldots, n-1$ and $s_n^* = m + M$

$$E(\tau, \mathbf{t}^*) = (2n-2)I\left(\frac{1-m-M}{2(n-1)}\right) + I(m) + I(M).$$

The theorem is proved.

References

1. Adams R, Fournier J (2003) Sobolev spaces. Elsevier Science
2. Anastassiou G (1995) Ostrowski type inequalities. Proc AMS 123:3775–3791. https://doi.org/10.2307/2161906
3. Anastassiou G (2003) Fuzzy Ostrowski type inequalities. Comput Appl Math 22(2):279–292. https://doi.org/10.1590/S0101-82052003000200007
4. Anastassiou G (2012) Ostrowski and landau inequalities for Banach space valued functions. Math Comput Model 55:312–329. https://doi.org/10.1016/j.mcm.2011.08.003
5. Arestov V (1996) Approximation of unbounded operators by bounded operators and related extremal problems. Russ Math Surv 51(6):1093. https://doi.org/10.1070/RM1996v051n06ABEH003001

6. Arnold B (1987) Majorization and the lorenz order: a brief introduction. Lecture notes in statistics, vol 43. Springer
7. Aseev SM (1986) Quasilinear operators and their application in the theory of multivalued mappings. Proc Steklov Inst Math 167:23–52
8. Babenko V (1976) Asymptotically sharp bounds for the best quadrature formulas for several classes of functions. Math Notes 19(3):187–193. https://doi.org/10.1007/BF01437850
9. Babenko V (1976) Faithful asymptotics of remainders optimal for some glasses of functions with cubic weight formulas. Math Notes 20(4):887–890. https://doi.org/10.1007/BF01098908
10. Babenko V (1977) On the optimal error bound for cubature formulas on certain classes of continuous functions. Anal Math 3(1):3–9
11. Babenko V, Babenko V, Kovalenko O (2020) Optimal recovery of monotone operators in partially ordered L-spaces. Numer Func Anal Opt 41(11):1373–1397. https://doi.org/10.1080/01630563.2020.1775251
12. Babenko V, Babenko V, Kovalenko O (2021) Fixed point theorems in hausdorff M-distance spaces (2021). https://doi.org/10.48550/arXiv.2111.13625. Arxiv.org/abs/HausdDistSpaces
13. Babenko V, Babenko V, Kovalenko O (2021) Fixed point theorems in M-distance spaces (2021). https://doi.org/10.48550/arXiv.2103.13914. Arxiv.org/abs/2103.13914
14. Babenko V, Babenko V, Kovalenko O (2023) Korneichuk-Stechkin lemma, Ostrowski and Landau inequalities, and optimal recovery problems for L-space valued functions. Numer Func Anal Opt 44(12):1309–1341. https://doi.org/10.1080/01630563.2023.2246540
15. Babenko V, Babenko V, Kovalenko O (2024) Ostrowski-type inequalities in abstract distance spaces. Res Math 32(2):9–20. https://doi.org/10.15421/242416
16. Babenko V, Babenko V, Kovalenko O, Parfinovych N (2022) General form of (λ, φ)-additive operators on spaces of L-space-valued functions. Res Math **30**(1), 3–9 (2022). https://doi.org/10.15421/242201
17. Babenko V, Babenko V, Kovalenko O, Parfinovych N (2023) Nagy type inequalities in metric measure spaces and some applications. Carpathian Math Publ 15(2):563–575 (2023). https://doi.org/10.15330/cmp.15.2.563-575
18. Babenko V, Babenko V, Kovalenko O, Parfinovych N (2023) On Landau—Kolmogorov type inequalities for charges and their applications. Res Math 31(1):3–16. https://doi.org/10.15421/242301
19. Babenko V, Babenko V, Kovalenko O, Parfinovych N (2024) Some sharp Landau-Kolmogorov-Nagy-type inequalities in Sobolev spaces of multivariate functions. Ukr Math J 75:1525–1532. https://doi.org/10.1007/s11253-024-02275-1
20. Babenko V, Babenko V, Kovalenko O, Polishchuk M (2021) Optimal recovery of operators in function L-spaces. Anal Math 47:13–32. https://doi.org/10.1007/s10476-021-0065-y
21. Babenko V, Babenko V, Polischuk M (2016) On the optimal recovery of integrals of set-valued functions. Ukrainian Math J 67(9):1306–1315. https://doi.org/10.1007/s11253-016-1154-0
22. Babenko V, Babenko Y, Kovalenko O (2020) On multivariate Ostrowski type inequalities and their applications. Math Ineq Appl 23(2):569–583. https://doi.org/10.7153/mia-2020-23-47
23. Babenko V, Churilova M (2001) On inequalities of Kolmogorov type for derivatives of fractional order. Bull Dnepropetrovsk Univ Math 6:16–20 (in Russian)
24. Babenko V, Churilova M (2007) Kolmogorov type inequalities for hypersingular integrals with homogeneous characteristic. Banach J Math Anal 1(1):66–77. https://doi.org/10.15352/bjma/1240321556
25. Babenko V, Kofanov V, Pichugov S (1998) Inequalities of Kolmogorov type and some their applications in approximation theory. Rendiconti del Circolo Matematico di Palermo Serie II(Suppl. 52):223–237
26. Babenko V, Kofanov V, Pichugov S (2005) Inequalities of Nagy type for periodic functions. East J Approx 11(1):3–11

27. Babenko V, Korneichuk NP, Kofanov VA, Pichugov SA (2003) Inequalities for derivatives and their applications. Naukova Dumka. (in Russian)
28. Babenko V, Kovalenko O, Parfinovych N (2022) On approximation of hypersingular integral operators by bounded ones. J Math Anal Appl 513(2):126215. https://doi.org/10.1016/j.jmaa.2022.126215
29. Babenko V, Kovalenko O, Parfinovych N (2024) Kolmogorov-type inequalities for hypersingular integrals with homogeneous characteristics. Res Math **32**(2), 21–39 (2024). https://doi.org/10.15421/242417
30. Babenko V, Parfinovich N (2012) Inequalities of the Kolmogorov type for norms of Riesz derivatives of multivariate functions and some of their applications. J Math Sci 187:9–21. https://doi.org/10.1007/s10958-012-1045-3
31. Babenko V, Parfinovich N (2012) Kolmogorov type inequalities for norms of Riesz derivatives of multivariate functions and some applications. Proc. Steklov Inst. Math. 277(SUPPL. 1):9–20. https://doi.org/10.1134/S0081543812050033
32. Babenko V, Parfinovich N, Pichugov S (2014) Kolmogorov-type inequalities for norms of Riesz derivatives of functions of several variables with Laplacian bounded in L_∞ and related problems. Math Notes 95:3–14. https://doi.org/10.1134/S0001434614010015
33. Bagdasarov S (1998) Chebyshev Splines and Kolmogorov Inequalities. Operator theory Springer. https://doi.org/10.1007/978-3-0348-8808-0
34. Berezanski Y, Us G, G, SZ (2003) Functional analysis. Elsevier Science
35. Chernaya E (1995) Asymptotically exact estimation of the error of weighted cubature formulas optimal in some classes of continuous functions. Ukr Math J 47(10):1606–1618
36. Chernaya E (1995) On the optimization of weighted cubature formulae on certain classes of continuous functions. East J Approx 1:47–60
37. Clarkson JA, Adams CR (1933) On definitions of bounded variation for functions of two variables. Trans Amer Math Soc 35(4):824–854
38. Courant R, Hilbert D (2008) Methods of mathematical physics, vol 2. Wiley
39. Dragomir S (2001) On the Ostrowski's integral inequality for mappings with bounded variation and applications. Math Ineq Appl 4(1):33–40. https://doi.org/10.7153/mia-04-05
40. Dragomir S, Rassias T (2002) Ostrowski type inequalities and applications in numerical integration. Kluwer Academic Publishers, Dordrecht/Boston/London. https://doi.org/10.1007/978-94-017-2519-4
41. Dragomir SS (2017) Ostrowski type inequalities for Lebesgue integral: a survey of recent results. Australian J Math Anal Appl 14(1):1–287
42. Drozhzhina LV (1975) On quadrature formulas for random processes. Dopovidi Akad Nauk Ukrain RSR Ser A 9:775–777. (in Ukrainian)
43. Dyn N, Farkhi E, Mokhov A (2014) Approximation of set-valued functions: adaptation of classical approximation operators. World Scientific Publishing Company. https://doi.org/10.1142/p905
44. Federer H (1969) Geometric measure theory. Grundlehren der mathematischen Wissenschaften. Springer
45. Fink AM (1992) Bounds on the deviation of a function from its averages. Czechoslov Math J 42:289–310
46. Hadamard J (1914) Sur le module maximum d'une fonction et de ses derivees. C R Soc Math France 41:68–72
47. Hardy G, Littlewood J (1912) Contribution to the arithmetic theory of series. Proc London Math Soc 11(2):411–478
48. Janković S, Kadelburg Z, Radenović S (2011) On cone metric spaces: a survey. Nonlinear Anal Theory Methods Appl 74(7):2591–2601. https://doi.org/10.1016/j.na.2010.12.014

49. Karamata J (1932) Sur une inégalité relative aux fonctions convexes. Publ Math Univ Belgrade 1:145–148 (in French)
50. Karlin S (1968) Total positivity. No v 1 in total positivity. Stanford University Press (1968)
51. Kelley J (1964) General topology. The university series in higher Mathematics. D Van Nortrand Company (Canada) Ltd, Canada
52. Kirk W, Shahzad N (2014) Fixed point theory in distance spaces. Springer New York (2014). https://doi.org/10.1007/978-3-319-10927-5
53. Kofanov V, Popovich I (2020) Sharp Nagy type inequalities for the classes of functions with given quotient of the uniform norms of positive and negative parts of a function. Res Math 28(1):3–11. https://doi.org/10.15421/242001
54. Kolmogorov A (1938) Une generalisation de J.Hadamard entre les bornes superieures des derivees succesives d'une fonction. CR Acad Sci 36:764–765
55. Korneichuk N (1991) Exact constants in approximation theory. Encyclopedia of mathematics and its applications. Cambridge University Press. https://doi.org/10.1017/CBO9781107325791
56. Kovalenko O (2017) Ostrowski type inequalities for sets and functions of bounded variation. J Inequal Appl 151. https://doi.org/10.1186/s13660-017-1429-5
57. Kovalenko O (2020) On multidimensional Ostrowski-type inequalities. Ukr Math J 72:741–758. https://doi.org/10.1007/s11253-020-01814-w
58. Kovalenko O (2020) On optimal recovery of integrals of random processes. J Math Anal Appl 487(1):123949. https://doi.org/10.1016/j.jmaa.2020.123949
59. Kovalenko O (2024) On a general approach to some problems of approximation of operators. J Math Sci 279:67–76. https://doi.org/10.1007/s10958-024-06987-4
60. Kovalenko O (2024) On optimization of cubature formulae for Sobolev classes of functions defined on star domains. Mat Stud 61(1):84–96. https://doi.org/10.30970/ms.61.1.84-96
61. Landau E (1913) Einige Ungleichungen fur zweimal differenzierbare Funktionen. Proc London Math Soc 13:43–49
62. Lieb E, Loss M (2001) Analysis. CRM Proceedings and lecture notes. American Mathematical Society (2001)
63. Mitrinovic DS, Pecaric J, Fink AM (1994) Inequalities involving functions and their integrals and derivatives. Kluwer Academic Publishers, Dordrecht
64. Motornyi VP, Babenko VF, Dovgoshei AA, Kuznetsova OI (2011) Approximation theory and harmonic analysis. Naukova Dumka, Kyiv
65. Nikolsky SM (1946) Fourier series of functions with a given modulus of continuity. Dokl Akad Nauk SSSR 52(3):191–194
66. Ostrowski A (1938) Uber die Absolutabweichung einer differentienbaren Funktionen von ihren Integralmittelwert. Comment Math Hel 10:226–227
67. Parfinovich NV (2015) Inequalities of Kolmogorov type for norms of hypersingular integrals with homogeneous characteristic of constant sign. Res Math 20:58–69 (in Russian)
68. Rockafellar RT (1970) Convex analysis. Princeton University Press. https://doi.org/10.1515/9781400873173
69. Samko S, Kilbas A, Marichev OI (1993) Fractional integrals and derivatives: theory and applications. Gordon and Breach Science Publishers, Yveron
70. Smolyak SA (1965) On optimal restoration of functions and functionals of them. Candidate's Dissertation, Physical-Mathematical Sciences. Moscow State University. (in Russian)
71. Stechkin S (1967) Best approximation of linear operators. Math Notes 1(2):91–99
72. Stein E (1970) Singular integrals and differentiability properties of functions. Monographs in harmonic analysis. Princeton University Press
73. Stepanets AI (2018) Uniform Approximations by Trigonometric Polynomials. Walter de Gruyter GmbH & Co KG. https://doi.org/10.1515/9783110926033

74. SZ-Nagy B (1941) Über Integralungleichungen zwischen einer Function und ihrer Ableitung. Acta Sci Math 10:64–74
75. Vahrameev SA (1980) Applied mathematics and mathematical software of computers. M: MSU Publisher. (in Russian)
76. Vulich B (1967) Introduction to the theory of partially ordered spaces. Noordhoff
77. Zabrejko PP (1997) K-metric and K-normed linear spaces: survey. Collect Math 48(4-5-6):825–859

The Bojanov–Naidenov Problem for Differentiable Functions and the Erdös Problem for Polynomials and Splines

Abstract

It is well known that the problem to find a sharp constant in a Kolmogorov-type inequality for functions defined on the real axis, is equivalent to the extremal Kolmogorov problem to find the exact upper bound of the norm of an intermediate derivative of a function on the class of functions with restrictions on the norms of the function and its higher derivative. Despite a large number of works devoted to Kolmogorov-type inequalities, sharp constants for derivatives of arbitrary order are known only in a few cases. Therefore, the modification of the Kolmogorov problem considered by Boyanov and Naidyonov is interesting. In this modification, the norm of the intermediate derivative on the entire line is substituted by its norm on an arbitrary finite segment. In this chapter, the Boyanov-Naidyonov problem is solved on classes of functions with a given comparison function for norms of the positive and negative parts of the intermediate derivative of the function. In particular, this problem is solved on the Sobolev classes and on the spaces of trigonometric polynomials and polynomial splines. In addition, a solution to an analogue of the Erdös problem is obtained; we characterize a polynomial (spline) with a given uniform norm that has maximal possible total length of the arcs of the graph of its positive (negative) part on a given segment.

2.1 Introduction

Let $G = \mathbb{R}$ or $G = [\alpha, \beta]$. Consider the spaces $L_p(G)$, $0 < p \leq \infty$, of all Lebesgue-measurable functions such that $\|x\|_{L_p(G)} < \infty$, where

$$\|x\|_{L_p(G)} := \begin{cases} \left(\int_G |x(t)|^p \, dt\right)^{1/p}, & \text{if } 0 < p < \infty; \\ \operatorname*{ess\,sup}_{t \in G} |x(t)|, & \text{if } p = \infty. \end{cases}$$

For $r \in \mathbb{N}$ and $p, s \in (0, \infty]$, by $L^r_{p,s}$ we denote the space of all functions $x \in L_p(\mathbb{R})$ with locally absolutely continuous derivatives up to the $(r-1)$-th order inclusively and such that $x^{(r)} \in L_s(\mathbb{R})$. We write $\|x\|_p$ instead of $\|x\|_{L_p(\mathbb{R})}$ and L^r_∞ instead of $L^r_{\infty,\infty}$.

It is known (see e.g., [3, p. 47]) that the problem to determine the sharp constant C in the Kolmogorov–Nagy-type inequality

$$\|x^{(k)}\|_q \leq C \|x\|_p^\alpha \|x^{(r)}\|_s^{1-\alpha} \tag{2.1}$$

on the class $L^r_{p,s}$, where $\alpha = \frac{r-k+1/q-1/s}{r+1/p-1/s}$, $q, p, s \geq 1$, and the parameters $r \in \mathbb{N}$ and $k \in \mathbb{N}_0 := \mathbb{N} \cup \{0\}$, $k < r$ satisfy the condition $\alpha \leq (r-k)/r$, is equivalent to the extremal problem

$$\|x^{(k)}\|_q \to \sup \tag{2.2}$$

on the class of functions $x \in L^r_{p,s}$ with the following restrictions:

$$\|x^{(r)}\|_s \leq A_r, \quad \|x\|_p \leq A_0, \tag{2.3}$$

where A_0 and A_r are given positive numbers.

There are numerous works devoted to this class of problems (for a detailed bibliography, see [1, 3, 16]). Note that the question of coincidence of the sharp constants in inequalities of type (2.1) for periodic functions and the same inequalities for non-periodic functions on the axis was investigated in [2]. Despite a large number of works devoted to inequalities of the form (2.1), a sharp constant C in this inequality is known for all $r \in \mathbb{N}$ and all $k < r$ only in a few cases. For this reason it is of interest to analyze the Bojanov–Naidenov modification of problem (2.2) with restrictions (2.3) proposed in [4].

We say that $f \in L^1_\infty$ is a comparison function for $x \in L^1_\infty$, if $\|x_\pm\|_\infty \leq \|f_\pm\|_\infty$ and the equality $x(\xi) = f(\eta), \xi, \eta \in \mathbb{R}$, yields the inequality

$$|x'(\xi)| \leq |f'(\eta)|,$$

provided that the indicated derivatives exist.

We say that an odd 2ω-periodic function $\varphi \in L^1_\infty$ is an S-function, if it has the following properties:

1. φ is even with respect to $\omega/2$;
2. $|\varphi|$ is concave on $[0, \omega]$;
3. $|\varphi|$ is strictly monotone on $[0, \omega/2]$.

2.1 Introduction

For $k = 0, 1, 2, \ldots$ and an S-function $\varphi \in L_\infty^{k+1}$, by S_φ^k we denote the class of functions $x \in L_\infty^{k+1}$ such that $\varphi^{(i)}$ is a comparison function for $x^{(i)}$, $i = 0, 1, \ldots, k$. As examples of the classes S_φ^k, we can mention the Sobolev classes

$$\{x \in L_\infty^r : \|x^{(r)}\|_\infty \leq A_r, \|x\|_\infty \leq A_0\},$$

bounded subsets of the spaces T_n (of trigonometric polynomials of degree $\leq n$), and the spaces $S_{n,r}$ (of splines of order r with defect 1 and nodes at the points $l\pi/n, l \in \mathbb{Z}$).

For an arbitrary segment $[\alpha, \beta] \subset \mathbb{R}$, in [4], Bojanov and Naidenov solved the following problem

$$\int_\alpha^\beta \Phi(|x^{(k)}(t)|)dt \to \sup, k = 1, 2, \ldots$$

on the class S_φ^k, where Φ is a continuously differentiable function on $[0, \infty)$ such that $\Phi(t)/t$ is non-decreasing and $\Phi(0) = 0$. As a result, they solved the Erdös problem to characterize a trigonometric polynomial with fixed uniform norm whose graph has the maximal length on a given segment $[\alpha, \beta] \subset \mathbb{R}$, see [5]. For continuous splines on the axis, this problem was solved in [10].

By W we denote the class of continuous, nonnegative, and convex functions Φ defined on $[0, \infty)$ and such that $\Phi(0) = 0$. For $p > 0$, we set [17]

$$L(x)_p := \sup \left\{ \left(\int_a^b |x(t)|^p \, dt \right)^{\frac{1}{p}} : a, b \in \mathbb{R}, |x(t)| > 0, t \in (a, b) \right\}. \tag{2.4}$$

Note that $L(x)_\infty = \|x\|_\infty$ and $L(x')_1 \leq 2\|x\|_\infty$.

In [6–8] the Bojanov–Naidenov problem was also solved for $k = 0$, namely the extremal problem

$$\int_\alpha^\beta \Phi(|x(t)|^p)dt \to \sup, \Phi \in W, p > 0,$$

was solved on the class of functions S_φ^0 satisfying the condition $L(x)_p \leq L(\varphi)_p$. As a result, we obtained a solution of the problem

$$\int_\alpha^\beta \Phi(|x^{(k)}(t)|)dt \to \sup, \Phi \in W, k = 1, 2, \ldots$$

on the classes of functions $x \in S_\varphi^k$.

The Bojanov–Naidenov problem and the Kolmogorov–Nagy-type inequalities for functions with asymmetric restrictions imposed on the higher derivative were studied in [9, 12]. Among other works devoted to related problems, we can mention [11, 13].

In this chapter, we solve the problem (see Theorem 2.1)

$$\int_a^b \Phi(x_\pm^p(t))dt \to \sup, \Phi \in W, p > 0, \qquad (2.5)$$

on the class of pairs (x, I) of functions $x \in S_\varphi^0$ and segments $I = [a, b]$ such that $L(x)_p \le L(\varphi)_p$ and the following condition is satisfied:

$$\mu\left(\operatorname{supp}_{[a,b]} x_\pm\right) \le \mu, \mu > 0. \qquad (2.6)$$

In addition, we also solve (Theorem 2.2) the problem

$$\int_a^b \Phi(x_\pm^{(k)}(t))dt \to \sup, \Phi \in W, k = 1, 2, \ldots \qquad (2.7)$$

on the class of pairs (x, I) of functions $x \in S_\varphi^k$ and segments $I = [a, b]$ for which the following condition is satisfied:

$$\mu\left(\operatorname{supp}_{[a,b]} x_\pm^{(k)}\right) \le \mu, \mu > 0, \qquad (2.8)$$

where

$$\operatorname{supp}_{[a,b]} x := \{t \in [a, b] : |x(t)| > 0\}.$$

In particular, problems (2.5) and (2.7) with restrictions (2.6) and (2.8) are solved, respectively, on the classes

$$\Omega_p^r(A_0, A_r) := \{x \in L_\infty^r : \|x^{(r)}\|_\infty \le A_r, L(x)_p \le A_0\}$$

(Theorem 2.2) and on bounded subsets of the spaces T_n and $S_{n,r}$ (Theorems 2.4 and 2.5).

In addition, we obtain a solution (Theorem 2.6) of an analogue of the Erdös problem of characterization of a pair (x, I) formed by a polynomial $T \in T_n$ with given uniform norm and a segment I whose support measure $\mu\left(\operatorname{supp}_I T'_\pm\right)$ is bounded by a given number and is such that the total length of arcs of the graph of positive (negative) part of the polynomial $T_\pm$ is maximal on the segment I. A similar problem is solved by the same theorem for splines from the set

$$\tilde{S}_{n,r} := \{s(\cdot + \tau) : s \in S_{n,r}, \tau \in \mathbb{R}\}.$$

The main results of this chapter are contained in [14].

2.2 Auxiliary Statements

Note that if a function $x \in S_\varphi^0$ satisfies the condition $L(x)_p < \infty$ with some $p > 0$, $|x(t)| > 0$ for $t \in (a, b)$, and moreover, $a = -\infty$ or $b = +\infty$, then $x(t) \to 0$, as $t \to -\infty$ or $t \to +\infty$. In this case, we assume that $x(-\infty) = 0$ or $x(+\infty) = 0$.

For a summable function x on the segment $[a, b]$, by $r(x, t)$ we denote the permutation of the function $|x|$ (see, e.g., [15, Sect. 1.3]). Moreover, we set $r(x, t) = 0$ for $t > b - a$.

Lemma 2.1 *Suppose that φ is a function with period 2ω, $p > 0$, $\Phi \in W$, and a function $x \in S_\varphi^0$ satisfies*

$$L(x)_p \leq L(\varphi)_p, \tag{2.9}$$

where the quantity $L(x)_p$ is given by equality (2.4).

If a (finite or infinite) interval $(a_\pm, b_\pm) \subset \mathbb{R}$ and a segment $[A_\pm, B_\pm] \subset \mathbb{R}$ are such that

$$x(a_\pm) = x(b_\pm) = 0, \; x_\pm(t) > 0, \; t \in (a_\pm, b_\pm), \tag{2.10}$$

and

$$\varphi(A_\pm) = \varphi(B_\pm) = 0, \; \varphi_\pm(t) > 0, \; t \in (A_\pm, B_\pm), \tag{2.11}$$

then, for any $\xi > 0$ and any function $\Phi \in W$, the following inequalities are true

$$\int_{a_\pm}^{a_\pm + \xi} \Phi\left(\overline{x}_\pm^p(t)\right) dt \leq \int_{A_\pm}^{A_\pm + \xi} \Phi\left(\overline{\varphi}_\pm^p(t)\right) dt \tag{2.12}$$

and

$$\int_{b_\pm - \xi}^{b_\pm} \Phi\left(\overline{x}_\pm^p(t)\right) dt \leq \int_{B_\pm - \xi}^{B_\pm} \Phi\left(\overline{\varphi}_\pm^p(t)\right) dt, \tag{2.13}$$

where $\overline{x}_\pm$ is the restriction of $x_\pm$ to $(a_\pm, b_\pm)$ and $\overline{\varphi}_\pm$ is the restriction of $\varphi_\pm$ to $[A_\pm, B_\pm]$. Moreover, outside the corresponding intervals, the functions $\overline{x}_\pm$ and $\overline{\varphi}_\pm$ are set to be equal to zero.

In addition, if

$$b_\pm - a_\pm \leq B_\pm - A_\pm, \tag{2.14}$$

then for any segment $[\alpha_\pm, \beta_\pm] \subset [A_\pm, B_\pm]$ such that

$$\beta_\pm - \alpha_\pm = b_\pm - a_\pm, \tag{2.15}$$

the following inequality is true:

$$\int_{a_\pm}^{b_\pm} \Phi\left(x_\pm^p(t)\right) dt \leq \int_{\alpha_\pm}^{\beta_\pm} \Phi\left(\varphi_\pm^p(t)\right) dt, \; \Phi \in W. \tag{2.16}$$

Proof We fix a function x and segments $(a_\pm, b_\pm)$ and $[A_\pm, B_\pm]$ satisfying the conditions of the lemma. We now establish inequality (2.12); inequality (2.13) can be proved similarly.

We first establish the inequality

$$\int_0^\xi r^p(\overline{x}_\pm, t)dt \leq \int_0^\xi r^p(\overline{\varphi}_\pm, t)dt, \xi > 0. \tag{2.17}$$

To do this, we first show that the difference

$$\delta_\pm(t) := r(\overline{x}_\pm, t) - r(\overline{\varphi}_\pm, t)$$

changes its sign (from minus to plus) on $[0, \infty)$ at most once. To prove this, we note that

$$\delta_\pm(0) \leq \|x_\pm\|_\infty - \|\varphi\|_\infty \leq 0 \tag{2.18}$$

because $x \in S_\varphi^0$. In view of this inequality and relations (2.10) and (2.11), for any $z_\pm \in [0, \|\overline{x}_\pm\|_{L_\infty[a_\pm, b_\pm]})$, there exist points

$$t_i^\pm \in [a_\pm, b_\pm], i = 1, \ldots, m, m \geq 2, y_j^\pm \in [A_\pm, B_\pm], j = 1, 2,$$

such that

$$z_\pm = \overline{x}_\pm(t_i^\pm) = \overline{\varphi}_\pm(y_j^\pm). \tag{2.19}$$

In view of the inclusion $x \in S_\varphi^0$, the following inequality

$$|\overline{x}_\pm'(t_i^\pm)| \leq |\overline{\varphi}_\pm'(y_j^\pm)|. \tag{2.20}$$

holds for points $t_i^\pm$ and $y_j^\pm$ satisfying relation (2.19). Thus if the points $\theta_1^\pm, \theta_2^\pm > 0$ are chosen such that

$$z_\pm = r(\overline{x}_\pm, \theta_1^\pm) = r(\overline{\varphi}_\pm, \theta_2^\pm),$$

then, by the theorem on the derivative of permutation (see, e.g., [15, Proposition 1.3.2]), in view of inequality (2.20), we get

$$|r'(\overline{x}_\pm, \theta_1^\pm)| = \left[\sum_{i=1}^m |\overline{x}_\pm'(t_i^\pm)|^{-1}\right]^{-1} \leq \left[\sum_{j=1}^2 |\overline{\varphi}_\pm'(y_j^\pm)|^{-1}\right]^{-1} = |r'(\overline{\varphi}_\pm, \theta_2^\pm)|.$$

By virtue of (2.18), this implies that the difference $\delta^\pm(t) := r(\overline{x}_\pm, t) - r(\overline{\varphi}_\pm, t)$ changes its sign (from minus to plus) on $[0, \infty)$ at most once. The same is also true for the difference

$$\delta_p^\pm(t) := r^p(\overline{x}_\pm, t) - r^p(\overline{\varphi}_\pm, t).$$

Consider an integral

$$I_p^\pm(\xi) := \int_0^\xi \delta_p^\pm(t)dt, \xi \geq 0.$$

2.2 Auxiliary Statements

It is clear that $I_p^\pm(0) = 0$ and, in view of condition (2.9), for $\xi \geq \max\{b_\pm - a_\pm, B_\pm - A_\pm\}$, we ge

$$I_p^\pm(\xi) \leq L(x_\pm)_p - L(\varphi_\pm)_p \leq 0.$$

Moreover, the derivative $(I_p^\pm)'(t) = \delta_p^\pm(t)$ changes its sign (from minus to plus) at most once. Thus,

$$I_p^\pm(\xi) \leq 0$$

for all $\xi \geq 0$. Inequality (2.17) is true. By the Hardy–Littlewood–Polya theorem (see, e.g., [15, Theorem 1.3.11]), this inequality implies that

$$\int_{a_\pm}^{b_\pm} \Phi\left(x_\pm^p(t)\right) dt \leq \int_{A_\pm}^{B_\pm} \Phi\left(\varphi_\pm^p(t)\right) dt, \; \Phi \in W. \tag{2.21}$$

We now establish inequality (2.12). Passing to the shifts of the functions $\overline{x}$ and $\overline{\varphi}$, we can assume that

$$a_\pm = A_\pm = 0. \tag{2.22}$$

In view of the inclusion $x \in S_\varphi^0$, the difference $\Delta^\pm(t) := \overline{x}_\pm(t) - \overline{\varphi}_\pm(t)$ changes its sign (from minus to plus) on $[0, \infty)$ at most once. Since the functions $f(t) = t^p$ and $\Phi \in W$ are monotonically increasing, the same is also true for the difference

$$\Delta_\Phi^\pm(t) := \Phi\left(\overline{x}_\pm^p(t)\right) - \Phi\left(\overline{\varphi}_\pm^p(t)\right).$$

We set

$$I_\Phi^\pm(\xi) := \int_0^\xi \Delta_\Phi^\pm(t) dt, \; \xi \geq 0.$$

It is clear that $I_\Phi^\pm(0) = 0$. Further, using inequality (2.21) and assumption (2.22), we obtain

$$I_\Phi^\pm(\xi) \leq \int_{a_\pm}^{b_\pm} \Phi\left(\overline{x}_\pm^p(t)\right) dt - \int_{A_\pm}^{B_\pm} \Phi\left(\overline{\varphi}_\pm^p(t)\right) dt \leq 0$$

for $\xi \geq \max\{b_\pm - a_\pm, B_\pm - A_\pm\}$. In addition, the derivative $(I_\Phi^\pm)'(t) = \Delta_\Phi^\pm(t)$ changes its sign (from minus to plus) on $[0, \infty)$ at most once. Thus,

$$I_\Phi^\pm(\xi) \leq 0 \text{ for all } \xi \geq 0.$$

In view of assumption (2.22), this is equivalent to inequality (2.12).

It remains to establish inequality (2.16) under conditions (2.14) and (2.15). Assume that the last two conditions are satisfied. Thus, passing, if necessary, to a shift of the function x, we can assume that

$$a_\pm = \alpha_\pm, \, b_\pm = \beta_\pm. \tag{2.23}$$

Hence, using the inclusion $x \in S_\varphi^0$ and condition (2.10), we arrive at the inequality

$$x_\pm(t) \leq \varphi_\pm(t), \, t \in [a_\pm, b_\pm].$$

In view of assumption (2.23), this directly yields inequality (2.16).

In the proof of Lemma 2.1, we have established inequality (2.21). Thus, the following corollary is true:

Corollary 2.1 *Under the conditions of Lemma 2.1, for any function* $\Phi \in W$, *the inequality*

$$\int_{a_\pm}^{b_\pm} \Phi\left(x_\pm^p(t)\right) dt \leq \int_{A_\pm}^{B_\pm} \Phi\left(\varphi_\pm^p(t)\right) dt = \int_0^{2\omega} \Phi\left(\varphi_\pm^p(t)\right) dt \tag{2.24}$$

is true.

Lemma 2.2 *Suppose that* φ *is an S-function with period* 2ω, $p > 0$, $\Phi \in W$, $[a, b] \subset \mathbb{R}$. *If the function* $x \in S_\varphi^0$ *satisfies the condition*

$$L(x)_p \leq L(\varphi)_p, \tag{2.25}$$

where the quantity $L(x)_p$ *is given by equality* (2.4), *and one of the requirements*

$$\delta_\pm := \mu\left(\mathrm{supp}_{[a,b]} x_\pm\right) \leq \omega, \tag{2.26}$$

then, for any function $\Phi \in W$, *the following inequality is true:*

$$\int_a^b \Phi\left(x_\pm^p(t)\right) dt \leq \int_{m^\pm - \Theta^\pm}^{m^\pm + \Theta^\pm} \Phi\left(\varphi_\pm^p(t)\right) dt. \tag{2.27}$$

Here, $m^\pm$ *are points of local maximum of the functions* $\varphi_\pm$ *and the numbers* $\Theta^\pm > 0$ *are such that*

$$\varphi(m^\pm - \Theta^\pm) = \varphi(m^\pm + \Theta^\pm), \tag{2.28}$$

and, moreover,

$$2\Theta^\pm = \delta_\pm. \tag{2.29}$$

Proof We fix a function $x \in S_\varphi^0$ and a segment $[a, b]$ satisfying the conditions of Lemma 2.1. Inequality (2.27) is established for x_+; for x_-, the proof is similar. Suppose that the segment $[a, b]$ satisfies the corresponding requirement (2.26). Assume that

2.2 Auxiliary Statements

$$x_+(a) > 0, x_+(b) > 0 \qquad (2.30)$$

[if at least one of these inequalities is not true, then the proof of inequality (2.27) is simplified].

Assume that the function x does not have zeros on (a, b). Since $L(x)_p < \infty$ by condition (2.25), there exists a (finite or infinite) interval (c, d) such that $(a, b) \subset (c, d)$ and, in addition,

$$x_+(c) = x_+(d) = 0, x_+(t) > 0, t \in (c, d).$$

By $\overline{x}_+$ we denote the restriction of x_+ to (c, d). Moreover, by $\overline{\varphi}_+$ we denote the restriction of φ_+ to $[0, 2\omega]$. Applying inequality (2.24) to the interval (c, d), we arrive at the estimate

$$\int_c^d \Phi\left(\overline{x}_+^p(t)\right) dt \leq \int_0^{2\omega} \Phi\left(\overline{\varphi}_+^p(t)\right) dt,$$

which can be rewritten in the form

$$\int_0^{d-c} \Phi\left(r^p(\overline{x}_+, t)\right) dt \leq \int_0^{2\omega} \Phi\left(r^p(\overline{\varphi}_+, t)\right) dt. \qquad (2.31)$$

As in the proof of Lemma 2.1, we can show that the difference

$$\delta_\Phi(t) := \Phi\left(r^p(\overline{x}_+, t)\right) - \Phi\left(r^p(\overline{\varphi}_+, t)\right)$$

changes its sign (from minus to plus) on $[0, \infty)$ at most once. In view of this fact and inequality (2.31), we get the following inequality:

$$\int_0^\xi \Phi\left(r^p(\overline{x}_+, t)\right) dt \leq \int_0^\xi \Phi\left(r^p(\overline{\varphi}_+, t)\right) dt, \xi > 0.$$

It is clear that this inequality also holds if $\overline{x}_+$ is the restriction of x_+ to (a, b). For the same restriction $\overline{x}_+$ we conclude that

$$\int_a^b \Phi\left((\overline{x}_+^p(t))\right) dt = \int_0^{b-a} \Phi\left(r^p(\overline{x}_+, t)\right) dt \leq \int_0^{b-a} \Phi\left(r^p(\overline{\varphi}_+, t)\right) dt$$

$$= \int_{m^+-\Theta^+}^{m^++\Theta^+} \Phi\left(\varphi_+^p(t)\right) dt,$$

where m^+ is a point of local maximum of the spline φ_+ and $\Theta^+ > 0$ satisfies conditions (2.28) and (2.29). Moreover, $\delta_+ = b - a$. Thus, in the case where x does not have zeros on (a, b), inequality (2.27) is true.

We now assume that x has zeros on (a, b). We set

$$a' := \inf\{t \in (a, b) : x_+(t) = 0\}, \, b' := \sup\{t \in (a, b) : x_+(t) = 0\}.$$

In view of (2.30), the support $\mathrm{supp}_{[a,b]} x_+$ has the form

$$\mathrm{supp}_{[a,b]} x_+ = (a, a') \cup (b', b) \cup \bigcup_k (a_k, b_k), \tag{2.32}$$

where $(a_k, b_k) \subset (a', b')$. Moreover,

$$x(a_k) = x(b_k) = 0, \, x_+(t) > 0, \, t \in (a_k, b_k)$$

(the set of these intervals (a_k, b_k) can be empty). In view of relation (2.26), assumption (2.30), and the definitions of the numbers a' and b', we obtain

$$\delta_+ = (a' - a) + (b - b') + \sum_k (b_k - a_k) \le \omega. \tag{2.33}$$

Let A_+ and B_+ be two neighboring zeros of the function φ and, moreover, $\varphi_+(t) > 0$ for $t \in (A_+, B_+)$. In view of (2.25), we have $L(x)_p < \infty$. Hence, there exist (finite or infinite) intervals (α', a') and (b', β') such that

$$x_+(\alpha') = x_+(a') = 0, \, x_+(t) > 0, \, t \in (\alpha', a')$$

and

$$x_+(b') = x_+(\beta') = 0, \, x_+(t) > 0, \, t \in (b', \beta').$$

Applying inequalities (2.12) and (2.13) to the intervals (α', a') and (b', β') and the segment $[A_+, B_+]$, we find

$$\int_{b'}^{b} \Phi\left(x_+^p(t)\right) dt \le \int_{A_+}^{A_+ + \xi} \Phi\left(\varphi_+^p(t)\right) dt, \, \xi = b - b', \tag{2.34}$$

and

$$\int_{a}^{a'} \Phi\left(x_+^p(t)\right) dt \le \int_{B_+ - \eta}^{B_+} \Phi\left(\varphi_+^p(t)\right) dt, \, \eta = a' - a. \tag{2.35}$$

In view of (2.33), $\bar{x}_+$ in inequality (2.12) can be replaced by x_+, whereas $\bar{\varphi}_+$ can be replaced by φ_+. By virtue of (2.33), there exist mutually disjoint intervals (α_k, β_k) such that

$$(\alpha_k, \beta_k) \subset (A_+ + \xi, B_+ - \eta) \text{ and } \beta_k - \alpha_k = b_k - a_k.$$

According to relation (2.16), for these intervals, the following inequality is true:

$$\int_{a_k}^{b_k} \Phi\left(x_+^p(t)\right) dt \leq \int_{\alpha_k}^{\beta_k} \Phi\left(\varphi_+^p(t)\right) dt. \qquad (2.36)$$

Using estimates (2.34) to (2.36) and relation (2.32), we get

$$\int_a^b \Phi\left(x_+^p(t)\right) dt = \int_a^{a'} \Phi\left(x_+^p(t)\right) dt + \int_{b'}^b \Phi\left(x_+^p(t)\right) dt + \sum_k \int_{a_k}^{b_k} \Phi\left(x_+^p(t)\right) dt$$

$$\leq \int_{A_+}^{A_++\xi} \Phi\left(\varphi_+^p(t)\right) dt + \int_{B_+-\eta}^{B_+} \Phi\left(\varphi_+^p(t)\right) dt + \sum_k \int_{\alpha_k}^{\beta_k} \Phi\left(\varphi_+^p(t)\right) dt.$$

Since $\beta_k - \alpha_k = b_k - a_k$, in view of (2.33), we can write

$$\xi + \eta + \sum_k (\beta_k - \alpha_k) = \delta_+.$$

Thus, the sum of integrals on the right-hand side of the obtained estimate does not exceed

$$\int_0^{\delta_+} r\left(\Phi\left(\varphi_+^p\right), t\right) dt = \int_{m^+-\Theta^+}^{m^++\Theta^+} \Phi\left(\varphi_+^p(t)\right) dt,$$

where m^+ is the point of local maximum of the function φ_{++} and $\Theta^+ > 0$ satisfies relations (2.28) and (2.29). Inequality (2.27) is proved.

Corollary 2.2 *Under the conditions of Lemma 2.2 and in the case where one of the assumptions $\mu\left(\mathrm{supp}_{[a,b]} x_\pm\right) \leq \omega$ is true, the corresponding inequality*

$$\int_a^b \Phi\left(x_\pm^p(t)\right) dt \leq \int_0^{2\omega} \Phi\left(\varphi_\pm^p(t)\right) dt \qquad (2.37)$$

holds.

2.3 The Bojanov–Naidenov Problem for the Classes of Functions with a Given Comparison Function

Let $p, \omega > 0$ and φ be an S-function with period 2ω. We set

$$L_\varphi(p, \omega) := \left\{ x \in S_\varphi^0 : L(x)_p \leq L(\varphi)_p \right\}, \qquad (2.38)$$

where the quantity $L(x)_p$ is given by equality (2.4). We fix a number $\mu > 0$ and introduce a class $L_\varphi^\pm(p, \omega, \mu)$ of pairs (x, I) of functions x and segments $I = [a, b]$ by the formula

$$L_\varphi^\pm(p, \omega, \mu) := \{(x, I) : x \in L_\varphi(p, \omega), \mu\left(\mathrm{supp}_I x_\pm\right) \leq \mu\}. \tag{2.39}$$

We rewrite the number μ in the form

$$\mu = n \cdot \omega + 2\Theta, n \in \mathbb{N} \cup \{0\}, \Theta \in [0, \omega/2). \tag{2.40}$$

Note that if the numbers $\tau^\pm \in \mathbb{R}$ and the segment $[A, B]$ are such that

$$B - A = 2n \cdot \omega + 2\Theta, \tag{2.41}$$

$$\varphi_\pm(A + \Theta + \tau^\pm) = \varphi_\pm(B - \Theta + \tau^\pm) = \|\varphi\|_\infty, \tag{2.42}$$

then $\left(\varphi(\cdot + \tau^\pm), [A, B]\right) \in L_\varphi^\pm(p, \omega, \mu)$.

Theorem 2.1 *Suppose that $p, \omega, \mu > 0$, and φ is an S-function with period 2ω. Then, for any function $\Phi \in W$,*

$$\sup\left\{\int_a^b \Phi\left(x_\pm^p(t)\right) dt : (x, [a, b]) \in L_\varphi^\pm(p, \omega, \mu)\right\} = \int_A^B \Phi\left(\varphi_\pm^p(t + \tau^\pm)\right) dt,$$

where the sets $L_\varphi^\pm(p, \omega, \mu)$, the numbers $\tau^\pm$, and the segment $[A, B]$ are given by relations (2.38) to (2.42).

Proof We fix an arbitrary pair $(x, I) \in L_\varphi^\pm(p, \omega, \mu)$ formed by a function x and a segment $I = [a, b]$. We prove the theorem for x_+; for x_- the proof is similar. To do this, we first establish the inequality

$$\mathcal{I} := \int_a^b \Phi\left(x_+^p(t)\right) dt \leq \int_A^B \Phi\left(\varphi_+^p(t + \tau^+)\right) dt := \mathcal{I}(\mu). \tag{2.43}$$

We first consider the case where $\mathrm{supp}_{[a, b]} x_+ = \mu$. Since μ satisfies relation (2.40), the segment $[a, b]$ can be rewritten in the form

$$[a, b] = \bigcup_{k=1}^n [\alpha_k, \beta_k] \cup [\alpha, \beta].$$

Moreover, the intervals (α_k, β_k) and (α, β) are mutually disjoint and

$$\mu(\mathrm{supp}_{[\alpha_k, \beta_k]} x_+) = \omega, \mu(\mathrm{supp}_{[\alpha, \beta]} x_+) = 2\Theta.$$

2.3 The Bojanov–Naidenov Problem for the Classes of Functions ...

Hence,

$$\int_a^b \Phi\left(x_+^p(t)\right) dt = \sum_{k=1}^n \int_{\alpha_k}^{\beta_k} \Phi\left(x_+^p(t)\right) dt + \int_\alpha^\beta \Phi\left(x_+^p(t)\right) dt.$$

To estimate the integrals on the right-hand side of this equality, we apply inequalities (2.37) and (2.27) and

$$\int_a^b \Phi\left(x_+^p(t)\right) dt \leq n \int_0^{2\omega} \Phi\left(\varphi_+^p(t)\right) dt + \int_{m^+-\Theta}^{m^++\Theta} \Phi\left(\varphi_+^p(t)\right) dt$$

$$= \int_A^B \Phi\left(\varphi_+^p(t+\tau^+)\right) dt,$$

where m^+ is the point of local maximum of the function φ_+, and the last equality in this sequence of relations follows from (2.41). Thus, inequality (2.43) is established in the case where $\mathrm{supp}_{[a,\,b]}x_+ = \mu$.

Now let $\mu_1 := \mathrm{supp}_{[a,\,b]}x_+ < \mu$.

Note that the number μ can be uniquely represented in the form (2.40). Hence, the segment $[A, B]$ and the number τ^+ are uniquely (up to a shift) determined by this number. Therefore, the integral $I(\mu)$ on the right-hand side of (2.43) is uniquely determined by the number μ. Moreover, it is clear that $I(\mu)$ does not decrease as a function of μ. Hence, repeating the reasoning used in the previous case, we obtain the following estimate for the integral I on the left-hand side of (2.43):

$$I \leq I(\mu_1) \leq I(\mu).$$

Thus, the proof of inequality (2.43) is completed.

Note that, for the pair $\left(\varphi(\cdot + \tau^\pm), [A, B]\right) \in L_\varphi^\pm(p, \omega, \mu)$ formed by a function $x(\cdot) = \varphi(\cdot + \tau^+)$ and a segment $[A, B]$ given by relations (2.40)–(2.42), inequality (2.43) turns into equality.

Let $k \in \mathbb{N}$, and let φ be an S-function with period 2ω such that $\varphi \in L_\infty^{k+1}$. Thus, $\varphi^{(i)}$ is a comparison function for $x^{(i)}$, $i = 0, 1, \ldots, k$. Therefore,

$$L(x^{(k)})_1 \leq 2\|x^{(k-1)}\|_\infty \leq 2\|\varphi^{(k-1)}\|_\infty = L(\varphi^{(k)})_1. \qquad (2.44)$$

Hence, $x^{(k)} \in S_{\varphi^{(k)}}(1, \omega)$. We fix a number $\mu > 0$ and introduce a class of pairs (x, I) of functions x and segments $I = [a, b]$ by the formula

$$S_{\varphi,\,k}^\pm(\omega, \mu) := \left\{(x, I) : x \in S_\varphi^k,\, \mu\left(\mathrm{supp}_I x_\pm^{(k)}\right) \leq \mu\right\}. \qquad (2.45)$$

Using these definitions and relation (2.44), we arrive at the implication

$$(x, I) \in S_{\varphi, k}^{\pm}(\omega, \mu) \implies (x^{(k)}, I) \in L_{\varphi^{(k)}}^{\pm}(1, \omega, \mu), \qquad (2.46)$$

where the sets $L_{\varphi}^{\pm}(p, \omega, \mu)$ are determined in (2.39).

We represent the number μ in the form

$$\mu = n \cdot \omega + 2\Theta, \, n \in \mathbb{N} \cup \{0\}, \, \Theta \in (0, \omega/2). \qquad (2.47)$$

Further, we choose the numbers $\tau_k^{\pm} \in \mathbb{R}$ and the segment $[A, B]$ such that

$$B - A = 2n \cdot \omega + 2\Theta, \qquad (2.48)$$

$$\varphi_{\pm}^{(k)}(A + \Theta + \tau_k^{\pm}) = \varphi_{\pm}^{(k)}(B - \Theta + \tau_k^{\pm}) = \left\|\varphi^{(k)}\right\|_{\infty}. \qquad (2.49)$$

Then $(\varphi(\cdot + \tau^{\pm}), [A, B]) \in S_{\varphi, k}^{\pm}(\omega, \mu)$.

Theorem 2.2 *Suppose that $k \in \mathbb{N}$, $\omega, \mu > 0$, and φ is an S-function with period 2ω such that $\varphi \in L_{\infty}^{k+1}$. Then, for any function $\Phi \in W$,*

$$\sup\left\{\int_a^b \Phi\left(x_{\pm}^{(k)}(t)\right) dt : (x, I) \in S_{\varphi, k}^{\pm}(\omega, \mu)\right\} = \int_A^B \Phi\left(\varphi_{\pm}^{(k)}(t + \tau_k^{\pm})\right) dt,$$

where the set $S_{\varphi, k}^{\pm}(\omega, \mu)$, the numbers $\tau_k^{\pm}$, and the segment $[A, B]$ are given by relations (2.45) to (2.49).

Proof In view of implication (2.46), if $(x, I) \in S_{\varphi, k}^{\pm}(\omega, \mu)$, then $(x^{(k)}, I) \in L_{\varphi^{(k)}}^{\pm}(1, \omega, \mu)$, where the set $L_{\varphi}^{\pm}(p, \omega, \mu)$ is given by (2.39). Thus, applying Theorem 2.1 to the class $L_{\varphi^{(k)}}^{\pm}(1, \omega, \mu)$, we arrive at the assertion of Theorem 2.2.

Setting $\Phi(t) = t^{q/p}$ in (2.1) and $\Phi(t) = t^q$ in Theorem 2.2, we get the following corollary.

Corollary 2.3 *Let $k \in \mathbb{N}$, $p, \omega, \mu > 0$, φ be an S-function with period 2ω, and suppose $\Phi \in W$. Then for any $q \geq p$*

$$\sup\left\{\int_a^b x_{\pm}^q(t) dt : (x, I) \in L_{\varphi}^{\pm}(p, \omega, \mu)\right\} = \int_A^B \varphi_{\pm}^q(t + \tau^{\pm}) dt,$$

where the sets $L_{\varphi}^{\pm}(p, \omega, \mu)$, the numbers $\tau^{\pm}$, and the segment $[A, B]$ are given by relations (2.38) to (2.42).

In addition, if $k \in \mathbb{N}$, $\mu > 0$, and $\varphi \in L_\infty^{k+1}$, then, for any $q \geq 1$,

$$\sup\left\{\int_a^b \left(x_\pm^{(k)}(t)\right)^q dt : (x, I) \in S_{\varphi,k}^\pm(\omega, \mu)\right\} = \int_A^B \left(\varphi_\pm^{(k)}(t + \tau_k^\pm)\right)^q dt,$$

where the sets $S_{\varphi,k}^\pm(\omega, \mu)$, the numbers $\tau_k^\pm$, and the segment $[A, B]$ are given by relations (2.45) *to* (2.49).

2.4 The Bojanov–Naidenov Problem for Sobolev Classes

By $\varphi_r(t)$, $r \in \mathbb{N}$, we denote a shift of the r-th 2π-periodic integral with zero mean value over a period of the function $\varphi_0(t) = \operatorname{sgn} \sin t$. This quantity satisfies the condition $\varphi_r(0) = 0$. For $\lambda > 0$, we set $\varphi_{\lambda,r}(t) := \lambda^{-r} \varphi_r(\lambda t)$.

Let A_r, A_0, $p > 0$. We choose $\lambda > 0$ such that

$$A_0 = A_r L(\varphi_{\lambda,r})_p, \tag{2.50}$$

where the quantity $L(x)_p$ is given by equality (2.4), and set

$$\varphi(t) := A_r \varphi_{\lambda,r}(t). \tag{2.51}$$

It is clear that φ is an S-function with period $2\pi/\lambda$; moreover, $\|\varphi^{(r)}\|_\infty = A_r$, $L(\varphi)_p = A_0$. Consider the class of functions

$$\Omega_p^r(A_0, A_r) := \{x \in L_\infty^r : \|x^{(r)}\|_\infty \leq A_r, L(x)_p \leq A_0\}. \tag{2.52}$$

Lemma 2.3 ([8]) *Suppose that $r \in \mathbb{N}$, A_0, A_r, $p > 0$. Then, for any $k = 0, 1, \ldots, r - 1$*

$$\Omega_p^r(A_0, A_r) \subset S_\varphi^k,$$

where the function φ is defined by equality (2.51) *and the number λ is given by equality* (2.50).

Let $r \in \mathbb{N}$, $k = 0, 1, \ldots, r - 1$; $\mu > 0$. Consider the set of pairs (x, I) of functions x and segments $I = [\alpha, \beta]$ given by the formula

$$\Omega_p^{r,k}(A_0, A_r)_\pm := \{(x, I) : x \in \Omega_p^r(A_0, A_r), \mu\left(\operatorname{supp}_I x_\pm^{(k)}\right) \leq \mu\}. \tag{2.53}$$

We represent the number μ in the form

$$\mu = n \cdot \frac{\pi}{\lambda} + 2\Theta, n \in \mathbb{N} \cup \{0\}, \Theta \in (0, \pi/(2\lambda)). \tag{2.54}$$

Further, we choose numbers $\tau^{\pm} \in \mathbb{R}$ and a segment $[A, B]$ such that

$$B - A = 2n \cdot \frac{\pi}{\lambda} + 2\Theta, \tag{2.55}$$

$$\left(\varphi_{\lambda, r-k}\right)_{\pm} (A + \Theta + \tau^{\pm}) = \left(\varphi_{\lambda, r-k}\right)_{\pm} (B - \Theta + \tau_k^{\pm}) = \left\|\varphi_{\lambda, r-k}\right\|_{\infty}. \tag{2.56}$$

Thus, $(\varphi_{\lambda, r}(\cdot + \tau^{\pm}), [A, B]) \in \Omega_p^{r,k}(A_0, A_r)_{\pm}$.

Theorems 2.1 and 2.2 and Lemma 2.3 imply the following statement.

Theorem 2.3 *Suppose that* $r \in \mathbb{N}$, $A_0, A_r, p > 0$, $\Phi \in W$. *Then*

$$\sup \left\{ \int_{\alpha}^{\beta} \Phi(x_{\pm}^p(t)) dt : (x, [\alpha, \beta]) \in \Omega_p^{r,0}(A_0, A_r)_{\pm} \right\}$$

$$= \int_A^B \Phi((A_r \varphi_{\lambda, r})_{\pm}^p (t + \tau^{\pm})) dt.$$

At the same time, if $k \in \mathbb{N}$, $k < r$, *then*

$$\sup \left\{ \int_{\alpha}^{\beta} \Phi(x_{\pm}^{(k)}(t)) dt : (x, [\alpha, \beta]) \in \Omega_p^{r,k}(A_0, A_r)_{\pm} \right\}$$

$$= \int_A^B \Phi(A_r \left(\varphi_{\lambda, r-k}\right)_{\pm} (t + \tau^{\pm})) dt,$$

where the classes $\Omega_p^{r,k}(A_0, A_r)_{\pm}$, *the numbers* λ, $\tau^{\pm}$, *and the segment* $[A, B]$ *are given by* (2.52) *to* (2.56).

Setting $\Phi(t) = t^{q/p}$, $q \geq p$, in the first relation of Theorem 2.3 and $\Phi(t) = t^q$, $q \geq 1$, in the second relation, we obtain, as in Corollary 2.3, sharp estimates for the norms $\|x_{\pm}^{(k)}\|_{L_q[\alpha, \beta]}$, $k = 0, 1, \ldots, r-1$, in the classes $\Omega_p^{r,k}(A_0, A_r)_{\pm}$.

2.5 The Bojanov–Naidenov Problem for Trigonometric Polynomials

By T_n we denote the space of trigonometric polynomials of degree at most n. For $A_0, p > 0$, we se

$$T_n(A_0, p) := \{T \in T_n : L(T)_p \leq A_0 L(\sin n(\cdot))_p\},$$

where the quantity $L(x)_p$ is given by equality (2.4).

2.5 The Bojanov–Naidenov Problem for Trigonometric Polynomials

Lemma 2.4 ([8]) *Suppose that* $n \in \mathbb{N}$, $A_0, p > 0$. *Then for any* $k = 0, 1, \ldots$,

$$T_n(A_0, p) \subset S_\varphi^k,$$

where $\varphi(t) = A_0 \sin nt$.

Let $k \in \mathbb{N} \cup \{0\}$ and $\mu > 0$. We introduce a set of pairs (T, I) formed by polynomials T and the segment $I = [\alpha, \beta]$ by the formula

$$T_{n,k}^\pm(A_0, p, \mu) := \{(T, I) : T \in T_n(A_0, p), \mu\left(\operatorname{supp}_I T_\pm^{(k)}\right) \leq \mu\}. \tag{2.57}$$

The number μ is represented in the form

$$\mu = m \cdot \frac{\pi}{n} + 2\Theta, \, m \in \mathbb{N} \cup \{0\}, \, \Theta \in (0, \pi/(2n)). \tag{2.58}$$

Further, we choose the numbers $\tau^\pm \in \mathbb{R}$ and the segment $[A, B]$ such that

$$B - A = 2m \cdot \frac{\pi}{n} + 2\Theta, \tag{2.59}$$

$$\left(\sin n \left(A + \Theta + \tau^\pm\right)\right)_\pm = \left(\sin n \left(B - \Theta + \tau^\pm\right)\right)_\pm = 1. \tag{2.60}$$

By Theorems 2.1 and 2.2 and Lemma 2.4, we obtain the following statement.

Theorem 2.4 *Suppose that* $A_0, p, \mu > 0$, $\Phi \in W$. *Then*

$$\sup\left\{\int_\alpha^\beta \Phi(T_\pm^p(t))dt : (T, [\alpha, \beta]) \in T_{n,0}^\pm(A_0, p, \mu)\right\}$$

$$= \int_A^B \Phi\left((A_0 \sin n(t + \tau^\pm))_\pm^p\right) dt$$

and, for any $k \in \mathbb{N}$,

$$\sup\left\{\int_\alpha^\beta \Phi(T_\pm^{(k)}(t))dt : (T, [\alpha, \beta]) \in T_{n,k}^\pm(A_0, p, \mu)\right\}$$

$$= \int_A^B \Phi\left(n^k A_0 \left(\sin n(t + \tau^\pm)\right)_\pm\right) dt,$$

where the classes $T_{n,k}^\pm(A_0, p, \mu)$, *the numbers* $\tau^\pm$, *and the segment* $[A, B]$ *are given by* (2.57) *to* (2.60).

2.6 The Bojanov–Naidenov Problem for Splines

By $S_{n,r}$ we denote the space of 2π-periodic polynomial splines of order r with defect 1 and nodes at the points $k\pi/n$, $k \in \mathbb{Z}$. For $A_0, p > 0$, we set

$$S_{n,r}(A_0, p) := \{s(\cdot + \tau) : s \in S_{n,r}, L(s)_p \leq A_0 L(\varphi_{n,r})_p, \tau \in \mathbb{R}\},$$

where the quantity $L(x)_p$ is given by equality (2.4).

Lemma 2.5 ([8]) *Suppose that $r, n \in \mathbb{N}$ and $A_0, p > 0$. Then for any $k = 0, 1, \ldots, r - 1$,*

$$S_{n,r}(A_0, p) \subset S_\varphi^k,$$

where $\varphi(t) = A_0 \varphi_{n,r}(t)$.

Let $r, n \in \mathbb{N}, k = 0, 1, \ldots, r - 1, \mu > 0$. We consider the set of pairs (s, I) of splines s and segments $I = [\alpha, \beta]$ given by the formula

$$S_{n,r}^k(A_0, p, \mu)_\pm := \{(s, I) : s \in S_{n,r}(A_0, p), \mu\left(\mathrm{supp}_I s_\pm^{(k)}\right) \leq \mu\}. \tag{2.61}$$

We rewrite the number μ in the form

$$\mu = m \cdot \frac{\pi}{n} + 2\Theta, m \in \mathbb{N} \cup \{0\}, \Theta \in (0, \pi/(2n)). \tag{2.62}$$

Further, we choose the numbers $\tau^\pm$ and the segment $[A, B]$ such that

$$B - A = 2m \cdot \frac{\pi}{n} + 2\Theta, \tag{2.63}$$

$$\left(\varphi_{n,r-k}\right)_\pm \left(A + \Theta + \tau^\pm\right) = \left(\varphi_{n,r-k}\right)_\pm \left(B - \Theta + \tau^\pm\right) = \|\varphi_{n,r-k}\|_\infty. \tag{2.64}$$

Using Theorems 2.1 and 2.2 and Lemma 2.5, we obtain the following statement.

Theorem 2.5 *Suppose that $r, n \in \mathbb{N}, A_0, p, \mu > 0$ and $\Phi \in W$. Then*

$$\sup\left\{\int_\alpha^\beta \Phi(s_\pm^p(t))dt : (s, [\alpha, \beta]) \in S_{n,r}^0(A_0, p, \mu)_\pm\right\}$$

$$= \int_A^B \Phi((A_0 \varphi_{n,r})_\pm^p(t + \tau^\pm))dt,$$

and for any $k = 1, 2, \ldots, r - 1$

$$\sup\left\{\int_\alpha^\beta \Phi(s_\pm^{(k)}(t))dt : s \in S_{n,r}^k(A_0, p, \mu)_\pm\right\} = \int_\alpha^\beta \Phi(A_0(\varphi_{n,r-k})_\pm(t+\tau^\pm)|)dt,$$

where the classes $S_{n,r}^k(A_0, p, \mu)_\pm$, the numbers $\tau^\pm$, and the segment $[A, B]$ are given by (2.61) to (2.64).

2.7 The Erdös Problem for Spaces of Trigonometric Polynomials and Splines

In [4], Bojanov and Naidenov solved the Erdös problem [5] of characterization of a trigonometric polynomial $T \in T_n$ with fixed uniform norm whose graph has the maximal length on a given segment $[\alpha, \beta] \subset \mathbb{R}$.

In the next theorem, we solve a similar problem of characterization of a pair (T, I) formed by a polynomial $T \in T_n$ with given uniform norm and a segment I whose support measure $\mu\left(\text{supp}_I T'_\pm\right)$ is bounded by a given number, and is such that the total length of arcs of the graph of positive (negative) part of the polynomial T on the segment I is maximal. In this theorem, the same problem is solved for splines from the set

$$\tilde{S}_{n,r} := \{s(\cdot + \tau) : s \in S_{n,r}, \tau \in \mathbb{R}\}.$$

It is known that the length of an arc $l[a, b]$ of the graph of a function $x \in L^1[a, b]$ is given by the formula $l[a, b] = \int_a^b \sqrt{1 + x'(t)^2} dt$.

It is clear that, for the function $\Phi_0(t) = \sqrt{1+t^2}$, the inclusion $\Phi_0 \in W$ is true. Setting $\Phi = \Phi_0$, $k = 1$, $p = \infty$ in Theorems 2.4 and 2.5, we obtain the following assertion:

Theorem 2.6 *Suppose that $n \in \mathbb{N}$, $M, \mu > 0$ and μ has the form*

$$\mu = m \cdot \frac{\pi}{n} + 2\Theta, m \in \mathbb{N} \cup \{0\}, \Theta \in (0, \pi/(2n)).$$

Among all pairs (x, I) of polynomials $x \in T_n$ with given uniform norm M and segments I from the family

$$S := \{I \subset \mathbb{R} : \mu\left(\text{supp}_I x'_\pm\right) \leq \mu\},$$

the maximal total length of arcs of the graph of positive (negative) part $x_\pm$ on the segment I has the polynomial $x(t) = M \sin n(t + \tau^\pm)$ on the segment $[A, B]$ such that

$$B - A = 2m \cdot \frac{\pi}{n} + 2\Theta, \tag{2.65}$$

$$\left(\sin n\left(A + \Theta + \tau^\pm\right)\right)_\pm = \left(\sin n\left(B - \Theta + \tau^\pm\right)\right)_\pm = 1.$$

Among all pairs (x, I) of shifts of the splines $x \in \tilde{S}_{n,r}$ with given uniform norm M and segments I of the family S, the maximal total length of arcs of the graph of positive (negative) *part $x_\pm$ on the segment I is observed for the shift of the spline $x(t) = \frac{M}{\|\varphi_{n,r}\|_\infty}\varphi_{n,r}(t + \tau^\pm)$ on the segment $[A, B]$ such that equality* (2.65) *is true and, in addition,*

$$\left(\varphi_{n,r-1}\right)_\pm \left(A + \Theta + \tau^\pm\right) = \left(\varphi_{n,r-1}\right)_\pm \left(B - \Theta + \tau^\pm\right) = \|\varphi_{n,r-1}\|_\infty.$$

References

1. Babenko V (2000) Investigations of Dnepropetrovsk mathematicians related to inequalities for derivatives of periodic functions and their applications. Ukr Math J 52:8–28. https://doi.org/10.1007/BF02514133
2. Babenko V, Kofanov V, Pichugov S (2003) Comparison of exact constants in inequalities for derivatives of functions defined on the real axis and a circle. Ukr Math J 55:699–711. https://doi.org/10.1023/B:UKMA.0000010250.39603.d4
3. Babenko V, Korneichuk N, Kofanov V, Pichugov S (2003) Inequalities for derivatives and their applications. Naukova Dumka, Kyiv (In Russian)
4. Bojanov B, Naidenov N (1999) An extension of the Landau-Kolmogorov inequality. Solution of a problem of Erdos. J Anal Math 78(5):263–280
5. Erdös P (1994) Open problems. In: Bojanov B (ed) Open problems in approximation theory. SCT Publications, Singapore, pp 238–242
6. Kofanov V (2009) On some extremal problems of different metrics for differentiable functions on the axis. Ukr Math J 61(6):908–922. https://doi.org/10.1007/s11253-009-0254-5
7. Kofanov V (2010) Some extremal problems of various metrics and sharp inequalities of Nagy-Kolmogorov type. East J Approx 16(4):313–334
8. Kofanov V (2011) Sharp upper bounds of norms of functions and their derivatives on classes of functions with given comparison function. Ukr Math J 63(7):1118–1135. https://doi.org/10.1007/s11253-011-0567-z
9. Kofanov V (2012) Inequalities for derivatives of functions on an axis with nonsymmetrically bounded higher derivatives. Ukr Math J 64(5):721–736. https://doi.org/10.1007/s11253-012-0674-5
10. Kofanov V (2014) Inequalities for nonperiodic splines on the real axis and their derivatives. Ukr Math J 66(2):242–252. https://doi.org/10.1007/s11253-014-0926-7
11. Kofanov V (2015) Inequalities of different metrics for differentiable periodic functions. Ukr Math J 67(2):230–242. https://doi.org/10.1007/s11253-015-1076-2
12. Kofanov V (2019) Bojanov–Naidenov problem for functions with asymmetric restrictions for the higher derivative. Ukr Math J 71(3):419–434. https://doi.org/10.1007/s11253-019-01655-2
13. Kofanov V (2019) Problem for differentiable functions on the real line and the inequalities of various metrics. Ukr Math J 71(6):896–911. https://doi.org/10.1007/s11253-019-01687-8
14. Kofanov V (2023) Bojanov–Naidenov problem for differentiable functions and the Erdos problem for polynomials and splines. Ukr Math J 75(2):206–224. https://doi.org/10.1007/s11253-023-02194-7
15. Korneichuk N, Babenko V, Ligun A (1992) Extremal properties of polynomials and splines. Naukova Dumka, Kyiv (In Russian)
16. Kwong M, Zettl A (1992) Norm inequalities for derivatives and differences. Springer, Berlin
17. Pinkus A, Shisha O (1982) Variations on the chebyshev and L^q theories of best approximation. J Approx Theory 35(2):148–168. https://doi.org/10.1016/0021-9045(82)90033-8

Remez-Type Inequalities

3

Abstract

Remez-type inequalities play an important role in approximation theory. This topic was initiated in the work of Remez in 1936, in which he found a sharp constant in an inequality of this type for algebraic polynomials. At the end of the 20th century, a new surge of works on this topic was observed. The efforts of many mathematicians aimed at finding the sharp constant in the Remez-type inequality for trigonometric polynomials. Only in 2019 this problem was solved in the work of Tikhonov and Yuditski. In the author's works, Remez-type inequalities were extended to wider classes of functions. In this chapter, sharp Remez-type inequalities for functions with a given comparison function are obtained in various metrics. As a result, such type of inequalities were proved for functions from the Sobolev classes, for trigonometric polynomials and polynomial splines with a given ratio of norms of their positive and negative parts.

3.1 Introduction

By I_d we denote a circle realized in the form of a segment $[0, d]$ whose ends are identified. For the sake of brevity, we write $\|x\|_p$ instead of $\|x\|_{L_p(I_{2\pi})}$.

For $r \in \mathbb{N}$, $G = \mathbb{R}$ or $G = I_d$, by $L_\infty^r(G)$ we denote the set of all functions $x \in L_\infty(G)$ with locally absolutely continuous derivatives up to the $(r-1)$-th order satisfying the condition $x^{(r)} \in L_\infty(G)$.

By $\varphi_r(t)$ we denote the shift of the r-th 2π-periodic integral of the function $\varphi_0(t) = \operatorname{sgn} \sin t$ with zero mean value over the period such that $\varphi_r(0) = 0$. For $\lambda > 0$, we set $\varphi_{\lambda,r}(t) := \lambda^{-r}\varphi_r(\lambda t)$.

The following theorem was proved in [1]:

Theorem A. *Suppose that $r \in \mathbb{N}$ and $q > p > 0$. Then for any function $x \in L_\infty^r(I_{2\pi})$ that has zeroes, the following sharp on class $L_\infty^r(I_{2\pi})$ inequality holds.*

$$\|x\|_q \leq \sup_{c \in [0, K_r]} \frac{\|\varphi_r + c\|_q}{\|\varphi_r + c\|_p^\alpha} \|x\|_p^\alpha \|x^{(r)}\|_\infty^{1-\alpha}, \tag{3.1}$$

where $\alpha = \frac{r+1/q}{r+1/p}$, $K_r := \|\varphi_r\|_\infty$ is the Favard constant.

In the proof of inequality (3.1) in [1] it was established that if for a given function $x \in L_\infty^r(I_{2\pi})$ that has zeroes, the number $c \in [-K_r, K_r]$ is chosen to guarantee that the condition

$$\frac{\|x_+\|_p}{\|x_-\|_p} = \frac{\|(\varphi_r + c)_+\|_p}{\|(\varphi_r + c)_-\|_p}$$

is satisfied, then the inequality

$$\|x_\pm\|_q \leq \frac{\|(\varphi_r + c)_\pm\|_q}{\|(\varphi_r + c)_\pm\|_p^\alpha} \|x_\pm\|_p^\alpha \|x^{(r)}\|_\infty^{1-\alpha} \tag{3.2}$$

is true.

An analog of inequality (3.1) in which the L_q-norm of a periodic function is estimated via its local L_p-norm was established in [8]. Sufficient conditions under which the least upper bound in inequality (3.1) is attained for $c = 0$ were established in [10].

In this chapter we generalize inequalities (3.1) and (3.2) to the classes of functions with given comparison function. Moreover, these generalizations contain the "Remez effect". We now present necessary definitions.

Let us remind that a function $f \in L_\infty^1(\mathbb{R})$ is called a comparison function for a function $x \in L_\infty^1(\mathbb{R})$, if there exists $c \in \mathbb{R}$ such that

$$\min_{t \in \mathbb{R}} f(t) + c \leq x(t) \leq \max_{t \in \mathbb{R}} f(t) + c, \, t \in \mathbb{R},$$

and the equality $x(\xi) = f(\eta) + c$, where $\xi, \eta \in \mathbb{R}$, yields the inequality $|x'(\xi)| \leq |f'(\eta)|$ provided that the indicated derivatives exist. An odd 2ω-periodic function $\varphi \in L_\infty^1(I_{2\omega})$ is called an S-function if it has the following properties: φ is even with respect to $\omega/2$, $|\varphi|$ is convex upward on $[0, \omega]$ and strictly monotone on $[0, \omega/2]$. For a 2ω-periodic S-function φ, by $S_\varphi(\omega)$ we denote the class of functions $x \in L_\infty^1(I_d)$ for which φ is a comparison function. Note that the classes $S_\varphi(\omega)$ were considered in [3, 9].

An important role in the approximation theory is played by the Remez-type inequalities

$$\|T\|_{L_\infty(I_{2\pi})} \leq C(n, \beta) \|T\|_{L_\infty(I_{2\pi} \setminus B)} \tag{3.3}$$

on the class T_n, where B is an arbitrary Lebesgue-measurable set $B \subset I_{2\pi}$, $\mu B \leq \beta$.

The foundations of this direction were laid by Remez [20] who determined the sharp constant $C(n, \beta)$ in an inequality of the form (3.3) for algebraic polynomials. In inequality (3.3) for trigonometric polynomials, two-sided estimates for the sharp constants $C(n, \beta)$

were established in a series of works. Moreover, the asymptotic behaviors of the constants $C(n, \beta)$ as $\beta \to 2\pi$ [7] and as $\beta \to 0$ [19] are known. For the bibliography in this field, see [4, 6, 7, 19]. In [19], the inequality

$$\|T\|_{L_\infty(I_{2\pi})} \leq \left(1 + 2\tan^2 \frac{n\beta}{4m}\right) \|T\|_{L_\infty(I_{2\pi} \setminus B)} \tag{3.4}$$

was proved for any polynomial $T \in T_n$ with the minimal period $2\pi/m$ and any Lebesgue-measurable set $B \subset I_{2\pi}$, $\mu B \leq \beta$, where $\beta \in (0, 2\pi m/n)$. The equality in (3.4) is attained for the polynomial

$$T(t) = \cos nx + \frac{1}{2}(1 - \cos \beta/2).$$

Recently, a sharp constant for the Remez-type inequality (3.3) for trigonometric polynomials has been found in [22].

In [11], the result obtained in [22] was generalized to the classes $S_\varphi(\omega)$. As a consequence, an analog of inequality (3.4) for polynomial splines and functions from the classes $L_\infty^r(I_d)$ was obtained. In [5, 12–14, 16] some sharp Remez-type inequalities of different metrics and Kolmogorov–Remez-type inequalities were proved for the classes $S_\varphi(\omega)$ and, in particular, for the differentiable periodic functions, trigonometric polynomials, and splines.

In the present chapter, we prove sharp Remez-type inequalities of different metrics for the functions $x \in S_\varphi(\omega)$ with given ratio of the L_p-norms of their positive and negative parts. As a consequence, we prove these inequalities for functions from the classes $L_\infty^r(I_{2\pi})$, trigonometric polynomials, and polynomial splines with given ratio of the L_p-norms of their positive and negative parts.

The main results of this chapter are published in [15].

3.2   Classes $S_\varphi(\omega)$

Theorem 3.1 *Suppose that* $q, p > 0$, $q \geq p$, φ *is an S-function with period* 2ω, *and* $\beta \in [0, 2\omega)$. *If for a d-periodic function* $x \in S_\varphi(\omega)$ *with zeroes there exists* $c \in x[-\|\varphi\|_\infty, \|\varphi\|_\infty]$ *satisfying the condition*

$$\|x_\pm\|_{L_p(I_d)} = \|(\varphi + c)_\pm\|_{L_p(I_{2\omega})}, \tag{3.5}$$

then for any Lebesgue-measurable set $B \subset I_d$, $\mu B \leq \beta$, *the following inequality is true:*

$$\|x\|_{L_q(I_d)} \leq \frac{\|\varphi + c\|_{L_q(I_{2\omega})}}{\|\varphi + c\|_{L_p(I_{2\omega} \setminus B_{y(\beta)})}} \|x\|_{L_p(I_d \setminus B)}, \tag{3.6}$$

where

$$B_y := \{t \in [0, 2\omega] : |\varphi(t) + c| > y\},$$

and, moreover, $y = y(\beta)$ *is chosen such that* $\mu B_{y(\beta)} = \beta$.

For any fixed $c \in [-\|\varphi\|_\infty, \|\varphi\|_\infty]$, inequality (3.6) is sharp on the class of functions $x \in S_\varphi(\omega)$ with zeroes satisfying condition (3.5). Equality in (3.6) is attained for the function $x(t) = \varphi(t) + c$ and the set $B = B_{y(\beta)}$.

We prove Theorem 3.1 by a series of lemmas, which are also used in the proofs of other theorems. We set
$$E_0(x)_\infty := \inf_{a \in \mathbb{R}} \|x - a\|_\infty.$$

Lemma 3.1 *Under the conditions of Theorem 3.1,*
$$\|x_\pm\|_\infty \leq \|(\varphi + c)_\pm\|_\infty, \tag{3.7}$$

and, in addition,
$$d \geq 2\omega. \tag{3.8}$$

Proof We fix a function $x \in S_\varphi(\omega)$ and a number $c \in [-\|\varphi\|_\infty, \|\varphi\|_\infty]$ satisfying the conditions of Theorem 3.1. Assume that inequality (3.7) is not true for the function x. Since φ is the comparison function for the function x, we have $E_0(x)_\infty \leq E_0(\varphi)_\infty$. Hence, the assumption made above means that exactly one inequality (3.7) is not true. Thus, let
$$\|x_+\|_\infty \leq \|(\varphi + c)_+\|_\infty, \|x_-\|_\infty > \|(\varphi + c)_-\|_\infty.$$

Then there exists $a > 0$ such that
$$\|(x + a)_+\|_\infty \leq \|(\varphi + c)_+\|_\infty, \|(x + a)_-\|_\infty = \|(\varphi + c)_-\|_\infty. \tag{3.9}$$

It is clear that $x + a \in S_\varphi(\omega)$. By m we denote the point of minimum of the function $\varphi + c$ and assume that $t_1(t_2)$ is the left (right) zero of this function nearest to m. In view of the second relation in (3.9), there exists a shift $x(\cdot + \tau)$ of the function x such that
$$x(m + \tau) + a = \varphi(m) + c.$$

In addition, since $\varphi + c$ is the comparison function for the function x, we get
$$x(t + \tau) + a \leq \varphi(t) + c < 0, t \in (t_1, t_2).$$

In view of $a > 0$, this yields the estimate
$$\|x_-\|_{L_p(I_d)} > \|(x + a)_-\|_{L_p(I_d)} \geq \|(\varphi + c)_-\|_{L_p(I_{2\omega})},$$

which contradicts to condition (3.5). Thus inequality (3.7) is proved. Relation (3.8) directly follows from (3.5) and (3.7) in view of the inclusion $x \in S_\varphi(\omega)$.

3.2 Classes $S_\varphi(\omega)$

For $f \in L_1[a, b]$, by $r(f, t)$, $t \in [0, b-a]$, we denote the permutation of the function $|f|$ (see, e.g., [18, §1.3]) and set $r(f, t) = 0$ for $t > b - a$.

Lemma 3.2 *Under the conditions of Theorem 3.1*

$$\int_0^\xi r^p(\bar{x}_\pm, t) dt \leq \int_0^\xi r^p(\bar{\varphi}_\pm, t) dt, \xi > 0, \tag{3.10}$$

where $\bar{x}$ is the restriction of x to I_d and $\bar{\varphi}$ is the restriction of $\varphi + c$ to $I_{2\omega}$. In particular

$$\|x_\pm\|_{L_q(I_d)} \leq \|(\varphi + c)_\pm\|_{L_q(I_{2\omega})}. \tag{3.11}$$

Proof To prove (3.10), we note that, in view of (3.7), for any $y_\pm \in [0, \|\bar{x}_\pm\|_\infty)$, there exist points

$$t_i^\pm \in I_d, i = 1, 2, \ldots, m, m \geq 2, y_j^\pm \in I_{2\omega}, j = 1, 2,$$

such that

$$y_\pm = \bar{x}_\pm(t_i^\pm) = \bar{\varphi}_\pm(y_j^\pm).$$

Since $\varphi + c$ is a comparison function for x, we obtain

$$|\bar{x}'_\pm(t_i^\pm)| \leq |\bar{\varphi}'_\pm(y_j^\pm)|.$$

We now show that if the points $\theta_1^\pm \in [0, d]$ and $\theta_2^\pm \in [0, 2\omega]$ satisfy the condition

$$y_\pm = r(\bar{x}_\pm, \theta_1^\pm) = r(\bar{\varphi}_\pm, \theta_2^\pm),$$

then

$$|r'(\bar{x}_\pm, \theta_1^\pm)| \leq |r'(\bar{\varphi}_\pm, \theta_2^\pm)|.$$

Indeed, this directly follows from the theorem on the derivative of permutation (see, e.g., [18, Proposition 1.3.2]). According to this theorem, we get

$$|r'(\bar{x}_\pm, \theta_1^\pm)| = \left[\sum_{i=1}^m |\bar{x}'_\pm(t_i)|^{-1}\right]^{-1} \leq \left[\sum_{j=1}^2 |\bar{\varphi}'_\pm(y_j^\pm)|^{-1}\right]^{-1} = |r'(\bar{\varphi}_\pm, \theta_2^\pm)|.$$

Using the relation

$$r(\bar{x}_\pm, 0) = \|\bar{x}_\pm\|_\infty \leq \|\bar{\varphi}_\pm\|_\infty = r(\bar{\varphi}_\pm, 0),$$

which follows from (3.7), and the fact that the L_∞-norm is preserved by permutations, we conclude that the difference

$$\Delta^\pm(t) := r(\bar{x}_\pm, t) - r(\bar{\varphi}_\pm, t)$$

changes sign on $[0, \infty)$ at most once (from minus to plus). The same is also true for the difference
$$\Delta_p^{\pm}(t) := r^p(\bar{x}_\pm, t) - r^p(\bar{\varphi}_\pm, t).$$
We set
$$I_\pm(\xi) := \int_0^\xi \Delta_p^\pm(t) dt.$$
Hence $I_\pm(0) = 0$. Since permutations preserve the L_p-norm, in view of (3.5) and (3.8), we get
$$I(d) = \|\bar{x}_\pm\|_{L_p(I_d)} - \|\bar{\varphi}_\pm\|_{L_p(I_{2\omega})} = 0.$$
Moreover, $I'_\pm(\xi) = \Delta_p^\pm(\xi)$ changes sign (from minus to plus) at most once. Thus, $I(\xi) \le 0$, $\xi > 0$, which is equivalent to (3.10). In view of the Hardy–Littlewood–Polya theorem (see e.g., [18, Theorem 1.3.1]), inequality (3.10) yields inequality (3.11).

Lemma 3.3 *Under the conditions of Theorem 3.1*
$$\|x\|_{L_p(I_d \setminus B)} \ge \|\varphi + c\|_{L_p(I_{2\omega} \setminus B_{y(\beta)})}. \tag{3.12}$$

Proof As above, let $\bar{x}$ be the restriction of x to I_d and let $\bar{\varphi}$ be the restriction of $\varphi + c$ to $I_{2\omega}$. For any measurable set $B \subset I_d$, $\mu B \le \beta$, we have, in view of the well-known property
$$\int_B |x(t)|^p dt \le \int_0^\beta r^p(\bar{x}, t) dt. \tag{3.13}$$

Further, since permutations preserve the L_p-norm, we get
$$\|x\|^p_{L_p(I_d \setminus B)} = \int_{I_d} |x(t)|^p dt - \int_B |x(t)|^p dt \ge \int_0^d r^p(\bar{x}, t) dt - \int_0^\beta r^p(\bar{x}, t) dt.$$

Using (3.5) and the inequality
$$\int_0^\xi r^p(\bar{x}, t) dt \le \int_0^\xi r^p(\bar{\varphi}, t) dt, \xi > 0,$$

which follows from (3.10) according to Proposition 1.3.6 in [18], we obtain
$$\|x\|^p_{L_p(I_d \setminus B)} \ge \int_0^{2\omega} r^p(\bar{\varphi}, t) dt - \int_0^\beta r^p(\bar{\varphi}, t) dt = \int_\beta^{2\omega} r^p(\bar{\varphi}, t) dt = \int_{I_{2\omega} \setminus B_{y(\beta)}} |\varphi(t)|^p dt.$$

This yields (3.12).

Finally, we return to the proof of Theorem 3.1.

Proof We fix a d-periodic function $x \in S_\varphi(\omega)$ that has zeroes and satisfies conditions (3.5) with some $c \in [-\|\varphi\|_\infty, \|\varphi\|_\infty]$. By Lemmas 3.2 and 3.3, this function admits estimates (3.11) and (3.12), which directly imply inequality (3.6). It is clear that this inequality is sharp.

3.3 Classes $L_\infty^r(I_{2\pi})$

Recall that the symbol $\varphi_r(t)$, $r \in \mathbb{N}$, denotes a shift of the r-th 2π-periodic integral with zero mean value over the period of the function $\varphi_0(t) = \operatorname{sgn} \sin t$ satisfying the condition $\varphi_r(0) = 0$. It is clear that the spline $\varphi_{\lambda,r}(t) := \lambda^{-r} \varphi_r(\lambda t)$, $\lambda > 0$ is an S-function with period $2\pi/\lambda$.

For $r \in \mathbb{N}$, $p > 0$ and $f_p \in [0, \infty]$, we consider a class

$$f_p L_\infty^r(I_{2\pi}) := \left\{ x \in L_\infty^r(I_{2\pi}) : \frac{\|x_+\|_p}{\|x_-\|_p} = f_p \right\}.$$

It is clear that, for given p and f_p, there exists a unique number $c \in [-K_r, K_r]$ for which

$$\varphi_r + c \in f_p L_\infty^r(I_{2\pi}). \tag{3.14}$$

Theorem 3.2 *Suppose that $r \in \mathbb{N}$, $p, q > 0$, $q \geq p$, $f_p \in [0, \infty]$, $\beta \in [0, 2\pi)$. For any function $x \in f_p L_\infty^r(I_{2\pi})$ with zeroes, and any measurable set $B \subset I_{2\pi}$ such that $\mu B \leq \beta/\lambda$, where λ is chosen from the condition*

$$\|x\|_p = \|\varphi_{\lambda,r} + \lambda^{-r} c\|_{L_p(I_{2\pi/\lambda})} \cdot \|x^{(r)}\|_\infty, \tag{3.15}$$

and the number c satisfies condition (3.14), the following inequality is true:

$$\|x\|_q \leq \frac{\|\varphi_r + c\|_q}{\|\varphi_r + c\|_{L_p(I_{2\pi} \setminus B_{y(\beta)})}^\alpha} \|x\|_{L_p(I_{2\pi} \setminus B)}^\alpha \cdot \|x^{(r)}\|_\infty^{1-\alpha}, \tag{3.16}$$

where

$$\alpha = \frac{r + 1/q}{r + 1/p}, \quad B_y := \{t \in I_{2\pi} : |\varphi_r(t) + c| > y\},$$

and, in addition, $y = y(\beta)$ is chosen such that $\mu B_{y(\beta)} = \beta$.

Inequality (3.16) is sharp in the class of all pairs (x, B) formed by a function $x \in f_p L_\infty^r(I_{2\pi})$ that has zeroes, and a measurable set $B \subset I_{2\pi}$ for which $\mu B \leq \beta/\lambda$, where λ

satisfies condition (3.15). *The equality in* (3.16) *is attained for the pair* $(x, B_{y(\beta)})$, *where* $x(t) = \varphi_r(t) + c$.

Proof We fix a function $x \in f_p L^r_\infty(I_{2\pi})$ satisfying the conditions of the theorem. Since inequality (3.16) is homogeneous, we can assume that

$$\|x^{(r)}\|_\infty = 1. \tag{3.17}$$

Thus, in view of (3.14), (3.15) and the definition of the class $f_p L^r_\infty(I_{2\pi})$, we get

$$\|x_\pm\|_p = \|(\varphi_{\lambda,r} + \lambda^{-r} c)_\pm\|_{L_p(I_{2\pi/\lambda})}. \tag{3.18}$$

For functions $x \in f_p L^r_\infty(I_{2\pi})$ satisfying this condition, inequality (3.2) holds

$$\|x_\pm\|_q \leq \frac{\|(\varphi_r + c)_\pm\|_q}{\|(\varphi_r + c)_\pm\|_p^\alpha} \|x_\pm\|_p^\alpha \|x^{(r)}\|_\infty^{1-\alpha}.$$

Using this inequality, relations (3.17) and (3.18), and the following obvious equality

$$\|(\varphi_{\lambda,r} + \lambda^{-r} c)_\pm\|_{L_p(I_{2\pi/\lambda})} = \lambda^{-(r+1/p)} \|(\varphi_r + c)_\pm\|_p, \ p > 0,$$

we arrive at the estimate

$$\|x_\pm\|_q \leq \|(\varphi_{\lambda,r} + \lambda^{-r} c)_\pm\|_{L_q(I_{2\pi/\lambda})}. \tag{3.19}$$

In particular, in view of (3.17) and (3.19) ($q = \infty$), the function x satisfies the conditions of the Kolmogorov comparison theorem [17].

According to this theorem, the spline $\varphi(t) = \varphi_{\lambda,r}(t)$ is a comparison function for the function x i.e., $x \in S_\varphi(\frac{\pi}{\lambda})$. Hence, in view of (3.18), the function x satisfies all conditions of Theorem 3.1. By virtue of this theorem, for $q \geq p$ and an arbitrary measurable set $B \subset I_{2\pi}$, $\mu B \leq \beta/\lambda$, the inequality

$$\|x\|_q \leq \frac{\|\varphi_{\lambda,r} + \lambda^{-r} c\|_{L_q(I_{2\pi/\lambda})}}{\|\varphi_{\lambda,r} + \lambda^{-r} c\|_{L_p\left(I_{2\pi/\lambda} \setminus \frac{B_{y(\beta)}}{\lambda}\right)}} \|x\|_{L_p(I_{2\pi} \setminus B)}$$

is true. It follows from the last inequality (for $q = p$) and conditions (3.15) and (3.17) that

$$\|x\|_{L_p(I_{2\pi} \setminus B)} \geq \|\varphi_{\lambda,r} + \lambda^{-r} c\|_{L_p\left(I_{2\pi/\lambda} \setminus \frac{B_{y(\beta)}}{\lambda}\right)}.$$

Combining the obtained lower estimate with inequality (3.19), in view of the obvious relation

$$\|\varphi_{\lambda,r} + \lambda^{-r} c\|_{L_p\left(I_{2\pi/\lambda} \setminus \frac{B_{y(\beta)}}{\lambda}\right)} = \lambda^{-(r+1/p)} \|\varphi_r + c\|_{L_p(I_{2\pi} \setminus B_{y(\beta)})}$$

3.3 Classes $L^r_\infty(I_{2\pi})$

and the definition $\alpha = \frac{r+1/q}{r+1/p}$, we obtain

$$\frac{\|x\|_q}{\|x\|^\alpha_{L_p(I_{2\pi}\setminus B)}} \le \frac{\|\varphi_{\lambda,r} + \lambda^{-r}c\|_{L_q(I_{2\pi/\lambda})}}{\|\varphi_{\lambda,r} + \lambda^{-r}c\|^\alpha_{L_p\left(I_{2\pi/\lambda}\setminus \frac{B_{y(\beta)}}{\lambda}\right)}} = \frac{\|\varphi_r + c\|_q}{\|\varphi_r + c\|^\alpha_{L_p(I_{2\pi}\setminus B_{y(\beta)})}}.$$

By virtue of (3.17), this estimate yields (3.16). Thus, it is clear that inequality (3.16) is sharp.

Corollary 3.1 *Suppose that* $r \in \mathbb{N}$, $p, q > 0$, $q \ge p$, $\alpha = \frac{r+1/q}{r+1/p}$, $\beta \in [0, 2\pi)$, *and the number* $\bar{c} \in [0, K_r]$ *realizes the upper bound*

$$\sup_{c \in [0, K_r]} \frac{\|\varphi_r + c\|_q}{\|\varphi_r + c\|^\alpha_{L_p(I_{2\pi}\setminus B^c_{y(\beta)})}},$$

where $B^c_y := \{t \in I_{2\pi} : |\varphi_r(t) + c| > y\}$, *and, moreover,* $y = y(\beta)$ *such that* $\mu B^c_{y(\beta)} = \beta$.
Then for any function $x \in L^r_\infty(I_{2\pi})$ *with zeroes and an arbitrary measurable set* $B \subset I_{2\pi}$, $\mu B \le \beta/\lambda$, *where* λ *is chosen to guarantee that*

$$\|x\|_p = \|\varphi_{\lambda,r} + \lambda^{-r}c\|_{L_p(I_{2\pi/\lambda})} \cdot \|x^{(r)}\|_\infty, \tag{3.20}$$

and c satisfies the condition

$$\|x_+\|_p \cdot \|x_-\|_p^{-1} = \|(\varphi_r + c)_+\|_p \cdot \|(\varphi_r + c)_-\|_p^{-1},$$

the following inequality is true:

$$\|x\|_q \le \frac{\|\varphi_r + \bar{c}\|_q}{\|\varphi_r + \bar{c}\|^\alpha_{L_p(I_{2\pi}\setminus B^{\bar{c}}_{y(\beta)})}} \|x\|^\alpha_{L_p(I_{2\pi}\setminus B)} \cdot \|x^{(r)}\|^{1-\alpha}_\infty. \tag{3.21}$$

Inequality (3.21) *is sharp on the class of all pairs* (x, B) *formed by a function* $x \in L^r_\infty(I_{2\pi})$ *with zeroes and a measurable set* $B \subset I_{2\pi}$ *such that* $\mu B \le \beta/\lambda$, *where* λ *satisfies condition* (3.20). *Equality in* (3.21) *is attained for the pair* $(x, B^{\bar{c}}_{y(\beta)})$, *where* $x(t) = \varphi_r(t) + \bar{c}$.

We note that:

1. For $\beta = 0$, Theorem 3.2 and Corollary 3.1 were proved in [1].
2. For functions $x \in L^r_\infty(I_{2\pi})$ satisfying the condition $\|x_+\|_p = \|x_-\|_p$, the constant in inequality (3.16) is equal to zero.
3. For functions of constant sign $x \in L^r_\infty(I_{2\pi})$ with zeroes, inequality (3.16) turns into the inequality for the best one-sided approximations by the constant

$$E^\pm_0(x)_{L_s(G)} := \inf_{c \in \mathbb{R}} \{\|x - c\|_{L_s(G)} : \forall t \in G, \pm(x(t) - c)_\pm \ge 0\}, \tag{3.22}$$

i.e., the norms $\|x\|_q$ and $\|x\|_{L_p(I_{2\pi}\setminus B)}$ in inequality (3.16) for these functions are replaced by $E_0^{\pm}(x)_q$ and $E_0^{\pm}(x)_{L_p(I_{2\pi}\setminus B)}$ respectively. Moreover, the constant c in this inequality is replaced by the Favard constant K_r.

3.4 Classes of Trigonometric Polynomials

Recall that T_n is the space of trigonometric polynomials of degree at most n. For $p > 0$, $f_p \in [0, \infty]$, we set

$$f_p T_n := \left\{ T \in T_n : \frac{\|T_+\|_p}{\|T_-\|_p} = f_p \right\}.$$

Theorem 3.3 *Suppose that $n, m \in \mathbb{N}$, $p, q > 0$, $q \geq p$, $f_p \in [0, \infty]$. If a trigonometric polynomial $T \in f_p T_n$ with the minimal period $2\pi/m$ has zeroes, then for any measurable set $B \subset I_{2\pi}$, $\mu B \leq \frac{m}{n}\beta$, $\beta \in [0, 2\pi)$, the following inequality is true:*

$$\|T\|_q \leq \left(\frac{n}{m}\right)^{\frac{1}{p}-\frac{1}{q}} \frac{\|\sin(\cdot) + c\|_q}{\|\sin(\cdot) + c\|_{L_p(I_{2\pi}\setminus B_{y(\beta)})}} \|T\|_{L_p(I_{2\pi}\setminus B)}, \tag{3.23}$$

where the number $c \in [-1, 1]$ satisfies the condition

$$\sin(\cdot) + c \in f_p T_n, \tag{3.24}$$

$B_y := \{t \in I_{2\pi} : |\sin t + c| > y\}$, *and $y = y(\beta)$ is such that $\mu B_{y(\beta)} = \beta$. Inequality (3.23) is sharp in the following sense:*

$$\sup_{(n,m)\in N_{n,m}} \sup_{(T,B)\in P_n^m} \frac{\|T\|_q}{(n/m)^{1/p-1/q} \|T\|_{L_p(I_{2\pi}\setminus B)}} = \frac{\|\sin(\cdot) + c\|_q}{\|\sin(\cdot) + c\|_{L_p(I_{2\pi}\setminus B_{y(\beta)})}}, \tag{3.25}$$

where $N_{n,m}$ is the set of pairs (n, m) of natural numbers such that $m \leq n$ and P_n^m is the set of pairs (T, B) formed by a polynomial $T \in f_p T_n$ with zeroes and the minimal period $2\pi/m$, and a measurable set $B \subset I_{2\pi}$, $\mu B \leq \frac{m}{n}\beta$.

Proof We fix a polynomial $T \in f_p T_n$ satisfying the conditions of Theorem 3.3. For the sake of brevity, we set $\varphi(t) := \sin nt$, $\psi(t) := \varphi(t) + c$, $t \in \mathbb{R}$. In view of the homogeneity of inequality (3.23), we can assume that

$$\|T\|_{L_p(I_{2\pi/m})} = \|\psi\|_{L_p(I_{2\pi/n})}. \tag{3.26}$$

3.4 Classes of Trigonometric Polynomials

In view of condition (3.24) and the definition of the class $f_p T_n$, this yields the equality

$$\|T_\pm\|_{L_p(I_{2\pi/m})} = \|\psi_\pm\|_{L_p(I_{2\pi/n})}. \tag{3.27}$$

We now show that

$$\|T_\pm\|_\infty \le \|\psi_\pm\|_\infty. \tag{3.28}$$

Indeed, assume the contrary i.e., that there exists $\gamma \in (0, 1)$ such that

$$\|\gamma T_\pm\|_\infty \le \|\psi_\pm\|_\infty.$$

Moreover, one of these inequalities turns into equality. Thus let

$$\|\gamma T_+\|_\infty \le \|\psi_+\|_\infty, \|\gamma T_-\|_\infty = \|\psi_-\|_\infty.$$

Then the polynomial ψ is a comparison function for the polynomial γT (see the proof of [2, Theorem 8.1.1]). Let m be a point of minimum of the function ψ and let $t_1(t_2)$ be the nearest (to m) from the left (resp. right) zero of this function. Passing, if necessary, to the shift of the polynomial γT, we can assume that

$$\|\gamma T_-\|_\infty = -\gamma T(m).$$

Since ψ is a comparison function for the polynomial γT, we get

$$\gamma T(t) \le \psi(t) < 0, t \in (t_1, t_2).$$

This yields the estimate

$$\|T_-\|_{L_p(2\pi/m)} > \|\gamma T_-\|_{L_p(2\pi/m)} \ge \|\psi_-\|_{L_p(2\pi/n)},$$

which contradicts to (3.27). Thus inequality (3.28) is proved.

This inequality and the proof of [2, Theorem 8.1.1] imply that $\varphi(t) = \sin nt$ is a comparison function for the polynomial $T(t)$ i.e., $T \in S_\varphi(\frac{\pi}{n})$. Hence, in view of (3.26), the polynomial T satisfies all conditions of Theorem 3.1 and, therefore, also the conditions of Lemmas 3.1, 3.2 and 3.3.

Next we establish the inequality

$$\|T\|_q \le \left(\frac{m}{n}\right)^{1/q} \|\sin(\cdot) + c\|_q. \tag{3.29}$$

By virtue of inequality (3.11), we obtain

$$\|T\|_{L_q(I_{2\pi/m})} \le \|\varphi + c\|_{L_q(I_{2\pi/n})}.$$

This immediately yields (3.29), because the polynomial T is $2\pi/m$-periodic and the function φ is $2\pi/n$-periodic.

We now prove the inequality

$$\|T\|_{L_p(I_{2\pi}\setminus B)} \geq \left(\frac{m}{n}\right)^{1/p} \|\sin(\cdot)+c\|_{L_p(I_{2\pi}\setminus B_{y(\beta)})} \qquad (3.30)$$

for any measurable set $B \subset I_{2\pi}$, $\mu B \leq \frac{m}{n}\beta$.

Let $\bar{T}$ be the restriction of the polynomial T to $I_{2\pi/m}$ and let $\bar{\varphi}$ be the restriction of $\varphi+c$ to $I_{2\pi/n}$. Using inequality (3.13), in view of the fact that permutation preserves the L_p-norm, we get

$$\|T\|_{L_p(I_{2\pi}\setminus B)}^p = \int_0^{2\pi} |T(t)|^p dt - \int_B |T(t)|^p dt \geq \int_0^{2\pi} r^p(T,t)dt - \int_0^{\frac{m}{n}\beta} r^p(T,t)dt$$

$$= m\left[\int_0^{2\pi/m} r^p(\bar{T},t)dt - \int_0^{\beta/n} r^p(\bar{T},t)dt\right].$$

Thus, by virtue of (3.26) and the inequality

$$\int_0^\xi r^p(\bar{T},t)dt \leq \int_0^\xi r^p(\bar{\varphi},t)dt, \xi > 0,$$

which follows from (3.10), according to [18, Proposition 1.3.6], we arrive at the following lower estimate:

$$\|T\|_{L_p(I_{2\pi}\setminus B)}^p \geq m\left[\int_0^{2\pi/n} r^p(\bar{\varphi},t)dt - \int_0^{\beta/n} r^p(\bar{\varphi},t)dt\right] = m\int_{\beta/n}^{2\pi/n} r^p(\bar{\varphi},t)dt$$

$$= \frac{m}{n}\int_\beta^{2\pi} r^p(\varphi+c,t)dt = \frac{m}{n}\int_{I_{2\pi}\setminus B_y(n)} |\varphi(t)+c|^p dt$$

$$= \frac{m}{n}\|\sin(\cdot)+c\|_{L_p(I_{2\pi}\setminus B_{y(\beta)})}^p,$$

where $B_y(n) := \{t \in I_{2\pi} : |\sin nt + c| > y\}$, and $y = y(\beta)$ is such that $\mu B_y(n) = \beta$. The obtained estimate yields inequality (3.30). Combining (3.29) and (3.30), we arrive at inequality (3.23). It is clear that (3.23) is sharp in the sense of (3.25).

Corollary 3.2 *Suppose that $n, m \in \mathbb{N}$, $q, p > 0$, $q \geq p$, $\beta \in [0, 2\pi)$, and the number $\bar{c} \in [0,1]$ realizes the upper bound*

$$\sup_{c\in[0,1]} \frac{\|\sin(\cdot)+c\|_q}{\|\sin(\cdot)+c\|_{L_p(I_{2\pi}\setminus B_{y(\beta)}^c)}},$$

where $B_y^c := \{t \in I_{2\pi} : |\sin t + c| > y\}$ and $y = y(\beta)$ is such that $\mu B_{y(\beta)}^c = \beta$. Then for any trigonometric polynomial $T \in T_n$ with zeroes and the minimal period $2\pi/m$ and any measurable set $B \subset I_{2\pi}$, $\mu B \leq \frac{m}{n}\beta$, the following inequality is true:

$$\|T\|_q \leq \left(\frac{n}{m}\right)^{\frac{1}{p}-\frac{1}{q}} \frac{\|\sin(\cdot) + \bar{c}\|_q}{\|\sin(\cdot) + \bar{c}\|_{L_p(I_{2\pi} \setminus B_{y(\beta)}^{\bar{c}})}} \|T\|_{L_p(I_{2\pi} \setminus B)}. \quad (3.31)$$

Inequality (3.31) is sharp in the following sense:

$$\sup_{(n,m) \in N_{n,m}} \sup_{(T,B) \in Q_n^m} \frac{\|T\|_q}{(n/m)^{1/p-1/q} \|T\|_{L_p(I_{2\pi} \setminus B)}} = \frac{\|\sin(\cdot) + \bar{c}\|_q}{\|\sin(\cdot) + \bar{c}\|_{L_p(I_{2\pi} \setminus B_{y(\beta)}^{\bar{c}})}},$$

where $N_{n,m}$ is the set of pairs (n, m) of natural numbers such that $m \leq n$ and Q_n^m is the set of pairs (T, B) formed by a polynomial $T \in T_n$ with zeroes and the minimal period $2\pi/m$, and a measurable set $B \subset I_{2\pi}$, $\mu B \leq \frac{m}{n}\beta$.

We note that:

1. For $\beta = 0$ and $m = 1$, Theorem 3.3 and Corollary 3.2 were proved in [1].
2. For polynomials $T \in T_n$ satisfying the condition $\|T_+\|_p = \|T_-\|_p$, the constant c in inequality (3.23) is equal to zero.
3. For sign-preserving polynomials $T \in T_n$ that have zeroes, inequality (3.23) turns into the equality for the best one-sided approximations by a constant [see (3.22)] i.e., the norms $\|T\|_q$ and $\|T\|_{L_p(I_{2\pi} \setminus B)}$ in inequality(3.23) for these polynomials should be replaced by $E_0^{\pm}(T)_q$ and $E_0^{\pm}(T)_{L_p(I_{2\pi} \setminus B)}$) respectively. Moreover, the constant c in this inequality is equal to 1.

3.5 Classes of Splines

Recall that $S_{n,r}$ is a space of 2π-periodic splines of order r with defect 1 and nodes at the points $k\pi/n$, $k \in \mathbb{Z}$. For $p > 0$, $f_p \in [0, \infty]$, we set

$$f_p S_{n,r} := \left\{ s \in S_{n,r} : \frac{\|s_+\|_p}{\|s_-\|_p} = f_p \right\}.$$

Theorem 3.4 *Suppose that $n, m \in \mathbb{N}$ $p, q > 0$, $q \geq p$, $f_p \in [0, \infty]$.*

If a spline $s \in f_p S_{n,r}$ with the minimal period $2\pi/m$ has zeroes, then for any measurable set $B \subset I_{2\pi}$, $\mu B \leq \frac{m}{n}\beta$, $\beta \in [0, 2\pi)$, the following inequality is true:

$$\|s\|_q \leq \left(\frac{n}{m}\right)^{\frac{1}{p}-\frac{1}{q}} \frac{\|\varphi_r + c\|_q}{\|\varphi_r + c\|_{L_p(I_{2\pi} \setminus B_{y(\beta)})}} \|s\|_{L_p(I_{2\pi} \setminus B)}, \quad (3.32)$$

where $c \in [-K_r, K_r]$ satisfies the condition

$$\varphi_{n,r} + n^{-r}c \in f_p S_{n,r}, \qquad (3.33)$$

$B_y := \{t \in I_{2\pi} : |\varphi_r(t) + c| > y\}$, and $y = y(\beta)$ is such that $\mu B_{y(\beta)} = \beta$.
Inequality (3.32) is sharp in the following sense:

$$\sup_{(n,m)\in N_{n,m}} \sup_{(s,B)\in S_n^m} \frac{\|s\|_q}{(n/m)^{1/p-1/q}\|s\|_{L_p(I_{2\pi}\setminus B)}} = \frac{\|\varphi_r + c\|_q}{\|\varphi_r + c\|_{L_p(I_{2\pi}\setminus B_{y(\beta)})}}, \qquad (3.34)$$

where $N_{n,m}$ is the set of pairs (n, m) of natural numbers such that $m \leq n$, and S_n^m is the set of pairs (s, B) formed by a spline $s \in f_p S_{n,r}$ with zeroes and the minimal period $2\pi/m$, and a measurable set $B \subset I_{2\pi}$, $\mu B \leq \frac{m}{n}\beta$.

Proof We fix a spline $s \in f_p S_{n,r}$ satisfying the conditions of Theorem 3.4. For the sake of brevity, we set $\varphi(t) := \varphi_{n,r}(t)$, $\psi(t) := \varphi_{n,r}(t) + n^{-r}c$, $t \in \mathbb{R}$. In view of the homogeneity of inequality (3.32), we can assume that

$$\|s\|_{L_p(I_{2\pi/m})} = \|\psi\|_{L_p(I_{2\pi/n})}. \qquad (3.35)$$

Thus, in view of (3.33) and the definition of the class $f_p S_{n,r}$, we arrive at the equality

$$\|s_\pm\|_{L_p(I_{2\pi/m})} = \|\psi_\pm\|_{L_p(I_{2\pi/n})}. \qquad (3.36)$$

Next we show that

$$\|s_\pm\|_\infty \leq \|\psi_\pm\|_\infty. \qquad (3.37)$$

Assume the contrary i.e., that there exists $\gamma \in (0, 1)$ such that $\|\gamma s_\pm\|_\infty \leq \|\psi_\pm\|_\infty$ and, in addition, that one of these inequalities turns into the equality; e.g., that

$$\|\gamma s_+\|_\infty \leq \|\psi_+\|_\infty, \|\gamma s_-\|_\infty = \|\psi_-\|_\infty.$$

Then

$$E_0(\gamma s)_\infty \leq E_0(\psi)_\infty = \|\varphi_{n,r}\|_\infty$$

and, by virtue of the Tikhomirov inequality [21]

$$\left\|s^{(r)}\right\|_\infty \leq \frac{E_0(s)_\infty}{\|\varphi_{n,r}\|_\infty},$$

where $E_0(x)_\infty$ is the best uniform approximation of the function x by constants, we arrive at the inequality

$$\left\|\gamma s^{(r)}\right\|_\infty \leq 1.$$

3.5 Classes of Splines

Thus, the spline γs satisfies the conditions of the Kolmogorov comparison theorem [17]. By this theorem, the spline φ is a comparison function for the spline γs. Let m be the point of minimum of the function ψ and let $t_1(t_2)$ be the left (resp. right) nearest (to m) zero of this function. Passing, if necessary, to a shift of the spline γs, we can assume that

$$\|\gamma s_-\|_\infty = -\gamma s(m).$$

Since the spline ψ is a comparison function for the spline γs, we get

$$\gamma s(t) \leq \psi(t) < 0, t \in (t_1, t_2).$$

This yields the estimate

$$\|s_-\|_{L_p(2\pi/m)} > \|\gamma s_-\|_{L_p(2\pi/m)} \geq \|\psi_-\|_{L_p(2\pi/n)},$$

which contradicts (3.36). Thus, inequality (3.37) is proved. Using inequality (3.37), we obtain

$$E_0(s)_\infty \leq E_0(\psi)_\infty = \|\varphi_{n,r}\|_\infty.$$

Applying the Tikhomirov inequality, we obtain

$$\left\|s^{(r)}\right\|_\infty \leq \frac{E_0(s)_\infty}{\|\varphi_{n,r}\|_\infty} \leq 1.$$

Therefore, the spline s satisfies the conditions of the Kolmogorov comparison theorem [17]. According to this theorem, the spline φ is a comparison function for the spline s. Hence, $s \in S_\varphi(\frac{\pi}{n})$ and, in view of (3.36), the spline s satisfies the conditions of Theorem 3.1 and, therefore, also the conditions of Lemmas 3.1, 3.2 and 3.3.

We prove the inequality

$$\|s\|_q \leq n^{-r} \left(\frac{m}{n}\right)^{1/q} \|\varphi + c\|_q. \tag{3.38}$$

Indeed, by virtue of inequality (3.11), we get

$$\|s\|_{L_q(I_{2\pi/m})} \leq \|\varphi_{n,r} + n^{-r}c\|_{L_q(I_{2\pi/n})}.$$

This directly yields (3.38) because the spline s is $2\pi/m$-periodic and the spline $\varphi_{n,r}$ is $2\pi/n$-periodic.

We now prove the inequality

$$\|s\|_{L_q(I_{2\pi \setminus B})} \geq n^{-r} \left(\frac{m}{n}\right)^{1/p} \|\varphi_r + c\|_{L_q(I_{2\pi \setminus B_{y(\beta)}})} \tag{3.39}$$

for any measurable set $B \subset I_{2\pi}$, $\mu B \leq \frac{m}{n}\beta$. Let $\bar{s}$ be the restriction of the spline s to $I_{2\pi/m}$ and let $\bar{\psi}$ be the restriction of the spline ψ to $I_{2\pi/n}$. As in the proof of Theorem 3.3, using inequality (3.13) and taking into account the fact that permutations preserve the L_p-norm, we obtain

$$\|s\|_{L_p(I_{2\pi}\setminus B)}^p \geq m \left[\int_0^{2\pi/m} r^p(\bar{s}, t)dt - \int_0^{\beta/n} r^p(\bar{s}, t)dt \right].$$

Further, using (3.35) and the inequality

$$\int_0^\xi r^p(\bar{s}, t)dt \leq \int_0^\xi r^p(\bar{\psi}, t)dt, \xi > 0,$$

which follows from (3.10) according to [18, Proposition 1.3.6], as in the proof of Theorem 3.3, we obtain the following lower bound

$$\|s\|_{L_p(I_{2\pi}\setminus B)}^p \geq m \left[\int_0^{2\pi/n} r^p(\bar{\psi}, t)dt - \int_0^{\beta/n} r^p(\bar{\psi}, t)dt \right] = m \int_{\beta/n}^{2\pi/n} r^p(\bar{\psi}, t)dt$$

$$= \frac{m}{n} \int_\beta^{2\pi} r^p(\psi, t)dt = \frac{m}{n} n^{-rp} \int_{I_{2\pi}\setminus B_{y(\beta)}(n)} |\varphi_r(nt) + c|^p dt$$

$$= n^{-rp} \frac{m}{n} \|(\varphi_r + c)\|_{L_p(I_{2\pi}\setminus B_{y(\beta)}))}^p,$$

where $B_{y(\beta)}(n) := \{t \in I_{2\pi} : |\varphi_r(nt) + c| > y\}$, and $y = y(\beta)$ is such that $\mu B_{y(\beta)}(n) = \beta$.

The obtained lower bound is equivalent to (3.39). Inequality (3.32) directly follows from (3.38) and (3.39). It is clear that inequality (3.32) is sharp in the sense of (3.34).

Corollary 3.3 *Suppose that $n, m \in \mathbb{N}$, $m \leq n$, $q, p > 0$, $q \geq p$, $\beta \in [0, 2\pi)$, and the number $\bar{c} \in [0, K_r]$ realizes the upper bound*

$$\sup_{c \in [0, K_r]} \frac{\|\varphi_r + c\|_q}{\|\varphi_r + c\|_{L_p(I_{2\pi}\setminus B_{y(\beta)}^c)}},$$

where $B_y^c := \{t \in I_{2\pi} : |\varphi_r(t) + c| > y\}$, and $y = y(\beta)$ is such that $\mu B_{y(\beta)}^c = \beta$. Then for any spline $s \in S_{n,r}$ with zeroes and the minimal period $2\pi/m$, and an arbitrary measurable set $B \subset I_{2\pi}$, $\mu B \leq \frac{m}{n}\beta$, the following inequality is true:

$$\|s\|_q \leq \left(\frac{n}{m}\right)^{\frac{1}{p}-\frac{1}{q}} \frac{\|\varphi_r + \bar{c}\|_q}{\|\varphi_r + \bar{c}\|_{L_p(I_{2\pi}\setminus B_{y(\beta)}^{\bar{c}})}} \|s\|_{L_p(I_{2\pi}\setminus B)}. \tag{3.40}$$

Inequality (3.40) is sharp in the following sense:

$$\sup_{(n,m)\in N_{n,m}} \sup_{(s,B)\in \Sigma_n^m} \frac{\|s\|_q}{(n/m)^{1/p-1/q}\|s\|_{L_p(I_{2\pi}\setminus B)}} = \frac{\|\varphi_r + \bar{c}\|_q}{\|\varphi_r + \bar{c}\|_{L_p(I_{2\pi}\setminus B_{y(\beta)})}},$$

where $N_{n,m}$ is the set of pairs (n, m) of natural numbers such that $m \leq n$, and Σ_n^m is the set of pairs (s, B) formed by a spline $s \in S_{n,r}$ with zeroes and the minimal period $2\pi/m$, and a measurable set $B \subset I_{2\pi}$, $\mu B \leq \frac{m}{n}\beta$.

We note that:

1. For $\beta = 0$ and $m = 1$, Theorem 3.4 and Corollary 3.3 were obtained in [1].
2. For splines $s \in S_{n,r}$ satisfying the condition $\|s_+\|_p = \|s_-\|_p$, the constant c in inequality (3.32) is equal to zero.
3. For splines of constant sign $s \in S_{n,r}$ with zeroes, inequality (3.32) turns into the inequality for the best one-sided approximations by a constant i.e., the norms $\|s\|_q$ and $\|s\|_{L_p(I_{2\pi} \setminus B)}$ in inequality (3.32) for these splines should be replaced by $E_0^{\pm}(s)_q$ and $E_0^{\pm}(s)_{L_p(I_{2\pi} \setminus B)}$ respectively. Moreover, the constant c in this inequality is equal to the Favard constant K_r.

References

1. Babenko V, Kofanov V, Pichugov S (2002) Comparison of permutations and Kolmogorov-Nagy type inequalities for periodic functions. In: Approximation theory: a volume dedicated to Blagovest Sendov, Darba, Sofia, pp 24–53
2. Babenko V, Korneichuk N, Kofanov V, Pichugov S (2003) Inequalities for derivatives and their applications. Naukova Dumka, Kyiv (In Russian)
3. Bojanov B, Naidenov N (1999) An extension of the Landau-Kolmogorov inequality. Solution of a problem of Erdos. J Anal Math 78(5):263–280
4. Borwein P, Erd'elyi T (1995) Polynomials and polynomial inequalities. Springer, New York, NY. https://doi.org/10.1007/978-1-4612-0793-1
5. Gaidabura A, Kofanov V (2018) Sharp Remez-type inequalities of various metrics in the classes of functions with given comparison function. Ukr Math J 69(11):1710–1726. https://doi.org/10.1007/s11253-018-1465-4
6. Ganzburg M (2001) Polynomial inequalities on measurable sets and their applications. Constr Approx 17:275–306. https://doi.org/10.1007/s003650010020
7. Ganzburg M (2012) On a Remez-type inequality for trigonometric polynomials. J Approx Theory 164(9):1233–1237. https://doi.org/10.1016/j.jat.2012.05.006
8. Kofanov V (2009) On some extremal problems of different metrics for differentiable functions on the axis. Ukr Math J 61(6):908–922. https://doi.org/10.1007/s11253-009-0254-5
9. Kofanov V (2011) Sharp upper bounds of norms of functions and their derivatives on classes of functions with given comparison function. Ukr Math J 63(7):1118–1135. https://doi.org/10.1007/s11253-011-0567-z
10. Kofanov V (2015) Inequalities of different metrics for differentiable periodic functions. Ukr Math J 67(2):230–242. https://doi.org/10.1007/s11253-015-1076-2
11. Kofanov V (2016) Sharp Remez-type inequalities for differentiable periodic functions, polynomials, and splines. Ukr Math J 68(2):253–268. https://doi.org/10.1007/s11253-016-1222-5
12. Kofanov V (2017) Sharp Remez-type inequalities of different metrics for differentiable periodic functions, polynomials, and splines. Ukr Math J 69(2):205–223. https://doi.org/10.1007/s11253-017-1357-z

13. Kofanov V (2020) Sharp Kolmogorov-Remez-type inequalities for periodic functions of low smoothness. Ukr Math J 72(4):555–567. https://doi.org/10.1007/s11253-020-01800-2
14. Kofanov V (2021) On the relationship between sharp Kolmogorov-type inequalities and sharp Kolmogorov-Remez-type inequalities. Ukr Math J 73(4):592–600. https://doi.org/10.1007/s11253-021-01945-8
15. Kofanov V, Olexandrova T (2022) Sharp Remez type inequalities estimating the L_q-norm of a function via its L_p-norm. Ukr Math J 74(5):726–742. https://doi.org/10.1007/s11253-022-02097-z
16. Kofanov V, Popovich I (2020) Sharp Remez-type inequalities of various metrics with asymmetric restrictions imposed on the functions. Ukr Math J 72(7):1068–1079. https://doi.org/10.1007/s11253-020-01844-4
17. Kolmogorov AN (1991) Selected works I. Mathematics and mechanics, chap. On inequalities for suprema of successive derivatives of an arbitrary function on an infinite interval. Springer, Dordrecht, pp 277–290
18. Korneichuk N, Babenko V, Ligun A (1992) Extremal properties of polynomials and splines. Naukova Dumka, Kyiv (In Russian)
19. Nursultanov E, Tikhonov S (2013) Sharp Remez inequality for trigonometric polynomial. Constr Approx 38:101–132. https://doi.org/10.1007/s00365-012-9172-0
20. Remes E (1936) Sur une proprietie extremale des polynomes de Tchebychef. Zap Nauk-Doslid Inst Math Mek Kharkiv Math Tovar 4(13)(1):93–95
21. Tikhomirov VM (1960) Diameters of sets in function spaces and the theory of best approximations. Russ Math Surv 15(3):75–111. https://doi.org/10.1070/RM1960v015n03ABEH004093
22. Tikhonov S, Yuditski P (2020) Sharp Remez inequality. Constr Approx 52:233–246. https://doi.org/10.1007/s00365-019-09473-2

Restoration of the Noise Corrupted Optical Images with Their Simultaneous Contrast Enhancement

Abstract

In this chapter, we focus on the development of a variational approach for simultaneous contrast enhancement of color images and their denoising. With that in mind we propose a new variational model in Sobolev-Orlicz spaces with non-standard growth conditions of the objective functional and discuss its applications to the simultaneous fusion and denoising of each spectral channel for an input color images. The characteristic feature of the proposed model is the fact that we deal with a constrained minimization problem with a special objective functional that lives in variable Sobolev-Orlicz spaces. This functional contains a spatially variable exponent characterizing the growth conditions and it can be seen as a replacement for the standard 1-norm in TV regularization. We show that the proposed model allows to synthesize at a high level of accuracy noise- and blur-free color images, which were captured in extremely low light conditions.

A very promising approach to image quality enhancement is to reduce the influence from the noise and improve the perceptibility of objects in the scene by increasing the brightness difference between objects and their background. In recent years, many contrast enhancement techniques have been proposed for digital images. Some approaches allow to improve image contrast just in low light conditions [39, 44]. Other methods, called sharpening, focus on enforcing strong contours in order to remove the obtained blur, e.g., by Gaussian convolution [38]. However, this kind of enhancement concerns only strong image contours while the contrast enhancement attempts to modify gray level of objects not only in the contours neighborhood. In recent years, many different techniques have been proposed for reconstruction of noise-affected digital images and their contrast enhancement. We refer to [40], where the authors focus on the problem of contrast enhancement of natural images captured with a digital camera, and give a sufficiently complete overview of the existing methods with detailed analysis of all pros and cons.

In this chapter, we mainly focus on the development of a variational approach for simultaneous contrast enhancement of color images and their denoising. With that in mind we propose a new variational model in Sobolev-Orlicz spaces with non-standard growth conditions of the objective functional and discuss its applications to the simultaneous fusion and denoising of each spectral channel for an input color images. In contrast to [26], we don't provided the color image restoration using saturation-value Total Variation, but instead we are working just with the RGB-representation of color images. However, as follows from the results of numerical simulations, the proposed approach does not strongly modify the histogram of the original image. This enables the model to preserve the global lighting sensation and to show that the hue of the main objects does not drastically change with the illumination. One of the most important advantages of this approach is the fact that the proposed model allows to synthesize at a high level of accuracy noise- and blur-free color images, that were captured in extremely low light conditions. This situation is typical for the most of remote sensing problems. Indeed, the real-life satellite images frequently suffer from different types of noise, blur, and other atmosphere artifacts that can affect the radiation recovered by the sensors. As a result, such images lose their efficiency for the crop field monitoring problems and their utilization can lead to erroneous results and inferences.

The characteristic feature of the proposed model is that we deal with a constrained minimization problem with a special objective functional that lives in variable Sobolev-Orlicz spaces. This functional contains a spatially variable exponent characterizing the growth conditions and it can be seen as a replacement for the 1-norm in TV regularization. Moreover, the variable exponent, which is associated with non-standard growth, is unknown a priori and it depends on a particular function that belongs to the domain of objective functional.

The idea of using a spatially varying exponent in a TV-like regularization method for image denoising dates back as early as 1997 [7] and it has been put into practice in 2006 [11]. Both papers as well as some subsequent articles try to tackle variants of the problem

$$J(u) = \mathcal{D}(u) + \lambda \int_{\Omega} |\nabla u(x)|^{p(\nabla u(x))} dx \longrightarrow \inf, \qquad (4.1)$$

where the exponent depends directly on the image u, e.g.,

$$p(\nabla u) = 1 + \frac{a^2}{a^2 + |\nabla G_\sigma * u|^2}. \qquad (4.2)$$

Here, $(G_\sigma * v)(x)$ determines the convolution of function v with the 2-dimensional Gaussian filter kernel G_σ.

It has been demonstrated that this model possesses some favorable properties, particularly when edge preservation and effective noise suppression are primary goals in image reconstruction. Furthermore, this model has been introduced specifically to address the issue of staircasing [42], which refers to the regularizer's inclination towards piecewise constant functions. The appearance of the staircasing effect is a notable drawback of the classical TV model. However, the non-convex model (4.1) did not gain significant attention for a long period due to its high numerical complexity and the absence of a rigorous mathematical substantiation of its consistency. Only particular solutions to this problem have been derived for a smoothed version of the integrand, using a weak notion of solution (see, for instance, [45]).

A recently developed alternative variant is the TV-like method [35] (see also [1, 12]), which computes the variable exponent p in an offline step and keeps it as a fixed parameter in the final optimization problem This approach allows the exponent to vary based on spatial location, enabling users to locally select whether to preserve edges or smooth intensity variations. However, there are only two natural types of imaging problems where this approach can be applied:

- single-channel imaging where first the exponent is computed from the given data and then is applied as prior in the subsequent minimization problem;
- dual-channel imaging where the secondary channel provides the exponent map that is used for regularization of the primary channel.

Thus, this circumstance imposes significant limitations from practical point of view, especially in the case of multi-spectral satellite noisy images, where different channels can differ drastically.

Our main purpose in this chapter is to describe a robust approach for the simultaneous contrast enhancement and denoising of non-smooth multispectral images using an energy functional with nonstandard growth following the recent results that were obtained in cooperation with C.D'Apice, R. Manzo, and A. Parisi (see [21]). In particular, we apply a special form of anisotropic diffusion tensor for the regularization term and a term which is inspired by the variational model of Bertalmio et al. [5]. Following this approach, we aim to increase the perceptibility of objects in the scene and the noise robustness of the proposed model albeit it makes such variational problem completely non-smooth, non-convex, and, hence, significantly more difficult from a minimization point of view.

We consider a variational problem for the energy functional with nonstandard growth $p(x)$, where the principle edge information for the contrast enhancement is mainly accumulated. Namely, for the simultaneous denoising and contrast enhancement of color images, we propose to solve the following constrained minimization problems

$$J_i(f_i^0) = \inf_{v \in \Xi_i} J_i(v), \quad i = 1, 2, 3 \tag{4.3}$$

for each spectral channel of an input image separately, where the objective functional is non-convex and it takes the form

$$J_i(v) = \int_\Omega |R_\eta \nabla v(x)|^{p(|\nabla v|)} dx + Q_i(v). \tag{4.4}$$

Here, $Q_i(v)$ stands for the fidelity term and its specific form is described in details in Sect. 4.2 together with the operator R_η. The principle point that should be emphasized is the fact that we do not predefine the exponent $p(x)$ a priori using for that the original image, but instead we associate this characteristic with the current state of their spectral channels. In fact, we take it as follows

$$p(\nabla u) = 1 + \frac{a^2}{a^2 + |\nabla u|^2}. \tag{4.5}$$

So, in contrast to the well-know approach, coming from the pioneering papers [2, 10], the principle difference of the models (4.5) and (4.2) is that we do not apply in (4.5) any spatial regularization of gradient ∇u. Because of this, the model (4.3)–(4.4) becomes an ill-posed problem from the mathematical point of view and can produce many unexpected phenomena. In particular, to our best knowledge we have no results of existence and consistency of the optimization problem (4.3)–(4.4). To overcome this problem, we could apply some regularization of the variable exponent $p(x)$ in the form like (4.2).

However, it is well-known that optimization problem (4.3)–(4.4) with the spatially regularized gradient has several serious practical and theoretical difficulties. The first one is that the spatial regularization of gradient in the form (4.2) leads to the loss of accuracy in the case when the signal is noisy, with white noise (see, for instance, [10]). Then the noise introduces very large, in theory unbounded, oscillations of the gradient ∇u. As a result, the conditional smoothing introduced by the model will not help, since all these noise edges will be kept.

The second drawback of the model with the regularized gradient is the fact that the space-invariant Gaussian smoothing inside the divergent term tends to push the edges in u away from their original locations. We refer to [41] where this issue is studied in details. This effect, known as edge dislocation, can be detrimental especially in the context of the boundary detection problem and its application to the remote sensing and monitoring. So, our prime interest in this chapter is to study the optimization problem (4.3)–(4.4) without the space-invariant Gaussian smoothing of the variable exponent $p(x)$.

4.1 Preliminaries

Let us recall some useful notations. For vectors $\xi \in \mathbb{R}^2$ and $\eta \in \mathbb{R}^2$, $(\xi, \eta) = \xi^t \eta$ denotes the standard vector inner product in $\mathbb{R}^2$, where ^{t} stands for the transpose operator. The norm $|\xi|$ is the Euclidean norm given by $|\xi| = \sqrt{(\xi, \xi)}$. Let $\Omega \subset \mathbb{R}^2$ be a bounded open set with a Lipschitz boundary $\partial \Omega$ and nonzero Lebesgue measure. For any subset $E \subset \Omega$ we denote by $|E|$ its 2-dimensional Lebesgue measure $\mathcal{L}^2(E)$. Let $\overline{E}$ denote the closure of E, and ∂E stands for its boundary.

4.1 Preliminaries

4.1.1 Functional Spaces

For convenience of the reader, we collect here the basic facts on functional spaces that will be used in the sequel. Let X denote a real Banach space with norm $\|\cdot\|_X$, and let X' be its dual. Let $\langle \cdot, \cdot \rangle_{X';X}$ be the duality form on $X' \times X$. By $\rightharpoonup$ and $\stackrel{*}{\rightharpoonup}$ we denote the weak and weak* convergence in normed spaces, respectively. For given $1 \leq p \leq +\infty$, the space $L^p(\Omega; \mathbb{R}^2)$ is defined by

$$L^p(\Omega; \mathbb{R}^2) = \{f : \Omega \to \mathbb{R}^2 \; : \; \|f\|_{L^p(\Omega;\mathbb{R}^2)} < +\infty\},$$

where $\|f\|_{L^p(\Omega;\mathbb{R}^2)} = \left(\int_\Omega |f(x)|^p \, dx\right)^{1/p}$ for $1 \leq p < +\infty$. The inner product of two functions f and g in $L^p(\Omega; \mathbb{R}^2)$ with $p \in [1, \infty)$ is given by

$$(f, g)_{L^p(\Omega;\mathbb{R}^2)} = \int_\Omega (f(x), g(x)) \, dx = \int_\Omega \sum_{k=1}^2 f_k(x) g_k(x) \, dx.$$

We denote by $C_c^\infty(\mathbb{R}^2)$ a locally convex space of all infinitely differentiable functions with compact support in $\mathbb{R}^2$. We recall here some functional spaces that will be used throughout this paper. We define the Banach space $H^1(\Omega)$ as the closure of $C_c^\infty(\mathbb{R}^2)$ with respect to the norm

$$\|y\|_{H^1(\Omega)} = \left(\int_\Omega (y^2 + |\nabla y|^2) \, dx\right)^{1/2}.$$

We denote by $(H^1(\Omega))'$ the dual space of $H^1(\Omega)$. Hereinafter, $W^{1,1}(\Omega)$ stands for the Banach space of all functions $u \in L^1(\Omega)$ with respect to the norm

$$\|u\|_{W^{1,1}(\Omega)} = \|u\|_{L^1(\Omega)} + \|\nabla u\|_{L^1(\Omega)^2}.$$

Given a real Banach space X, we will denote by $C([0, T]; X)$ the space of all continuous functions from $[0, T]$ into X. We recall that a function $u : [0, T] \to X$ is said to be Lebesgue measurable if there exists a sequence $\{u_k\}_{k \in \mathbb{N}}$ of step functions (i.e., $u_k = \sum_{j=1}^{n_k} a_j^k \chi_{A_j^k}$ for a finite number n_k of Borel subsets $A_j^k \subset [0, T]$ and with $a_j^k \in X$) converging to u almost everywhere with respect to the Lebesgue measure in $[0, T]$.

Then for $1 \leq p < \infty$, $L^p(0, T; X)$ is the space of all measurable functions $u : [0, T] \to X$ such that

$$\|u\|_{L^p(0,T;X)} = \left(\int_0^T \|u(t)\|_X^p \, dt\right)^{\frac{1}{p}} < \infty,$$

while $L^\infty(0, T; X)$ is the space of measurable functions such that

$$\|u\|_{L^\infty(0,T;X)} = \operatorname*{ess\,sup}_{t \in [0,T]} \|u(t)\|_X < \infty.$$

A full presentation of this topic can be found in [23].

Let us recall that, for $1 \leq p \leq \infty$, $L^p(0, T; X)$ is a Banach space. Moreover, if X is separable and $1 \leq p < \infty$, then the dual space of $L^p(0, T; X)$ can be identified with $L^{p'}(0, T; X')$.

4.1.2 Basic Facts on the Lebesgue and Sobolev Spaces with Variable Exponents

Let $p: \Omega \to [p^-, p^+] \subset (1, +\infty)$, with $p^\pm = \text{const}$, be a given measurable function. Denote by $L^{p(\cdot)}(\Omega)$ the set of all measurable functions $f(x)$ on Ω such that $\int_\Omega |f(x)|^{p(x)} \, dx < \infty$. Then $L^{p(\cdot)}(\Omega)$ is a reflexive separable Banach space with respect to the Luxemburg norm

$$\|f\|_{L^{p(\cdot)}(\Omega)} = \inf \left\{ \lambda > 0 : \int_\Omega \left| \frac{f(x)}{\lambda} \right|^{p(x)} dx \leq 1 \right\}.$$

Moreover, in this case the set $C_0^\infty(\Omega)$ is dense in $L^{p(\cdot)}(\Omega)$. The relation between the modular $\int_\Omega |f(x)|^{p(x)} dx$ and the norm follows from the definition

$$\min \left\{ \|f\|_{L^{p(\cdot)}(\Omega)}^{p^-}, \|f\|_{L^{p(\cdot)}(\Omega)}^{p^+} \right\} \leq \int_\Omega |f(x)|^{p(x)} dx \leq \max \left\{ \|f\|_{L^{p(\cdot)}(\Omega)}^{p^-}, \|f\|_{L^{p(\cdot)}(\Omega)}^{p^+} \right\}.$$

If $p(\cdot) = \text{const} > 1$ then these inequalities transform into equalities. The following estimates are also well-known (see, for instance, [24, 46]): if $f \in L^{p(\cdot)}(\Omega)$ then

$$\|f\|_{L^{p^-}(\Omega)} \leq (1 + |\Omega|)^{1/p^-} \|f\|_{L^{p(\cdot)}(\Omega)}, \tag{4.6}$$

$$\|f\|_{L^{p(\cdot)}(\Omega)} \leq (1 + |\Omega|)^{1/(p^+)'} \|f\|_{L^{p^+}(\Omega)}, \tag{4.7}$$

$$(p^+)' = \frac{p^+}{p^+ - 1}, \quad \forall f \in L^{p^+}(\Omega),$$

$$\|f\|_{L^{p(\cdot)}(\Omega)}^{p^-} - 1 \leq \int_\Omega |f(x)|^{p(x)} dx \leq \|f\|_{L^{p(\cdot)}(\Omega)}^{p^+} + 1, \quad \forall f \in L^{p(\cdot)}(\Omega). \tag{4.8}$$

The next result can be viewed as an analogous of the Hölder inequality in Lebesgue spaces with variable exponents: If $f \in L^{p(\cdot)}(\Omega)$ and $g \in L^{p'(\cdot)}(\Omega)$ with

$$p(x) \in [p^-, p^+] \subset (1, +\infty), \quad p'(x) = \frac{p(x)}{p(x) - 1},$$

4.1 Preliminaries

then $fg \in L^1(\Omega)$ and

$$\int_\Omega (f, g)\, dx \leq \left(\frac{1}{p^-} + \frac{1}{(p')^-}\right) \|f\|_{L^{p(\cdot)}(\Omega)} \|g\|_{L^{p'(\cdot)}(\Omega)}$$

$$\leq 2 \|f\|_{L^{p(\cdot)}(\Omega)} \|g\|_{L^{p'(\cdot)}(\Omega)}. \quad (4.9)$$

Let $p_1(\cdot)$ and $p_2(\cdot)$ be measurable on Ω functions such that $p_i(x) \in [p_i^-, p_i^+] \subset (1, +\infty)$ a.e. in Ω. In case $p_1(x) \geq p_2(x)$ a.e. in Ω, the inclusion $L^{p_1(\cdot)}(\Omega) \subset L^{p_2(\cdot)}(\Omega)$ is continuous and

$$\|u\|_{L^{p_2(\cdot)}(\Omega)} \leq C \|u\|_{L^{p_1(\cdot)}(\Omega)}, \quad \forall u \in L^{p_1(\cdot)}(\Omega) \quad (4.10)$$

with a constant $C = C(|\Omega|, p_1^\pm, p_2^\pm)$.

The variable Sobolev space $W^{1,p(\cdot)}(\Omega)$ is defined as the set of functions

$$W^{1,p(\cdot)}(\Omega) := \left\{ u \in W^{1,1}(\Omega) \cap L^{p(\cdot)}(\Omega) : |\nabla u(x)|^{p(x)} \in L^1(\Omega) \right\}$$

equipped with the norm

$$\|u\|_{W^{1,p(\cdot)}(\Omega)} = \|u\|_{L^{p(\cdot)}(\Omega)} + \|\nabla u\|_{L^{p(\cdot)}(\Omega; \mathbb{R}^N)}. \quad (4.11)$$

Unlike classical Sobolev spaces, smooth functions are not necessarily dense in $W^{1,p(\cdot)}(\Omega)$. Therefore, we define $H^{1,p(\cdot)}(\Omega)$ as the closure of the set $C^\infty(\overline{\Omega})$ in $W^{1,p(\cdot)}(\Omega)$-norm.

Let $C^{log}(\overline{\Omega})$ be the set of functions continuous on $\overline{\Omega}$ with the logarithmic modulus of continuity, i.e.

$$|p(x_1) - p(x_2)| \leq \omega(|x_1 - x_2|),$$

where $\omega \geq 0$ satisfies the condition: $\limsup_{\tau \to 0^+} \omega(\tau) \log \dfrac{1}{\tau} = C < +\infty$, $C = $ const. It is well-known that for $p \in C^{log}(\overline{\Omega})$ the set $C^\infty(\overline{\Omega})$ is dense in $W^{1,p(\cdot)}(\Omega)$ and the space $W^{1,p(\cdot)}(\Omega)$ coincides with the closure of $C^\infty(\overline{\Omega})$ with respect to the norm (4.11), i.e. in this case $W^{1,p(\cdot)}(\Omega) = H^{1,p(\cdot)}(\Omega)$. In particular, if there exists $\delta \in (0, 1]$ such that $p \in C^{0,\delta}(\overline{\Omega})$, then the set $C^\infty(\overline{\Omega})$ is dense in $W^{1,p(\cdot)}(\Omega)$. Indeed, since

$$\lim_{t \to 0} |t|^\delta \log\left(\frac{1}{|t|}\right) = 0 \quad with\ \delta \in (0, 1],$$

it follows from the Hölder continuity of $p(\cdot)$ that

$$|p(x) - p(y)| \leq C|x - y|^\delta \leq \omega(|x - y|) \sup_{x, y \in \Omega} \left[|x - y|^\delta \log(|x - y|^{-1}) \right]$$

$$\leq C' \omega(|x - y|), \quad \forall x, y \in \Omega,$$

with $\omega(t) = C / \log(|t|^{-1})$.

Let $p(\cdot), q(\cdot) \in C(\overline{\Omega})$ be such that $p(x) \in [p^-, p^+] \subset (1, 2]$ and $q(x) < \frac{2p(x)}{2-p(x)}$ in $\overline{\Omega}$. Then the embedding $W^{1,p(\cdot)}(\Omega) \subset L^{q(\cdot)}(\Omega)$ is continuous and compact. Moreover, according to (4.10), we have a continuous embedding $W^{1,p(\cdot)}(\Omega) \subset W^{1,p^-}(\Omega)$.

For a more detailed presentation of the theory of these spaces, we refer to the monograph [24].

4.1.3 On the Dual Sobolev Space $H^{-1}(\Omega)$

Let $H_0^1(\Omega)$ be the standard Sobolev space, i.e. $H_0^1(\Omega)$ is the closure of $C_0^1(\Omega)$ with respect to the norm

$$\|u\|_{H_0^1(\Omega)} = \left(\int_\Omega |\nabla u(x)|^2 \, dx \right)^{\frac{1}{2}}.$$

It is well-known that for any $u^* \in H^{-1}(\Omega)$ there can be found a vector-function $g = [g_1, g_2]$ in $L^2(\Omega; \mathbb{R}^2)$ such that

$$\langle u^*, u \rangle_{H^{-1}(\Omega); H_0^1(\Omega)} = \int_\Omega (g, \nabla u)_{\mathbb{R}^2} \, dx = \int_\Omega \left[g_1 \frac{\partial u}{\partial x_1} + g_2 \frac{\partial u}{\partial x_2} \right] dx.$$

Therefore, it is clear now that

$$\|u^*\|_{H^{-1}(\Omega)} \le \sqrt{\int_\Omega \left(g_1^2(x) + g_2^2(x) \right) dx}. \quad (4.12)$$

On the other hand, due to the Lax-Milgram Theorem, the Dirichlet boundary value problem

$$-\Delta y = u^* \text{ in } \Omega, \quad y = 0 \text{ on } \partial \Omega, \quad (4.13)$$

has a unique solution $y = (-\Delta)^{-1} u^* \in H_0^1(\Omega)$ for each $u^* \in H^{-1}(\Omega)$. Moreover, in view of the energy equality

$$\int_\Omega (\nabla y, \nabla y)_{\mathbb{R}^2} \, dx = \|\nabla y\|_{L^2(\Omega; \mathbb{R}^2)}^2 = \|y\|_{H_0^1(\Omega)}^2 = \langle u^*, y \rangle_{H^{-1}(\Omega); H_0^1(\Omega)}, \quad (4.14)$$

which holds true for the weak solution of Dirichlet problems (4.13), we can deduce the following a priori estimate for the weak solution of Dirichlet problem (4.13)

$$\|y\|_{H_0^1(\Omega)} \equiv \|(-\Delta)^{-1} u^*\|_{H_0^1(\Omega)} \equiv \|\nabla (-\Delta)^{-1} u^*\|_{L^2(\Omega; \mathbb{R}^2)} \le \|u^*\|_{H^{-1}(\Omega)}.$$

Combining this result with (4.12), we get

4.1 Preliminaries

$$\|\nabla(-\Delta)^{-1}u^*\|_{L^2(\Omega;\mathbb{R}^2)} \leq \|u^*\|_{H^{-1}(\Omega)} \leq \sqrt{\int_\Omega \left(g_1^2(x) + g_2^2(x)\right) dx}$$

$$\stackrel{\text{by (4.14)}}{=} \sqrt{\int_\Omega |\nabla y|^2_{\mathbb{R}^2}\, dx} = \|\nabla y\|_{L^2(\Omega;\mathbb{R}^2)}$$

$$= \|y\|_{H_0^1(\Omega)} = \|\nabla(-\Delta)^{-1}u^*\|_{L^2(\Omega;\mathbb{R}^2)}.$$

Hence, the norm in $H^{-1}(\Omega)$ can be defined as follows

$$\|u^*\|_{H^{-1}(\Omega)} = \|\nabla(-\Delta)^{-1}u^*\|_{L^2(\Omega;\mathbb{R}^2)}. \tag{4.15}$$

4.1.4 Level Sets, Directional Gradients, and Texture Indexes

Let $u : \Omega \to \overline{\mathbb{R}}$ be a given function. Then for each $\lambda \in \mathbb{R}$ we can define the upper level set of u as follows

$$Z_\lambda(u) = \{u \geq \lambda\} := \{x \in \Omega : u(x) \geq \lambda\}.$$

In order to describe this set, we assume that $u \in W^{1,1}(\Omega)$. It was proven in [3] that if $u \in W^{1,1}(\Omega)$ then its upper level sets $Z_\lambda(u)$ are sets of finite perimeter. So, the boundaries of level sets can be described by a countable family of Jordan curves with finite length, i.e., by continuous maps from the circle into the plane $\mathbb{R}^2$ without crossing points. As a result, at almost all points of almost all level sets of $u \in W^{1,1}(\Omega)$ we may define a unit normal vector $\theta(x)$. This vector field formally satisfies the following relations

$$(\theta, \nabla u) = |\nabla u| \quad \text{and} \quad |\theta| \leq 1 \text{ a.e. in } \Omega.$$

In the sequel, we will refer to θ as the vector field of unit normals to the topographic map of a function u.

If $\theta \in L^\infty(\Omega, \mathbb{R}^2)$ is a vector field of unit normals to the topographic map of some function $u(\cdot)$, then for any function $v \in W^{1,1}(\Omega)$ we can define the so-called directional gradient of v following the rule (see [8, 9])

$$R_\eta \nabla v := \nabla v - \eta^2 (\theta, \nabla v)\, \theta, \tag{4.16}$$

where $\eta \in (0, 1)$ is a given threshold. Since, for each function $v \in W^{1,1}(\Omega)$, the expression $R_\eta \nabla v$ can be reduced to $(1 - \eta^2)\nabla v$ in those places of Ω where ∇v is collinear to the unit normal θ, and to ∇v if ∇v is orthogonal to θ, we have the following estimate

$$(1 - \eta^2)|\nabla v| \leq |R_\eta \nabla v| \leq |\nabla v| \quad \text{in } \Omega.$$

In what follows, with each function $u \in W^{1,1}(\Omega)$, we associate the so-called texture index $p(|\nabla u|)$ using the rule (see [18, 20, 22] for comparison)

$$p(s) = 1 + \delta + \frac{a^2(1-\delta)}{a^2 + s^2}, \quad \forall s \in [0, +\infty), \tag{4.17}$$

where $0 < \delta \ll 1$ is a given threshold. It is clear now that

$$p(|\nabla u|) \in [p^-, p^+] \subset (1, 2] \text{ a. e. in } \Omega \text{ with } p^- = 1 + \delta \text{ and } p^+ = 2$$

for each $u \in W^{1,1}(\Omega)$.

4.2 Statement of the Problem

In this section we present a novel variational problem for the denoising and contrast enhancement of non-smooth RGB images which can be viewed as an improved version of the variational model that has been recently proposed in [17].

Let $f = [f_1, f_2, f_3]^t \in L^2(\Omega; \mathbb{R}^3)$ be an original color image. Let $\theta = \theta(f_i) = [\theta_1, \theta_2]^t \in L^\infty(\Omega, \mathbb{R}^2)$ be a vector field of unit normals to the topographic map of the spectral channel f_i,

$$|\theta(x)| \leq 1 \text{ and } (\theta(x), \nabla f_i(x)) = |\nabla f_i(x)| \text{ a.e. in } \Omega.$$

In fact, this vector field can be defined by the rule $\theta(x) = \frac{\nabla U(t,x)}{|\nabla U(t,x)|}$ with $t > 0$ small enough, where $U(t, x)$ is a solution of the following initial-boundary value problem

$$\frac{\partial U}{\partial t} = \operatorname{div}\left(\frac{\nabla U}{|\nabla U| + \kappa}\right), \quad t \in (0, +\infty), \ x \in \Omega, \tag{4.18}$$

$$U(0, x) = f_i(x), \quad x \in \Omega, \tag{4.19}$$

$$\frac{\partial U(t, x)}{\partial \nu} = 0, \quad t \in (0, +\infty), \ x \in \partial\Omega \tag{4.20}$$

with a relaxed version of the $1D$-Laplace operator in the principle part of (4.18). Here, $\kappa > 0$ is a sufficiently small positive value.

Let $\eta \in (0, 1)$ be a given threshold. We define the linear operator $R_\eta : \mathbb{R}^2 \to \mathbb{R}^2$ as follows

$$R_\eta \nabla v := \nabla v - \eta^2 (\theta, \nabla v) \theta, \quad \forall v \in W^{1,1}(\Omega). \tag{4.21}$$

Since $R_\eta \nabla v$ reduces to $(1 - \eta^2)\nabla v$ in those regions where the gradient ∇v is co-linear to θ, and to ∇v, where ∇v is orthogonal to θ, this operator does not enforce gradients in the direction θ.

Remark 4.1 In the sequel, in order to reduce the number of parameters in the proposed model, we will set $\delta = \kappa$ in (4.18) and (4.17), and $\eta = 1 - \kappa$ in (4.16).

We also introduce the following set

4.2 Statement of the Problem

$$\Xi_i = \left\{ I \in H^{1,p(|\nabla I|)}(\Omega) \cap L^\infty(\Omega) \;\middle|\; \begin{array}{l} \gamma_{i,0} \leq I(x) \leq \gamma_{i,1} \quad \text{a.e. in } \Omega, \\ \gamma_{i,0} = \inf_{x \in \Omega} f_i(x), \\ \gamma_{i,1} = \sup_{x \in \Omega} f_i(x), \end{array} \right\}$$

where $p(\cdot)$ is given by (4.17), and $H^{1,p(|\nabla I|)}(\Omega)$ is the variable Sobolev space that can be defined as follows

$$H^{1,p(|\nabla I|)}(\Omega) := \mathrm{cl}_{\|\cdot\|_{W^{1,p(|\nabla I|)}(\Omega)}} C_0^\infty(\mathbb{R}^2). \tag{4.22}$$

Further, for a given gray-scale image $I \in L^2(\Omega)$, we define its average local contrast measure $D(I)$ as follows (for comparison, we refer to [40])

$$D(I) = \int_\Omega \int_\Omega W(x, y) \sqrt{\kappa^2 + |I(x) - I(y)|^2} \, dx dy,$$

where $\kappa > 0$ is the same parameter as in (4.18), and $W \in L^2(\Omega \times \Omega)$ is a symmetric non-negative kernel such that

$$\int_\Omega W(x, y) \, dx = 1, \quad \forall y \in \Omega.$$

A typical example of this function is the Gaussian kernel,

$$W(x, y) = \frac{1}{\sqrt{2\pi}\sigma} \exp\left(-\frac{|x-y|^2}{2\sigma^2}\right), \quad \sigma > 0.$$

As a result, the proposed variational approach for the contrast enhancement and denoising of color images can be stated as follows:

For each spectral channel f_i, $(i = 1, 2, 3)$, of a given image $f = [f_1, f_2, f_3]^t \in L^2(\Omega; \mathbb{R}^3)$, we generate a new one $f_i^0 \in L^2(\Omega)$ as a solution of the following constrained minimization problem

$$J_i(f_i^0) = \inf_{v \in \Xi_i} J_i(v), \tag{4.23}$$

where

$$J_i(v) = \int_\Omega |R_\eta \nabla v(x)|^{p(|\nabla v|)} \, dx + \frac{\mu}{2} \|v - f_i\|_{H^{-1}(\Omega)}^2$$

$$+ \frac{\lambda}{4} \left[D(v) - cD(f_i)\right]^2. \tag{4.24}$$

Here, $\lambda > 0$ and $\mu \in (0, 1)$ are tuning parameters. The parameter λ manages the trade-off between the fidelity term $\frac{\mu}{2} \|v - f_i\|_{H^{-1}(\Omega)}^2$ and the contrast term $\frac{\lambda}{4} \left[D(v) - cD(f_i)\right]^2$. As for the multiplier $c > 0$, we always suppose that $c > 1$ and it provides a control of the contrast level expected for the result.

Before proceeding further, we provide some qualitative analysis of the variational problem (4.23)–(4.24). To begin with, we notice that, for each feasible solution $v \in \Xi$, the following two-sides estimate

$$p^+ = 2 \geq p(|\nabla v|) > 1 + \delta =: p^-, \quad \text{for a.a. } x \in \Omega \qquad (4.25)$$

holds true with $0 < \delta \ll 1$. Moreover, since $\eta \in (0, 1)$ and $\eta \gg 0$, we see that

$$(1 - \eta^2)|\nabla v| \leq |R_\eta \nabla v| \leq |\nabla v| \quad \text{in } \Omega, \qquad (4.26)$$

$$\int_\Omega |R_\eta \nabla v(x)|^{p(|\nabla v|)} dx \geq \int_\Omega (1 - \eta^2)^{p(|\nabla v|)} |\nabla v(x)|^{p(|\nabla v|)} dx$$

$$\geq (1 - \eta^2)^2 \int_\Omega |\nabla v(x)|^{p(|\nabla v|)} dx, \qquad (4.27)$$

$$\forall v \in W^{1, p(|\nabla v|)}(\Omega).$$

As a result, $\forall v \in \Xi_i$, we have

$$\|v\|_{W^{1,p(|\nabla v|)}(\Omega)}^{p^-} = \left(\|v\|_{L^{p(|\nabla v|)}(\Omega)} + \|\nabla v\|_{L^{p(|\nabla v|)}(\Omega; \mathbb{R}^2)} \right)^{p^-}$$

$$\leq C \left(\|v\|_{L^{p(|\nabla v|)}(\Omega)}^{p^-} + \|\nabla v\|_{L^{p(|\nabla v|)}(\Omega; \mathbb{R}^2)}^{p^-} \right)$$

$$\stackrel{\text{by (4.8)}}{\leq} C \left(\int_\Omega |v(x)|^{p(|\nabla v|)} dx + \int_\Omega |\nabla v(x)|^{p(|\nabla v|)} dx + 2 \right)$$

$$\leq C \left(|\Omega| \gamma_{i,1}^2 + \int_\Omega |\nabla v(x)|^{p(|\nabla v|)} dx + 2 \right)$$

$$\stackrel{\text{by (4.27)}}{\leq} C \left(|\Omega| \gamma_{i,1}^2 + 2 + \frac{1}{(1 - \eta^2)^2} \int_\Omega |R_\eta \nabla v(x)|^{p(|\nabla v|)} dx \right)$$

$$\leq C \left(|\Omega| \gamma_{i,1}^2 + 2 + \frac{1}{(1 - \eta^2)^2} J_i(v) \right). \qquad (4.28)$$

Thus, the first term in the cost functional (4.24) can be considered as a regularizing term. As for the second term in (4.24), we make use of the following observation.

Remark 4.2 The model (4.24) is aimed not only to the contrast enhancement, but also to remove the additive noise in the so-called structured images, i.e. in images where the portion of high oscillatory edges is rather significant. In mostly cases the satellite images with crop fields typically contain many high oscillatory edges (boundaries of the crop locations). Moreover, "the portion of noise" in such images can be different from channel to channel. Because of this an important question is to separate pure noise from high oscillatory edges in each spectral channel. To handle this problem, Y. Meyer [37] suggested to replace the standard L^2-fidelity term $\frac{\mu}{2} \|v - f_i\|_{L^2(\Omega)}^2$, which is a typical component in the standard denoising models, by a weaker norm. As a plausible option of such weakening, Lieu and Vese [36] (see also Schönlieb [43]) have proposed to involve $H^{-1}(\Omega)$-norm instead of $\|\cdot\|_{L^2(\Omega)}$. Thus, from this point of view it is plausible to interpret the second term in (4.24) as a fidelity term.

4.2 Statement of the Problem

Before we move on to the existence issues, we make use of the following result concerning the lower semicontinuity property of the modular $\int_\Omega |f(x)|^{p(x)}\, dx$ with respect to the weak convergence in $L^{p_k(\cdot)}(\Omega)$. The proof of this assertion has been mainly inspired by the elegant proof of Lemma 1 in [13] (see also [47, Lemma 13.3] for comparison).

Proposition 4.3 *Let $\{p_k\}_{k\in\mathbb{N}} \subset [p^-, p^+]$ be a given sequence such that*

$$p_k(x) \to p(x) \quad \text{almost everywhere in } \Omega \text{ as } k \to \infty. \tag{4.29}$$

Let $\{v_k \in W^{1,p_k(\cdot)}(\Omega)\}_{k\in\mathbb{N}}$ be a sequence such that

$$\nabla v_k \rightharpoonup \nabla v \quad \text{weakly in } L^1(\Omega; \mathbb{R}^2), \tag{4.30}$$

$$\left\| |\nabla v_k(\cdot)|^{p_k(\cdot)} \right\|_{L^1(\Omega)} \le C \tag{4.31}$$

for some positive constant C not depending on k,

and let $R_\eta : \mathbb{R}^2 \to \mathbb{R}^2$ be the operator defined in (4.21) with some $\theta \in L^\infty(\Omega, \mathbb{R}^2)$. Then $\nabla v \in L^{p(\cdot)}(\Omega; \mathbb{R}^2)$ and

$$\liminf_{k\to\infty} \int_\Omega |R_\eta \nabla v_k(x)|^{p_k(x)}\, dx \ge \int_\Omega |R_\eta \nabla v(x)|^{p(x)}\, dx. \tag{4.32}$$

Proof By Young's inequality we have for $\xi, \zeta \in \mathbb{R}^2$ and $1 < p < \infty$,

$$(\xi, \zeta) \le |\xi||\zeta| \le |\xi|^p + \frac{|\zeta|^{p'}}{p' p^{p'/p}}, \quad \frac{1}{p} + \frac{1}{p'} = 1. \tag{4.33}$$

If now ζ is a function in $L^\infty(\Omega; \mathbb{R}^2)$ and we make $p = p_k$ in (4.33) and use the assumption $p^- \le p_k(x) \le p^+$ for all $k \in \mathbb{N}$, then we derive

$$\int_\Omega \left((R_\eta \nabla v_k, \zeta) - \frac{|\zeta|^{p_k'(x)}}{p_k'(x) p_k(x)^{p_k'(x)/p_k(x)}} \right) dx \le \int_\Omega |R_\eta \nabla v_k|^{p_k(x)}\, dx. \tag{4.34}$$

Using (4.21) and assumptions (4.29) and (4.30), we can pass to the limit in (4.34) as $k \to \infty$. As a result, we have

$$\int_\Omega \left((R_\eta \nabla v, \zeta) - \frac{|\zeta|^{p'(x)}}{p'(x) p(x)^{p'(x)/p(x)}} \right) dx$$

$$\le \liminf_{k\to\infty} \int_\Omega |R_\eta \nabla v_k|^{p_k(x)}\, dx := L. \tag{4.35}$$

Then we consider the following function:

$$\zeta := \frac{R_\eta \nabla v}{|R_\eta \nabla v|} p(x) |R_\eta \nabla v|_n^{\frac{1}{p'(x)-1}}, \text{ with } |R_\eta \nabla v|_n := \max\{|R_\eta \nabla v|, n\}, \ n > 0.$$

Inserting this function ζ into (4.35), we get

$$\int_\Omega \left(|R_\eta \nabla v| p(x) |R_\eta \nabla v|_n^{\frac{1}{p'(x)-1}} - |R_\eta \nabla v|_n^{\frac{p'(x)}{p'(x)-1}} \frac{p(x)}{p'(x)} \right) dx \leq L.$$

This implies

$$\int_\Omega |R_\eta \nabla v|_n^{\frac{1}{p'(x)-1}+1} dx \leq L.$$

Since $\frac{1}{p'(x)-1} + 1 = p(x)$, it follows that

$$\int_\Omega |R_\eta \nabla v|_n^{p(x)} dx \leq L. \tag{4.36}$$

To conclude the proof, it remains to notice that the announced inequality (4.32) follows by letting $n \to \infty$ in (4.36). As for the inclusion $\nabla v \in L^{p(\cdot)}(\Omega; \mathbb{R}^2)$ it is a direct consequence of assumption (4.31) and estimate (4.26).

Before proceeding to the existence issues, in the next section we provide a formal analysis of the optimality system for the problem (4.23)–(4.24).

4.3 Optimality Conditions

Let $f_i^0 \in \Xi_i$, with $i = 1, 2, 3$, be a point of local minimum in the problem (4.23)–(4.24), i.e., there exists a positive value $\tau > 0$ such that

$$J_i(f_i^0) - J_i(v) \leq 0, \quad \forall v \in \Xi \text{ s.t. } \|v - f_i^0\|_{W^{1,p^-}(\Omega)} < \tau. \tag{4.37}$$

For simplicity, we assume that the two-side inequality

$$\gamma_{i,1} < f_i^0(x) < \gamma_{i,2}$$

holds true almost everywhere in Ω. Then condition (4.37) can be rewritten as follows: for any smooth function $\varphi \in C^\infty(\overline{\Omega})$, the inequality

$$J_i(f_i^0) - J_i(f_i^0 + \sigma\varphi) \leq 0 \quad \text{for } \sigma \text{ small enough}$$

holds true. Hence, the scalar function

4.3 Optimality Conditions

$$\psi(\sigma) := J_i(f_i^0 + \sigma\varphi) = \int_\Omega |R_\eta\left(\nabla f_i^0(x) + \sigma\nabla\varphi(x)\right)|^{p(|\nabla f_i^0 + \sigma\nabla\varphi|)}\,dx$$
$$+ \frac{\mu}{2}\|f_i^0 + \sigma\varphi - f_i\|_{H^{-1}(\Omega)}^2 + \frac{\lambda}{4}\left[D(f_i^0 + \sigma\varphi) - cD(f_i)\right]^2$$

has a minimum at $\sigma = 0$.

Thus, to characterize the given feasible solution $f_i^0 \in \Xi_i$ to optimization problem (4.23)–(4.24), we make use of the Ferma's Theorem. To do so, we show that the objective functional $J_i(v)$ is Gâteaux differentiable at $v = f_i^0$, that is, there exists a linear bounded functional

$$D_G J_i(f_i^0) \in \left[H^{1,p[\nabla f_i^0]}(\Omega)\right]' = \mathcal{L}\left(H^{1,p[\nabla f_i^0]}(\Omega), \mathbb{R}\right)$$

such that

$$J_i\left(f_i^0 + \sigma h\right) = J_i\left(f_i^0\right) + \sigma D_G J_i(f_i^0)[h] + r_i(h, \sigma),$$
$$\forall h \in H^{1,p[\nabla f_i^0]}(\Omega),$$

where $|r_i(h, \sigma)| = o(|\sigma|)$ as $\sigma \to 0$. Then the condition $0 \in \mathrm{Argmin}\,\psi(\sigma)$ can be interpreted as

$$D_G J_i(f_i^0)[\varphi] = 0, \quad \forall \varphi \in C^\infty(\overline{\Omega}).$$

Keeping in mind the fact that the set of feasible solutions Ξ_i to the problem (4.23)–(4.24) has an empty topological interior, we begin with the following auxiliary results, where $F'(u)[h]$ stands for the directional derivative of a functional $F : X \to \mathbb{R}$ at the point $u \in X$ along a vector $h \in X$, i.e.,

$$F'(u)[h] = \lim_{\sigma \to 0} \frac{F(u + \sigma h) - F(u)}{\sigma}.$$

Proposition 4.4 *Let $f \in L^2(\Omega)$ be a given distribution and let*

$$F_1(u) = \frac{1}{2}\|u - f\|_{H^{-1}(\Omega)}^2, \quad \forall u \in L^2(\Omega).$$

Then

$$F_1'(u)[h] = \left((-\Delta)^{-1}(u - f), h\right)_{L^2(\Omega)}, \quad \forall h \in L^2(\Omega).$$

Proof The announced result immediately follows from the definition of the directional derivative and the following chain of transformations

$$F_1(u + \sigma h) - F_1(u) \stackrel{\text{by (4.15)}}{=} \frac{1}{2}\|\nabla(-\Delta)^{-1}(u + \sigma h - f)\|_{L^2(\Omega;\mathbb{R}^2)}$$
$$- \frac{1}{2}\|\nabla(-\Delta)^{-1}(u - f)\|_{L^2(\Omega;\mathbb{R}^2)}$$

$$= \sigma \left(\nabla(-\Delta)^{-1}(u-f), \nabla(-\Delta)^{-1}h\right)_{L^2(\Omega;\mathbb{R}^2)} + \sigma^2 \frac{1}{2}\|\nabla(-\Delta)^{-1}h\|^2_{L^2(\Omega;\mathbb{R}^2)}$$

$$= -\sigma \int_\Omega \operatorname{div}\left[\nabla(-\Delta)^{-1}(u-f)\right](-\Delta)^{-1}h\,dx + \sigma^2\frac{1}{2}\|h\|^2_{H^{-1}(\Omega)}$$

$$= \sigma \int_\Omega (-\Delta)(-\Delta)^{-1}(u-f)(-\Delta)^{-1}h\,dx + \sigma^2\frac{1}{2}\|h\|^2_{H^{-1}(\Omega)}$$

$$= \sigma \left((-\Delta)^{-1}(u-f), h\right)_{L^2(\Omega)} + o(\sigma), \quad \forall u \in L^2(\Omega).$$

Proposition 4.5 *Let $p : \Omega \to [p^-, p^+] \subset (1,2]$, with $p^\pm = \text{const}$, be a given exponent and let*

$$\widetilde{F}_2(u) = \int_\Omega |\nabla u(x)|^{p(x)}\,dx, \quad \forall u \in W^{1,p(\cdot)}(\Omega).$$

Then, for each $u \in W^{1,p(\cdot)}(\Omega)$, we have

$$\widetilde{F}'_2(u)[h] = \int_\Omega p(x)\left(|\nabla u(x)|^{p(x)-2}\nabla u(x), \nabla v(x)\right)\,dx, \qquad (4.38)$$

$$\forall h \in W^{1,p(\cdot)}(\Omega).$$

Proof Let $u, h \in W^{1,p(\cdot)}(\Omega)$ be given functions. We notice that

$$\frac{|\nabla u + \sigma\nabla h|^p - |\nabla u|^p}{\sigma} \to p\left(|\nabla u|^{p-2}\nabla u, \nabla h\right)$$

as $\sigma \to 0$ almost everywhere in Ω. Furthermore, by convexity,

$$|\xi|^p - |\eta|^p \le 2p\left(|\xi|^{p-1} + |\eta|^{p-1}\right)|\xi - \eta|,$$

we have

$$\left|\frac{1}{\sigma}\left(|\nabla u(x) + \sigma\nabla h(x)|^{p(x)} - |\nabla u(x)|^{p(x)}\right)\right|$$

$$\le 2p(x)\left(|\nabla u(x) + \sigma\nabla h(x)|^{p(x)-1} + |\nabla u(x)|^{p(x)-1}\right)|\nabla h(x)|$$

$$\le \text{const}\left(|\nabla u(x)|^{p(x)-1} + |\nabla h(x)|^{p(x)-1}\right)|\nabla h(x)|. \qquad (4.39)$$

Taking into account that

$$\int_\Omega |\nabla u(x)|^{p(x)-1}|\nabla v(x)|\,dx \le 2\||\nabla u(x)|^{p(x)-1}\|_{L^{p'(\cdot)}(\Omega)}\||\nabla h(x)|\|_{L^{p(\cdot)}(\Omega)}$$

$$\le 2\||\nabla u(x)|^{p(x)-1}\|_{L^{p'(\cdot)}(\Omega)}\|\nabla h(x)\|_{L^{p(\cdot)}(\Omega,\mathbb{R}^2)},$$

and

$$\int_\Omega |\nabla h(x)|^{p(x)}\,dx \overset{\text{by (4.8)}}{\le} \|\nabla h\|^2_{L^{p(\cdot)}(\Omega,\mathbb{R}^2)} + 1,$$

4.3 Optimality Conditions

we see that the right hand side of inequality (4.39) is an $L^1(\Omega)$-function. Therefore,

$$\lim_{\sigma \to 0} \frac{\widetilde{F}_2(u + \sigma h) - \widetilde{F}_2(u)}{\sigma} = \lim_{\sigma \to 0} \int_\Omega \frac{|\nabla u(x) + \sigma \nabla h(x)|^p - |\nabla u(x)|^p}{\sigma} \, dx$$
$$= \int_\Omega p(x) \left(|\nabla u(x)|^{p(x)-2} \nabla u(x), \nabla h(x) \right) dx$$

by the Lebesgue dominated convergence theorem. From this the representation (4.38) follows.

Proposition 4.6 *Let $p : \Omega \to [p^-, p^+] \subset (1, 2]$, with $p^\pm = \mathrm{const}$, be a given exponent and let*

$$F_2(u) = \int_\Omega |R_\eta \nabla u(x)|^{p(x)} \, dx, \quad \forall u \in W^{1, p(\cdot)}(\Omega),$$

where the linear operator $R_\eta : \mathbb{R}^2 \to \mathbb{R}^2$ is defined by the rule (4.21). Then, for each $u \in W^{1, p(\cdot)}(\Omega)$, we have

$$F_2'(u)[h] = \int_\Omega p(x) \left(|R_\eta \nabla u(x)|^{p(x)-2} R_\eta \nabla u(x), \nabla h(x) \right) dx$$
$$- \eta^2 \int_\Omega p(x) \left(|R_\eta \nabla u(x)|^{p(x)-2} R_\eta \nabla u(x), \theta(x) \right) (\theta(x), \nabla h(x)) \, dx,$$
$$\forall h \in W^{1, p(\cdot)}(\Omega). \quad (4.40)$$

Proof The representation (4.40) immediately follows from definition of the directional derivative and from Proposition 4.5.

Proposition 4.7 *Let $u \in \Xi_i$ be a given feasible solution, let*

$$p[\nabla u] := 1 + \delta + \frac{a^2(1 - \delta)}{a^2 + |\nabla u|^2},$$

and let the functional $F_3 : W^{1, 1+\delta}(\Omega) \to \mathbb{R}$ be defined as follows,

$$F_3(u) = \int_\Omega |R_\eta \nabla v(x)|^{p[\nabla u]} \, dx, \quad \forall u \in W^{1, 1+\delta}(\Omega),$$

where $v \in W^{1, p[\nabla u]}(\Omega)$ is a given function. Then, for each element $v \in W^{1, p[\nabla u]}(\Omega)$ and for all $h \in W^{1, 1+\delta}(\Omega)$, we have

$$F_3'(u)[h] = -\int_\Omega |R_\eta \nabla v(x)|^{p[\nabla u]} \times \frac{2a^2(1 - \delta) \log\left(|R_\eta \nabla v(x)|\right)}{\left(a^2 + |\nabla u|^2\right)^2} (\nabla u, \nabla h) \, dx. \quad (4.41)$$

Proof The representation (4.41) immediately follows from definition of the directional derivative.

Proposition 4.8 *Let $u \in \Xi$ be a feasible solution, and let*

$$F_4(u) = \frac{\lambda}{4}[D(u) - cD(f_i)]^2,$$

where $f_i \in L^2(\Omega)$ is a given spectral channel, $c = \text{const} > 1$, and

$$D(u) = \int_\Omega \int_\Omega W(x, y)\sqrt{\kappa^2 + |u(x) - u(y)|^2}\, dxdy.$$

Then the directional derivative of $F_4 : L^2(\Omega) \to \mathbb{R}$ at the given point u along a vector $h \in L^2(\Omega)$ takes the form

$$F_4'(u)[h] = \lambda\left(D(u) - cD(I_f)\right)$$
$$\times \int_\Omega \left(\int_\Omega W(x, y)\frac{u(x) - u(y)}{\sqrt{\kappa^2 + |u(x) - u(y)|^2}}\, dy\right) h(x)\, dx. \quad (4.42)$$

Proof The representation (4.42) immediately follows from definition of the directional derivative and the following chain of transformations

$$D(u + \sigma h) - D(u)$$
$$= \int_\Omega \int_\Omega W(x, y)\left(\sqrt{\kappa^2 + |u(x) - u(y) + \sigma(h(x) - h(y))|^2}\right.$$
$$\left. - \sqrt{\kappa^2 + |u(x) - u(y)|^2}\right)dxdy$$
$$= \int_\Omega \int_\Omega W(x, y)\left(|u(x) - u(y) + \sigma(h(x) - h(y))|^2 - |u(x) - u(y)|^2\right)$$
$$\times \frac{1}{\sqrt{\kappa^2 + |u(x) - u(y) + \sigma(h(x) - h(y))|^2} + \sqrt{\kappa^2 + |u(x) - u(y)|^2}}\, dxdy$$
$$= \sigma \int_\Omega \left(\int_\Omega [W(x, y) + W(y, x)]\frac{u(x) - u(y)}{\sqrt{\kappa^2 + |u(x) - u(y)|^2}}\, dy\right) h(x)\, dy$$
$$+ o(\sigma^2).$$

We are now able to show that the objective functional $J_i(v)$ is Gâteaux differentiable at $v = f_i^0$. With that in mind, we utilize the representation (4.24). As a result, we see that

4.3 Optimality Conditions

$$J_i'(f_i^0)[\varphi] = \int_\Omega p(|\nabla f_i^0|) \left(|R_\eta \nabla f_i^0(x)|^{p(|\nabla f_i^0|)-2} R_\eta \nabla f_i^0(x), \nabla \varphi(x)\right) dx$$

$$- \eta^2 \int_\Omega p(|\nabla f_i^0|) \left(|R_\eta \nabla f_i^0(x)|^{p(|\nabla f_i^0|)-2} R_\eta \nabla f_i^0(x), \theta(x)\right) (\theta(x), \nabla \varphi(x)) dx$$

$$- \int_\Omega |R_\eta \nabla f_i^0(x)|^{p(|\nabla f_i^0|)} \frac{2a^2(1-\delta) \log\left(|R_\eta \nabla f_i^0(x)|\right)}{\left(a^2 + |\nabla f_i^0(x))|^2\right)^2} (\nabla f_i^0, \nabla \varphi) dx$$

$$+ \lambda \left(D(f_i^0) - cD(f_i)\right)$$

$$\times \int_\Omega \left(\int_\Omega W(x,y) \frac{f_i^0(x) - f_i^0(y)}{\sqrt{\kappa^2 + |f_i^0(x) - f_i^0(y)|^2}} dy\right) \varphi(x) dx$$

$$+ \int_\Omega \left[(-\Delta)^{-1}(f_i^0 - f_i)\right] \varphi(x) dx = 0, \quad \forall \varphi \in C^\infty(\overline{\Omega}). \tag{4.43}$$

Thus, $J_i'(f_i^0) : C^\infty(\overline{\Omega}) \to \mathbb{R}$ is a linear functional.

Let us show that each term in (4.43) can be extended by continuity to the entire Sobolev space $H^{1,p(|\nabla f_i^0|)}(\Omega)$. To this end, it is enough to establish the existence of a constant $M > 0$ such that

$$\left|J_i'(f_i^0)[\varphi]\right| \leq M \|\varphi\|_{W^{1,p(|\nabla f_i^0|)}(\Omega)}, \quad \forall \varphi \in C^\infty(\overline{\Omega}). \tag{4.44}$$

Indeed, rewriting (4.43) in the form

$$J_i'(f_i^0)[\varphi] = S_1 + S_2 + S_3 + S_4 + S_5,$$

where the one-to-one correspondence to (4.43) is preserved, we see that

$$|S_1| \leq \|p\|_{L^\infty(\Omega)} \int_\Omega \left(|R_\eta \nabla f_i^0|^{p(|\nabla f_i^0|)-2} R_\eta \nabla f_i^0, \nabla \varphi\right) dx$$

$$\stackrel{\text{by (4.9)}}{\leq} 2p^+ \||R_\eta \nabla f_i^0|^{p(|\nabla f_i^0|)-2} R_\eta \nabla f_i^0\|_{L^{p'(|\nabla f_i^0|)}(\Omega;\mathbb{R}^2)} \|\nabla \varphi\|_{L^{p(|\nabla I^0|)}(\Omega;\mathbb{R}^2)}$$

$$\stackrel{\text{by (4.8)}}{\leq} 2p^+ \left(1 + \int_\Omega |R_\eta \nabla f_i^0|^{p(|\nabla f_i^0|)} dx\right)^{1/(p')^-} \|\varphi\|_{W^{1,p(|\nabla f_i^0|)}(\Omega)},$$

where $p^+ = 2$, $(p')^- = p^+/(p^+ - 1) = 2$, and

$$\int_\Omega |R_\eta \nabla f_i^0|^{p(|\nabla f_i^0|)} dx \stackrel{\text{by (4.8)}}{\leq} 1 + \|\nabla f_i^0\|_{L^{p(|\nabla f_i^0|)}(\Omega;\mathbb{R}^2)}^{p^+}$$

$$\leq 1 + \|f_i^0\|_{W^{1,p(|\nabla f_i^0|)}(\Omega)}^2 < +\infty$$

by the assumption $f_i^0 \in \Xi_i$. Thus, there exists a constant $M_1 > 0$ such that

$$|S_1| \leq M_1 \|\varphi\|_{W^{1,p(|\nabla f_i^0|)}(\Omega)}. \tag{4.45}$$

Arguing in a similar manner, it can be shown that a constant $M_2 > 0$ exists such that

$$|S_2| \leq M_2 \|\varphi\|_{W^{1,p(|\nabla f_i^0|)}(\Omega)}.$$

As for the third term in (4.43), we notice that

$$|\nabla f_i^0|^2 \frac{|\log(|\nabla f_i^0|)|}{(a^2 + |\nabla f_i^0|^2)^2} \leq \frac{|\log(|\nabla f_i^0|)|}{a^2 + |\nabla f_i^0|^2} < +\infty \quad \text{as } |\nabla f_i^0| \to \infty$$

by the L'Hôpital's rule. Using similar arguments, we see that

$$|\nabla f_i^0|^2 \frac{|\log(|\nabla f_i^0|)|}{(a^2 + |\nabla f_i^0|^2)^2} \leq \frac{1}{a^4} |\nabla f_i^0|^2 |\log(|\nabla f_i^0|)| < +\infty \quad \text{as } |\nabla f_i^0| \to 0.$$

Thus, we can deduce the existence a constant $M_2 > 0$ such that

$$|S_2| \leq 2a^2(1-\delta) \left\| |R_\eta \nabla f_i^0|^2 \frac{|\log(|R_\eta \nabla f_i^0|)|}{(a^2 + |\nabla f_i^0|^2)^2} \right\|_{L^\infty(\Omega)}$$
$$\times \int_\Omega \left(|R_\eta \nabla f_i^0|^{p(|\nabla f_i^0|)-2} R_\eta \nabla f_i^0, \nabla \varphi \right) dx$$
$$\leq M_3 \|\varphi\|_{W^{1,p(|\nabla f_i^0|)}(\Omega)}. \tag{4.46}$$

It remains to notice that in view of the obvious inclusions

$$\int_\Omega W(\cdot, y) \frac{f_i^0(\cdot) - f_i^0(y)}{\sqrt{\kappa^2 + |f_i^0(\cdot) - f_i^0(y)|^2}} \, dy \in L^2(\Omega),$$

$$\left[(-\Delta)^{-1} (f_i^0 - f_i) \right] \in L^2(\Omega),$$

the existence of constants M_3 and M_4 such that

$$|S_j| \leq M_j \|\varphi\|_{L^2(\Omega)} \leq M_j \|\varphi\|_{W^{1,p(|\nabla f_i^0|)}(\Omega)}, \quad j = 4, 5, \tag{4.47}$$

immediately follows from (4.43) and the Cauchy inequality.

Utilizing the estimates (4.45), (4.46), and (4.47), we finally arrive at the inequality (4.44) with $M = \max\{M_1, M_2, M_3, M_4, M_5\}$. Thus, the mapping $\varphi \mapsto J_i'(f_i^0)[\varphi]$ can be defined for all $\varphi \in H^{1,p(|\nabla f_i^0|)}(\Omega)$ using the density of $C^\infty(\overline{\Omega})$ in $H^{1,p(|\nabla f_i^0|)}(\Omega)$ (see (4.22)) and the standard rule

$$D_G J_i(f_i^0)[\varphi] = \lim_{k \to \infty} D_G J_i(f_i^0)[\varphi_k],$$

where $\{\varphi_k\}_{k \in \mathbb{N}} \subset C_c^\infty(\mathbb{R}^2)$ and $\varphi_k \to \varphi$ strongly in $H^{1,p(|\nabla f_i^0|)}(\Omega)$. Hence, the objective functional $J_i(v)$ is Gâteaux differentiable at $v = f_i^0$, and

4.4 Existence Issues and Regularization of the Original Optimization Problem

$$D_G J_i(f_i^0)[h] = J'(f_i^0)[h], \quad \forall h \in H^{1,p(|\nabla f_i^0|)}(\Omega).$$

In order to get the final relations for optimality conditions, it remains to observe that identity (4.43) implies the following equalities in the sense of distributions

$$-\operatorname{div}\left[p(x)|R_\eta \nabla f_i^0|^{p(|\nabla f_i^0|)-2} R_\eta \nabla f_i^0\right]$$
$$+\eta^2 \operatorname{div}\left[p(x)\left(|R_\eta \nabla f_i^0|^{p(|\nabla f_i^0|)-2} R_\eta \nabla f_i^0, \theta\right)\theta\right]$$
$$+2a^2(1-\delta)\operatorname{div}\left[|R_\eta \nabla f_i^0|^{p(|\nabla f_i^0|)} \frac{\log\left(|R_\eta \nabla f_i^0|\right)}{\left(a^2+|\nabla f_i^0|^2\right)^2} \nabla f_i^0\right]$$
$$+\lambda\left(D(f_i^0) - cD(f_i)\right) \int_\Omega W(x,y) \frac{f_i^0(x) - f_i^0(y)}{\sqrt{\kappa^2 + |f_i^0(x) - f_i^0(y)|^2}} dy$$
$$+\mu(-\Delta)^{-1}(f_i^0 - f_i) = 0, \quad \text{in } \Omega, \quad (4.48)$$
$$\left(|\nabla f_i^0|^{p(|\nabla f_i^0|)-2} \nabla f_i^0, \nu\right) = 0 \quad \text{on } \partial\Omega, \quad (4.49)$$

where ν denotes the unit outward normal to the boundary $\partial\Omega$.

4.4 Existence Issues and Regularization of the Original Optimization Problem

The main question we are going to discuss in this section is to find out whether the problem (4.23)–(4.24) admits at least one solution. With that in mind, we make use of the so-called indirect approach [16, 28] (for comparison, we refer to the recent publication [19]). The main idea of this approach is to show that the original minimization problem (4.23)–(4.24) can be efficiently approximated by a special family of optimization problems of a similar structure but with the spatial regularization of the exponent $p(|\nabla u|)$ in the form

$$p_\varepsilon(|\nabla u|) = 1 + \delta + \frac{a^2(1-\delta)}{a^2 + |(\nabla K_\varepsilon * u)(x)|^2}, \quad (4.50)$$

where $(\nabla K_\varepsilon * u)$ stands for the Steklov smoothing operator.

Let $K : \mathbb{R}^2 \to \mathbb{R}$ be a positive compactly supported function such that

$$K \in C_0^\infty(\mathbb{R}^2), \quad \int_{\mathbb{R}^2} K(x)\,dx = 1, \quad \text{and } K(x) = K(-x), \, \forall x \in \mathbb{R}^2.$$

For any $\varepsilon > 0$, we set $K_\varepsilon(x) = \varepsilon^{-2} K\left(\frac{x}{\varepsilon}\right)$. Then the following properties of the convolution

$$u_\varepsilon(x) := (K_\varepsilon * u)(x) = \int_\Omega K_\varepsilon(x-y)u(y)\,dy, \quad \forall u \in L^1(\Omega),$$

are well-known [25]:

(i) $u_\varepsilon \in C^\infty(\Omega)$ for all $\varepsilon > 0$;
(ii) $u_\varepsilon(x) \to u(x)$ almost everywhere in Ω;
(iii) If $u \in L^p(\Omega)$ with $1 \le p < \infty$, then $u_\varepsilon \to u$ in $L^p(\Omega)$.

We introduce the following family of approximating problems to the problem (4.23)–(4.24):

$$J_{i,\varepsilon}(f_{i,\varepsilon}^0) = \inf_{v \in \Xi_{i,\varepsilon}} J_{i,\varepsilon}(v), \quad i = 1, 2, 3, \tag{4.51}$$

where ε is a small parameter which varies within a strictly decreasing sequence of positive numbers converging to 0,

$$J_{i,\varepsilon}(v) = \int_\Omega |R_\eta \nabla v(x)|^{p_\varepsilon(|\nabla v|)} dx + \frac{\mu}{2} \|v - f_i\|_{H^{-1}(\Omega)}^2 + \frac{\lambda}{4} [D(v) - cD(f_i)]^2,$$

$$\Xi_{i,\varepsilon} = \left\{ I(x) \,\middle|\, \begin{array}{l} I \in H^{1,p_\varepsilon(|\nabla I|)}(\Omega) \cap L^\infty(\Omega), \\ \gamma_{i,0} \le I(x) \le \gamma_{i,1} \text{ a.e. in } \Omega, \\ \gamma_{i,0} = \inf_{x \in \Omega} f_i(x), \\ \gamma_{i,1} = \sup_{x \in \Omega} f_i(x), \end{array} \right\} \tag{4.52}$$

and $p_\varepsilon(|\nabla v|)$ is defined in (4.50).

Before proceeding further, we make use of a few technical results.

Lemma 4.9 *[17, Lemma 1] Let $\{v_k\}_{k \in \mathbb{N}} \subset L^\infty(\Omega)$ be a sequence of measurable functions such that $v_k(x) \to v(x)$ weakly-* in $L^\infty(\Omega)$ for some $v \in L^\infty(\Omega)$. Let*

$$\left\{ p_k = 1 + \delta + \frac{a^2(1-\delta)}{a^2 + |(\nabla K_\varepsilon * v_k)(x)|^2} \right\}_{k \in \mathbb{N}}$$

be the corresponding sequence of exponents. Then

$$p_{k,\varepsilon} \to p_\varepsilon = 1 + \delta + \frac{a^2(1-\delta)}{a^2 + |(\nabla K_\varepsilon * u)|^2} \quad \text{uniformly in } \overline{\Omega} \text{ as } k \to \infty,$$

$$1 + \delta + \frac{a^2(1-\delta)}{a^2 + \|K_\varepsilon\|_{C^1(\overline{\Omega - \Omega})}^2 \sup_{k \in \mathbb{N}} \|v_k\|_{L^1(\Omega)}^2} \le p_{k,\varepsilon}(x) \le 2, \ \forall k \in \mathbb{N},$$

where

$$\|K_\varepsilon\|_{C^1(\overline{\Omega-\Omega})} = \max_{\substack{z = x - y \\ x \in \overline{\Omega}, y \in \overline{\Omega}}} \Big[|K_\varepsilon(z)| + |\nabla K_\varepsilon(z)| \Big].$$

4.4 Existence Issues and Regularization of the Original Optimization Problem

Lemma 4.10 *[40, Proposition B.2] The mapping $v \mapsto \frac{\lambda}{4}[D(v) - cD(f_i)]^2$ is continuous from $L^2(\Omega)$ endowed with thee strong topology to $\mathbb{R}$ with pointwise convergence.*

Proposition 4.11 *[17] Let $\{p_{k,\varepsilon}\}_{k\in\mathbb{N}}$ be a sequence of exponents that satisfies all preconditions of Lemma 4.9. If a bounded sequence*

$$\left\{f_k \in L^{p_{k,\varepsilon}(\cdot)}(\Omega)\right\}_{k\in\mathbb{N}}$$

converges weakly in $L^{1+\delta}(\Omega)$ to f, then $f \in L^{p_\varepsilon(\cdot)}(\Omega)$, $f_k \rightharpoonup f$ in variable $L^{p_{k,\varepsilon}(\cdot)}(\Omega)$.

We are now in a position to prove the existence of minimizers for the proposed approximating problem (4.51)–(4.52).

Theorem 4.12 *Let Ω be an open bounded and connected sub-domain of $\mathbb{R}^2$ with a Lipschitz boundary. Let $f_i \in L^2(\Omega)$ be a given spectral channel of an image arguably contaminated by additive Gaussian noise with zero mean. Then, for each $\varepsilon > 0$, the minimization problem (4.51)–(4.52) admits at least one solution $f^0_{i,\varepsilon}$ in $W^{1,p^-}(\Omega) \cap L^\infty(\Omega)$ such that $I^0_{i,\varepsilon} \in H^{1,p[\nabla f^0_{i,\varepsilon}]}(\Omega)$.*

Proof To begin with, let us notice that, for each $\varepsilon > 0$, the indicated minimization problem is consistent, i.e. $J_{i,\varepsilon}(u) < +\infty$ for any $u \in \Xi_{i,\varepsilon}$. Since $\Xi_{i,\varepsilon} \neq \emptyset$ and $0 \leq J_{i,\varepsilon}(v) < +\infty$ for all $v \in \Xi_{i,\varepsilon}$, it follows that there exists a non-negative value $\zeta_\varepsilon \geq 0$ such that $\zeta_\varepsilon = \inf_{v \in \Xi_{i,\varepsilon}} J_{i,\varepsilon}(v)$. Let $\{v^\varepsilon_k\}_{k\in\mathbb{N}}$ be a minimizing sequence for (4.51)–(4.52), i.e.

$$\{v^\varepsilon_k\}_{k\in\mathbb{N}} \subset \Xi_{i,\varepsilon} \text{ and } \lim_{k\to\infty} J_{i,\varepsilon}\left(v^\varepsilon_k\right) = \zeta_\varepsilon.$$

Without lost of generality, we can suppose that $J_{i,\varepsilon}\left(v^\varepsilon_k\right) \leq \zeta_\varepsilon + 1$ for all $k \in \mathbb{N}$. From this and estimate (4.28), we deduce

$$\|v^\varepsilon_k\|^{p^-}_{W^{1,p_\varepsilon(|\nabla v^\varepsilon_k|)}} \leq C\left(|\Omega|\gamma^2_{i,1} + 2 + \frac{\zeta_\varepsilon + 1}{(1-\eta^2)^2}\right), \quad \forall k \in \mathbb{N}, \quad (4.53)$$

$$\|v^\varepsilon_k\|_{L^\infty(\Omega)} \leq \gamma_{i,1}, \quad \forall k \in \mathbb{N}.$$

Hence, in view of (4.6) and (4.25), the sequence $\{v^\varepsilon_k\}_{k\in\mathbb{N}}$ is bounded in $W^{1,p^-}(\Omega)$. Therefore, there exist a subsequence of $\{v^\varepsilon_k\}_{k\in\mathbb{N}}$, still denoted by the same index, and a vector function $f^0_{i,\varepsilon} \in W^{1,p^-}(\Omega)$ such that

$$v^\varepsilon_k \to f^0_{i,\varepsilon} \text{ strongly in } L^q(\Omega) \text{ for all } q \in [1, (p^-)^*) \text{ as } k \to \infty, \quad (4.54)$$

$$v^\varepsilon_k \overset{*}{\rightharpoonup} f^0_{i,\varepsilon} \text{ weakly-}* \text{ in } L^\infty(\Omega) \text{ as } k \to \infty, \quad (4.55)$$

$$v^\varepsilon_k \rightharpoonup f^0_{i,\varepsilon} \text{ weakly in } W^{1,p^-}(\Omega) \text{ as } k \to \infty, \quad (4.56)$$

where, by Sobolev embedding theorem,

$$(p^-)^* = \frac{2p^-}{2-p^-} = \frac{2+2\delta}{1-\delta} > 2.$$

In view of this and the smoothness of the kernel K_ε, we see that the operator

$$L^{p^-}(\Omega; \mathbb{R}^2) \ni \nabla v \mapsto p_\varepsilon(|\nabla v|) \in C(\overline{\Omega})$$

is compact (see Lemma 4.9). So, (4.55)–(4.56) imply that

$$p_\varepsilon(|\nabla v_k^\varepsilon|) \to p_\varepsilon(|\nabla f_{i,\varepsilon}^0|) \text{ in } C(\overline{\Omega}).$$

Passing then to a subsequence if necessary, we have (see Propositions 4.3 and 4.11):

$$v_k^\varepsilon(x) \to f_{i,\varepsilon}^0(x) \text{ a.e. in } \Omega. \tag{4.57}$$

$$v_k^\varepsilon \rightharpoonup f_{i,\varepsilon}^0 \text{ weakly in variable} L^{p_\varepsilon(|\nabla v_k^\varepsilon|)}(\Omega),$$

$$\nabla v_k^\varepsilon \rightharpoonup \nabla f_{i,\varepsilon}^0 \text{ weakly in variable } L^{p_\varepsilon(|\nabla v_k^\varepsilon|)}(\Omega; \mathbb{R}^2).$$

Hence, $f_{i,\varepsilon}^0 \in W^{1,p_\varepsilon(|\nabla f_{i,\varepsilon}^0|)}(\Omega)$.

Further we notice that, for each $k \in \mathbb{N}$, $\gamma_{i,0} \le v_k^\varepsilon(x) \le \gamma_{i,1}$ a.a. in Ω. Then it follows from (4.57) that the limit function $f_{i,\varepsilon}^0(x)$ is also subjected to the same restriction. Thus, $f_{i,\varepsilon}^0 \in W^{1,p_\varepsilon(|\nabla f_{i,\varepsilon}^0|)}(\Omega) \cap L^\infty(\Omega)$ is a feasible solution to minimization problem (4.51)–(4.52).

It remains to show that $f_{i,\varepsilon}^0$ is a minimizer of this problem. Indeed, taking into account the properties (4.53), (4.56) and the fact that $\theta \in L^\infty(\Omega, \mathbb{R}^2)$, we see that the sequence

$$\left\{ |R_\eta \nabla v_k^\varepsilon := \nabla v_k^\varepsilon - \eta^2 \left(\theta, \nabla v_k^\varepsilon\right) \theta \right| \in L^{p_\varepsilon(|\nabla v_k^\varepsilon|)}(\Omega; \mathbb{R}^2) \right\}_{k \in \mathbb{N}}$$

is bounded in variable space $L^{p_\varepsilon(|\nabla v_k^\varepsilon|)}(\Omega; \mathbb{R}^2)$ and weakly convergent to $|R_\eta \nabla f_{i,\varepsilon}^0|$ in $L^{p^-}(\Omega; \mathbb{R}^2)$. Hence, by Proposition 4.3, the following lower semicontinuous property

$$\liminf_{k \to \infty} \int_\Omega |R_\eta \nabla v_k^\varepsilon(x)|^{p_\varepsilon(|\nabla v_k^\varepsilon|)} \, dx \ge \int_\Omega |R_\eta \nabla f_{i,\varepsilon}^0(x)|^{p_\varepsilon(|\nabla f_{i,\varepsilon}^0|)} \, dx$$

holds true. Combining this relation with the following ones

$$\lim_{k \to \infty} \|v_k^\varepsilon - f_i\|_{H^{-1}(\Omega)}^2 = \|f_{i,\varepsilon}^0 - f_i\|_{H^{-1}(\Omega)}^2,$$

$$\lim_{k \to \infty} \left[D(v_k^\varepsilon) - cD(f_i)\right]^2 = \left[D(f_{i,\varepsilon}^0) - cD(f_i)\right]^2,$$

which are direct consequence of Lemma 4.10 and compactness of the embedding $L^2(\Omega) \subset H^{-1}(\Omega)$, we finally obtain

4.4 Existence Issues and Regularization of the Original Optimization Problem

$$\zeta_\varepsilon = \inf_{v \in \Xi_{i,\varepsilon}} J_{i,\varepsilon}(v) = \lim_{k \to \infty} J_{i,\varepsilon}\left(v_k^\varepsilon\right) = \liminf_{k \to \infty} J_{i,\varepsilon}\left(v_k^\varepsilon\right)$$

$$\geq \int_\Omega |R_\eta \nabla f_{i,\varepsilon}^0(x)|^{p_\varepsilon(|\nabla f_{i,\varepsilon}^0|)}\,dx + \frac{\mu}{2}\,\|f_{i,\varepsilon}^0 - f_i\|_{H^{-1}(\Omega)}^2$$

$$+ \frac{\lambda}{4}\left[D(f_{i,\varepsilon}^0) - cD(f_i)\right]^2 = J_{i,\varepsilon}(f_{i,\varepsilon}^0).$$

Thus, $f_{i,\varepsilon}^0$ is a minimizer to the problem (4.51)–(4.52).

Taking this existence result into account, we pass to the study of approximation properties of the problems (4.51)–(4.52). Namely, the main question we are going to discuss further is whether we can establish the convergence of minima of (4.51)–(4.52) to minima of (4.23)–(4.24) as ε tends to zero. In other words, our aim is to show that some optimal solutions to (4.23)–(4.24) can be approximated by the solutions of (4.51)–(4.52). To this end, we make use of some results of the variational convergence of minimization problems [29–31, 33, 34] and begin with some auxiliaries (see also [14, 15, 32] for other aspects of this concept).

Lemma 4.13 *Let $\{\varepsilon_k\}_{k \in \mathbb{N}}$ be a sequence of positive numbers converging to zero as $k \to \infty$. Let*

$$\{v_k\}_{k \in \mathbb{N}} \quad and \quad \left\{p_k := 1 + \delta + \frac{a^2(1-\delta)}{a^2 + |(\nabla K_{\varepsilon_k} \ast v_k)|^2}\right\}_{k \in \mathbb{N}}$$

be sequences such that

$$v_k \in \Xi_{i,\varepsilon_k}, \quad \forall k \in \mathbb{N}, \tag{4.58}$$

$$v_k(x) \to v(x) \quad a.e.\ in\ \Omega, \tag{4.59}$$

$$v_k \to v \quad strongly\ in\ L^2(\Omega), \tag{4.60}$$

$$\nabla v_k \rightharpoonup \nabla v \quad weakly\ in\ L^{p^-}(\Omega;\mathbb{R}^2), \tag{4.61}$$

$$\left\||\nabla v_k(\cdot)|^{p_k(\cdot)}\right\|_{L^1(\Omega)} \leq C \tag{4.62}$$

for some positive constant C not depending on k,

$$p_k(x) \to p(x) := 1 + \delta + \frac{a^2(1-\delta)}{a^2 + |\nabla v|^2} \quad a.e.\ in\ \Omega. \tag{4.63}$$

Then

$$v \in \Xi_i \quad and \quad J_i(v) \leq \liminf_{k \to \infty} J_{i,\varepsilon_k}(v_k), \quad \forall i = 1,2,3. \tag{4.64}$$

Proof The following relations

$$\lim_{k \to \infty} \|v_k - f_i\|_{H^{-1}(\Omega)}^2 = \|v - f_i\|_{H^{-1}(\Omega)}^2, \tag{4.65}$$

$$\lim_{k \to \infty} [D(v_k) - cD(f_i)]^2 = [D(v) - cD(f_i)]^2 \tag{4.66}$$

are a direct consequence of Lemma 4.10, compactness of the embedding $L^2(\Omega) \subset H^{-1}(\Omega)$, and condition (4.60). We also notice that, in view of representation

$$R_\eta \nabla v_k := \nabla v_k - \eta^2 (\theta, \nabla v_k) \theta, \quad \forall k \in \mathbb{N},$$

Proposition 4.3 and the initial assumptions (4.61)–(4.63) lead to the conclusion:

$$\nabla v \in L^{p(\cdot)}(\Omega; \mathbb{R}^2), \ \liminf_{k \to \infty} \int_\Omega |R_\eta \nabla v_k(x)|^{p_k(x)} \, dx \geq \int_\Omega |R_\eta \nabla v(x)|^{p(x)} \, dx.$$

As a result, combining the last inequality with (4.65)–(4.66), we arrive at the announced relation $(4.64)_2$.

It remains to show that v is a feasible solution to the problem (4.23)–(4.24), i.e., $v \in \Xi_i$. To this end, we take into account the inclusion $\nabla v \in L^{p(\cdot)}(\Omega; ; \mathbb{R}^2)$ established above and the fact that $v_k \in \Xi_{i,\varepsilon_k}$ for each $k \in \mathbb{N}$. Then it follows from (4.59) that $\gamma_{i,0} \leq v(x) \leq \gamma_{i,1}$ almost everywhere in Ω, and, therefore, $v \in \Xi_i$.

Lemma 4.14 *For each feasible solution $v \in \Xi_i$ to the original problem (4.23)–(4.24), there can be found a sequence $\{v_\varepsilon\}_{\varepsilon \to 0}$ such that*

$$v_\varepsilon \in \Xi_{i,\varepsilon}, \quad \forall \varepsilon \in (0, \varepsilon_0) \text{ with } \varepsilon_0 > 0 \text{ sufficiently small,} \tag{4.67}$$

$$v_\varepsilon(x) \to v(x) \ a.e. \text{ in } \Omega \text{ as } \varepsilon \to 0, \tag{4.68}$$

$$v_\varepsilon \to v \text{ strongly in } L^2(\Omega), \tag{4.69}$$

$$\nabla v_\varepsilon \to \nabla v \text{ strongly in } L^{p^-}(\Omega; \mathbb{R}^2), \tag{4.70}$$

$$\left\| |\nabla v_\varepsilon(\cdot)|^{p_\varepsilon(\cdot)} \right\|_{L^1(\Omega)} \leq C \tag{4.71}$$

for some positive constant C not depending on ε,

$$p_\varepsilon(x) := 1 + \delta + \frac{a^2(1-\delta)}{a^2 + |(\nabla K_\varepsilon * v_\varepsilon)|^2} \to \tag{4.72}$$

$$\to p(x) := 1 + \delta + \frac{a^2(1-\delta)}{a^2 + |\nabla v|^2} \ a.e. \text{ in } \Omega,$$

$$J_i(v) \geq \limsup_{\varepsilon \to 0} J_{i,\varepsilon}(v_\varepsilon). \tag{4.73}$$

Proof Let v be an arbitrary feasible solution to the problem (4.23)–(4.24). We define the sequence $\{v_\varepsilon\}_{\varepsilon \to 0}$ as a smooth mollification of v with the kernel K_ε, i.e.,

$$v_\varepsilon(x) := (K_\varepsilon * v)(x) = \int_\Omega K_\varepsilon(x-y) v(y) \, dy, \quad \forall x \in \Omega.$$

Then properties (4.68)–(4.70) are direct consequence of the Steklov smoothing procedure (see (i)–(iii)). Moreover, in view of (4.68) and the fact that $\gamma_{i,0} \leq v(x) \leq \gamma_{i,1}$ a.e. in Ω, we can suppose that the same restriction for v_ε

4.4 Existence Issues and Regularization of the Original Optimization Problem

$$\gamma_{i,0} \le v_\varepsilon(x) \le \gamma_{i,1} \quad \text{a.e. in } \Omega \qquad (4.74)$$

holds true with $\varepsilon > 0$ small enough.

Since $v_\varepsilon \to v$ strongly in $W^{1,p^-}(\Omega)$, we can suppose (without loss of generality) that $\nabla v_\varepsilon(x) \to \nabla v(x)$ almost everywhere in Ω. As a result, this implies the pointwise convergence (4.72). Hence,

$$|R_\eta \nabla v_\varepsilon(x)|^{p_\varepsilon(x)} \to |R_\eta \nabla v(x)|^{p(x)} \quad \text{a.e. in } \Omega.$$

From this and the fact that $|R_\eta \nabla v(x)|^{p(x)} \in L^1(\Omega)$, we deduce:

$$|R_\eta \nabla v_\varepsilon|^{p_\varepsilon(\cdot)} \to |R_\eta \nabla v|^{p(\cdot)} \quad \text{strongly in } L^1(\Omega). \qquad (4.75)$$

Thus, $\left\| |R_\eta \nabla v_\varepsilon(\cdot)|^{p_\varepsilon(\cdot)} \right\|_{L^1(\Omega)} \le C$ for some positive constant C not depending on ε. Hence, in view of estimates (4.26), we get: $\nabla v_\varepsilon \in L^{p_\varepsilon(\cdot)}(\Omega;;\mathbb{R}^2)$ for ε small enough. From this and (4.74), the assertion (4.67) follows. Moreover, the following equality

$$\lim_{\varepsilon \to 0} \int_\Omega |R_\eta \nabla v_\varepsilon|^{p_\varepsilon(\cdot)}\, dx = \int_\Omega |R_\eta \nabla v|^{p(\cdot)}\, dx \qquad (4.76)$$

immediately follows from (4.75).

It remains to observe that

$$\lim_{k \to \infty} \|v_k - f_i\|_{H^{-1}(\Omega)}^2 = \|v - f_i\|_{H^{-1}(\Omega)}^2, \qquad (4.77)$$

$$\lim_{k \to \infty} [D(v_k) - cD(f_i)]^2 = [D(v) - cD(f_i)]^2, \qquad (4.78)$$

by Lemma 4.10 and compactness of the embedding $L^2(\Omega) \subset H^{-1}(\Omega)$. As a result, we conclude from (4.76), (4.77), and (4.78) that, in fact, instead of the announced inequality (4.73), we have $J_i(v) = \lim_{\varepsilon \to 0} J_{i,\varepsilon}(v_\varepsilon)$. The proof is complete.

We are now in a position to prove the main result of this section.

Theorem 4.15 *Assume that original minimization problem (4.23)–(4.24) has a non-empty set of minimizers. Let $\left\{ f_{i,\varepsilon}^0 \in \Xi_{i,\varepsilon} \right\}_{\varepsilon > 0}$ be a sequence of solutions to the corresponding minimization problems (4.51)–(4.52). Let*

$$\left\{ p_\varepsilon := 1 + \delta + \frac{a^2(1-\delta)}{a^2 + \left| (\nabla K_\varepsilon * f_{i,\varepsilon}^0) \right|^2} \right\}_{\varepsilon > 0}$$

be the sequence of associated exponents. Assume that the sequence $\{p_\varepsilon\}_{\varepsilon > 0}$ is compact with respect to the pointwise convergence in Ω. Then there exists an element $f_i^0 \in \Xi$ such that, up to a subsequence,

$$f_{i,\varepsilon}^0(x) \to f_i^0(x) \text{ a.e. in } \Omega \text{ as } \varepsilon \to 0, \tag{4.79}$$

$$f_{i,\varepsilon}^0 \to f_i^0 \text{ strongly in } L^2(\Omega), \tag{4.80}$$

$$\nabla f_{i,\varepsilon}^0 \rightharpoonup \nabla f_i^0 \text{ weakly in } L^{p^-}(\Omega; \mathbb{R}^2), \tag{4.81}$$

$$\left\| |\nabla f_{i,\varepsilon}^0(\cdot)|^{p_\varepsilon(\cdot)} \right\|_{L^1(\Omega)} \leq C \tag{4.82}$$

for some positive constant C not depending on ε,

$$\inf_{v \in \Xi_{i,\varepsilon}} J_i(v) = J_i(f_i^0) = \lim_{\varepsilon \to 0} J_{i,\varepsilon}(f_{i,\varepsilon}^0) = \lim_{\varepsilon \to 0} \inf_{v \in \Xi_{i,\varepsilon}} J_{i,\varepsilon}(v_\varepsilon). \tag{4.83}$$

Proof First, we observe that a given sequence of minimizers for approximating problems (4.51)–(4.52) is compact with respect to the convergence (4.79)–(4.81). Indeed, for an arbitrary test function $\varphi \in C_c^\infty(\mathbb{R}^2)$, we have:

$$\varphi \in H^{1,p_\varepsilon(\cdot)}(\Omega), \quad \forall \varepsilon > 0.$$

Let's assume that this function satisfies the pointwise constraints $\gamma_{i,0} \leq \varphi(x) \leq \gamma_{i,1}$ in Ω. Then, $\varphi \in \Xi_{i,\varepsilon}$ for all $\varepsilon > 0$, and, therefore, we can suppose that

$$J_{i,\varepsilon}(f_{i,\varepsilon}^0) = \inf_{v \in \Xi_{i,\varepsilon}} J_{i,\varepsilon}(v_\varepsilon) \leq J_{i,\varepsilon}(\varphi) \leq \sup_{\varepsilon > 0} J_{i,\varepsilon}(\varphi) \leq C < +\infty \quad \forall \varepsilon > 0.$$

Hence,

$$\sup_{\varepsilon > 0} \int_\Omega |R_\eta \nabla f_{i,\varepsilon}^0(\cdot)|^{p_\varepsilon(\cdot)} \, dx \leq C \text{ and } \sup_{\varepsilon > 0} \|f_{i,\varepsilon}^0\|_{L^2(\Omega)} \leq \sqrt{|\Omega|} \gamma_{i,1}. \tag{4.84}$$

Combining this issue with estimates (4.26), we see that $\left\{ f_{i,\varepsilon}^0 \in \Xi_{i,\varepsilon} \right\}_{\varepsilon > 0}$ is a bounded sequence in $W^{1,p^-}(\Omega)$. Hence, there exist a subsequence $\left\{ f_{i,k}^0 \in \Xi_{i,\varepsilon_k} \right\}_{k \in \mathbb{N}}$ of $\left\{ f_{i,\varepsilon}^0 \in \Xi_{i,\varepsilon} \right\}_{\varepsilon > 0}$, and a function $f_i^0 \in W^{1,p^-}(\Omega)$ such that

$$\begin{aligned} f_{i,k}^0 &\to f_i^0 \text{ strongly in } L^q(\Omega) \text{ for all } q \in [1, (p^-)^*), \\ f_{i,k}^0 &\rightharpoonup f_i^0 \text{ weakly in } W^{1,p^-}(\Omega) \text{ as } k \to \infty, \end{aligned} \tag{4.85}$$

where, by Sobolev embedding theorem,

$$(p^-)^* = \frac{2p^-}{2 - p^-} = \frac{2 + 2\delta}{1 - \delta} > 2 + \delta.$$

From this, the conditions (4.79)–(4.81) follow, whereas (4.82) is a consequence of (4.26) and the boundedness property (4.84).

Thus, we may suppose that for the subsequence $\left\{ f_{i,k}^0 \in \Xi_{\varepsilon_k} \right\}_{k \in \mathbb{N}}$ all preconditions of Lemma 4.13 are fulfilled. Therefore, property (4.64) leads us to the conclusion that $f_i^0 \in \Xi_i$ and

4.4 Existence Issues and Regularization of the Original Optimization Problem

$$\liminf_{k\to\infty} \inf_{v\in \Xi_{i,\varepsilon_k}} J_{i,\varepsilon_k}(v) = \liminf_{k\to\infty} J_{i,\varepsilon_k}(f^0_{i,k}) \geq J_i(f^0_i)$$

$$\geq \inf_{v\in \Xi_i} J_i(v) = J_i(f^*_i), \qquad (4.86)$$

where $f^*_i \in \Xi$ is a minimizer for (4.23)–(4.24).

Then Lemma 4.14 implies the existence of a sequence $\left\{f^*_{i,\varepsilon} \in \Xi_{i,\varepsilon}\right\}_{\varepsilon>0}$ such that $f^*_{i,\varepsilon} \to f^*_i$ as $\varepsilon \to 0$ in the sense of relations (4.68)–(4.72), and

$$J_i(f^*_i) \geq \limsup_{\varepsilon\to 0} J_{i,\varepsilon}(f^*_{i,\varepsilon}).$$

Utilizing this fact, we have

$$\inf_{v\in \Xi_i} J_i(v) = J_i(f^*_i) \geq \limsup_{\varepsilon\to 0} J_{i,\varepsilon}(f^*_{i,\varepsilon}) \geq \limsup_{\varepsilon\to 0} \inf_{v\in \Xi_{i,\varepsilon}} J_{i,\varepsilon}(v)$$

$$\geq \limsup_{k\to\infty} \inf_{v\in \Xi_{i,\varepsilon_k}} J_{i,\varepsilon_k}(v) = \limsup_{k\to\infty} J_{i,\varepsilon_k}(f^0_{i,k}). \qquad (4.87)$$

From this and (4.86) we deduce that

$$\liminf_{k\to\infty} J_{i,\varepsilon_k}(f^0_{i,k}) \geq \limsup_{k\to\infty} J_{i,\varepsilon_k}(f^0_{i,k}).$$

Hence, we can combine (4.86) and (4.87) to get

$$J_i(f^0_i) = J_i(f^*_i) = \inf_{v\in \Xi_i} J_i(v) = \lim_{k\to\infty} \inf_{v\in \Xi_{i,\varepsilon_k}} J_{i,\varepsilon_k}(v). \qquad (4.88)$$

Using these relations and the fact that the problem (4.23)–(4.24) is solvable, we may suppose that $f^*_i = f^0_i$. Since equality (4.88) holds for all subsequences of $\left\{f^0_{i,\varepsilon}\right\}_{\varepsilon>0}$, which are convergent in the sense of relations (4.79)–(4.81), it follows that these limits coincide and, therefore, f^0_i is the limit of the whole sequence $\left\{f^0_{i,\varepsilon}\right\}_{\varepsilon>0}$. Then, using the same argument for the sequence of minimizers as we did it for the subsequence $\left\{f^0_{i,\varepsilon_k}\right\}_{k\in\mathbb{N}}$, we finally obtain

$$\liminf_{\varepsilon\to 0} \inf_{v\in \Xi_{i,\varepsilon}} J_{i,\varepsilon}(v) = \liminf_{\varepsilon\to 0} J_{i,\varepsilon}(f^0_{i,\varepsilon}) \geq J_i(f^0_i) = \inf_{v\in \Xi_i} J_i(v)$$

$$\geq \limsup_{\varepsilon\to 0} J_{i,\varepsilon}(f^*_{i,\varepsilon}) \geq \limsup_{\varepsilon\to 0} \inf_{v\in \Xi_{i,\varepsilon}} J_{i,\varepsilon}(v)$$

$$= \limsup_{\varepsilon\to 0} J_{i,\varepsilon}(f^0_{i,\varepsilon}),$$

and this concludes the proof.

Remark 4.16 It is worth focusing on a few principal issues related to Theorem 4.15. The first one is that, in practice, the assumption concerning solvability of the original optimization problem is not so restricted and, in principle, it can be omitted. Indeed, any digital color image $f = [f_1, f_2, f_3]^t$ is originally defined on some grid G. So, each of its spectral channel $f_i\big|_G$ can be associated with some real-valued matrix. Hence, we can always suppose that the exponent $p(|\nabla f_i|)\big|_G$ is the restriction on the same grid of some Lipschitz-continuous function $p(\cdot) : \Omega \to \mathbb{R}$. Then arguing as in the proof of Theorem 4.12, the solvability of the problem (4.23)–(4.24) can be easily established.

The second point, that should be emphasized here, is the assumption about compactness property of the sequence of associated exponents

$$\left\{ p_\varepsilon := 1 + \delta + \frac{a^2(1-\delta)}{a^2 + \left|(\nabla K_\varepsilon * f^0_{i,\varepsilon})\right|^2} \right\}_{\varepsilon > 0}$$

with respect to the pointwise convergence in Ω. Since this property is crucial in Theorem 4.15, we propose to consider it as an easy realized in practice criterion for the verification whether the approximating sequence $\left\{ f^0_{i,\varepsilon} \in \Xi_{i,\varepsilon} \right\}_{\varepsilon > 0}$ leads to some optimal solution of the original problem.

4.5 Numerical Results

To illustrate the implementation of the proposed model (4.23)–(4.23) to the simultaneous denoising and contrast enhancement of color images, we make use of the optimality conditions in the form of (4.49). In other words, we have dropped the two-side constraints $\gamma_{i,0} \leq \nu(x) \leq \gamma_{i,1}$ from the sets Ξ_i, and instead we control the fulfilment of this two-side constraint at each step of the numerical approximations (Figs. 4.1 and 4.2).

Since, in practical implementations, it is reasonable to define the solution of the problem (4.23)–(4.23) using a "gradient descent" strategy, we can start with some initial image $f = [f_1, f_2, f_3]^t \in L^2(\Omega; \mathbb{R}^3)$ and pass to the following system of three initial-boundary value problems for quasi-linear parabolic equations with Nuemann boundary conditions

$$\frac{\partial f^0_i}{\partial t} = \mathrm{div}\left[p(x)|R_\eta \nabla f^0_i|^{p(|\nabla f^0_i|)-2} R_\eta \nabla f^0_i \right]$$
$$- \eta^2 \mathrm{div}\left[p(x) \left(|R_\eta \nabla f^0_i|^{p(|\nabla f^0_i|)-2} R_\eta \nabla f^0_i, \theta \right) \theta \right]$$
$$- 2a^2(1-\delta) \mathrm{div}\left[|R_\eta \nabla f^0_i|^{p(|\nabla f^0_i|)} \frac{\log\left(|R_\eta \nabla f^0_i|\right)}{(a^2 + |\nabla f^0_i|^2)^2} \nabla f^0_i \right]$$
$$- \lambda \left(D(f^0_i) - cD(f_i) \right) \int_\Omega W(x,y) \frac{f^0_i(x) - f^0_i(y)}{\sqrt{\kappa^2 + |f^0_i(x) - f^0_i(y)|^2}} dy$$

4.5 Numerical Results

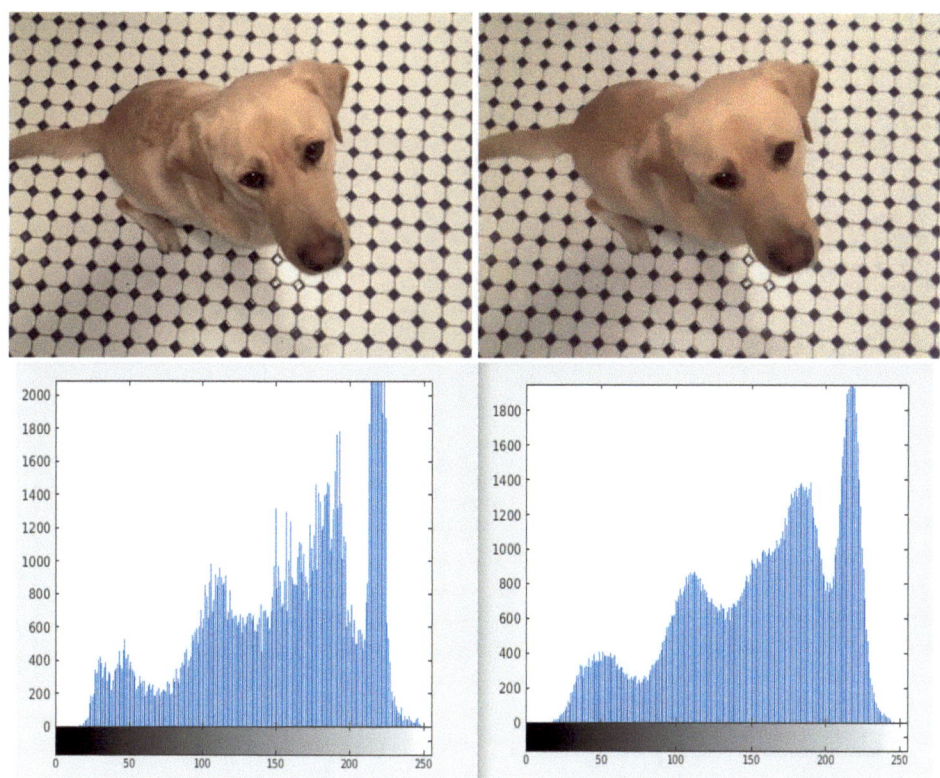

Fig. 4.1 Original image (left) and its smoothed version without contrasting ($\mu = 0$) (right)

$$-\mu(-\Delta)^{-1}(f_i^0 - f_i) = 0, \quad \text{in } (0, T) \times \Omega, \tag{4.89}$$

$$\left(|\nabla f_i^0|^{p(|\nabla f_i^0|)-2} \nabla f_i^0, \nu\right) = 0 \quad \text{on } (0, T) \times \partial\Omega, \tag{4.90}$$

$$f_i^0(0, \cdot) = f_i(\cdot), \quad i = 1, 2, 3, \quad \text{in } \Omega. \tag{4.91}$$

For numerical simulations, we set: $\delta = \kappa$ in (4.18) and (4.17), and $\eta = 1 - \kappa$ in (4.16), $\kappa = 0.001$, $\lambda = 0.1$, and $\mu = 2$. As for the noise estimator $a > 0$ in (4.17), we use the choice of Black et al. [6], i.e.

$$a = \frac{1.4826}{\sqrt{2}} MAD(\nabla f_i),$$

where MAD denotes the median absolute deviation of the corresponding spectral channel f_i of original image $f = [f_1, f_2, f_3]^t$ that can be computed as

$$MAD(\nabla f_i) = \text{median}\left[\Big||\nabla f_i| - \text{median}(|\nabla f_i|)\Big|\right]$$

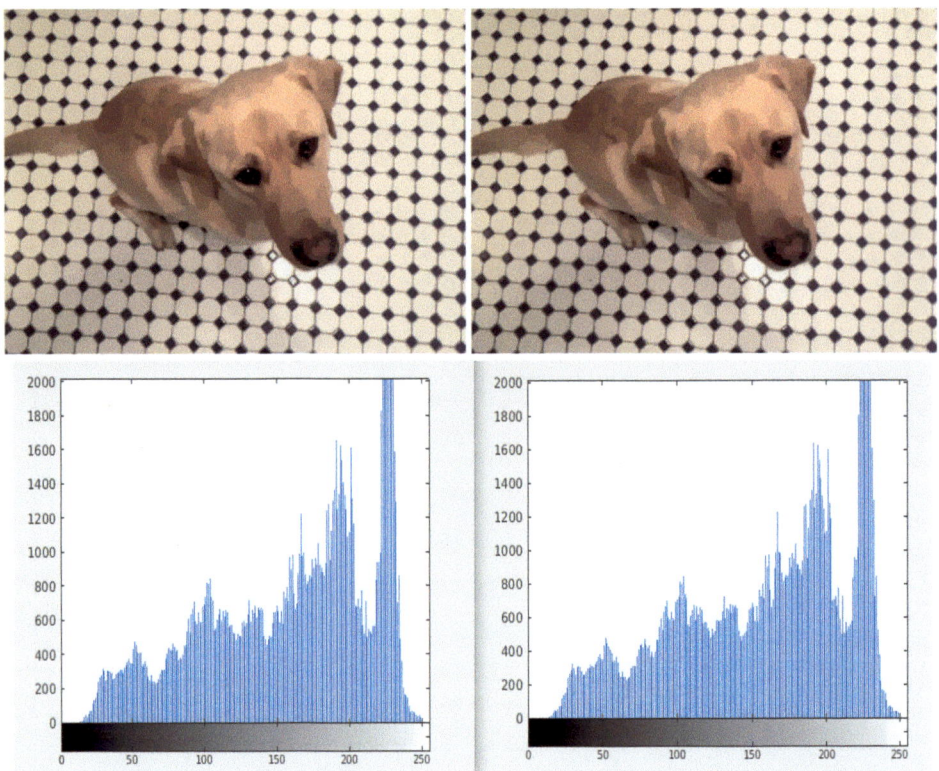

Fig. 4.2 Variants of contrast enhancement with the corresponding histograms (from the left to the right): $c = 2$ and $window = 5$, $c = 2$ and $window = 7$

and median ($|\nabla f_i|$) represents the median over the band f_i to the gradient amplitude (Fig. 4.5).

To guarantee the stability of the proposed algorithm, we make use of the following condition

$$2\left[\frac{1}{\kappa} + \lambda + \mu\right]\Delta t < 1.$$

There are numerous approaches to solve quasi-linear partial differential equations (see the references [4, 27] for various techniques). Since we are dealing with pixels in image processing, finite differences approaches and an explicit scheme of the forward Euler method are arguably the best options. The number of iterations for each spectral channel can be defined experimentally. We used 10^3-iterations. As for the size of the kernel $W(x, y)$ used for D, this size manages the scale of the contrast enhancement. In our experiments we used it equals to 3, 5, 7, 15, albeit it can be related to the size of the input image.

4.5 Numerical Results

The most expensive computation is the one of D and ∇D embedded in the computation of the right-hand side of the system (4.89). For acceleration of these computation, we can refer to [40], where the efficient Bernstein polynomials approximation has been proposed.

As follows from the result of numerical simulations (see Fig. 4.3, 4.4, 4.5, 4.6, 4.7, 4.8 and 4.9), parameters c, μ, λ, and the size of $window$ for the kernel $W(x, y)$ are crucial for the contrast enhancement and these parameters have to be tuned in dependence on the desired result. In particular, we observe that at a large scale ($window$) and low contrast level c, the proposed model is able to produce an image with more details, but with the same lighting sensation as the original one. In order to show how the choice of the parameters c, μ, λ, and $window$ affect the results of contrast enhancement, we supplied all images in Fig. 4.1, 4.2, 4.3, 4.4, 4.5, 4.6, 4.7, 4.8 and 4.9 by the histograms of their luma components which represent the perceptual brightness of the color images $I : \Omega \to \mathbb{R}^3$. To this end we used the following representation for the luma $Y_I(x) = \alpha_R I_R(x) + \alpha_G I_G(x) + \alpha_B I_B(x)$ with

$$\alpha_R = 0.299, \quad \alpha_G = 0.587, \quad \alpha_B = 0.114.$$

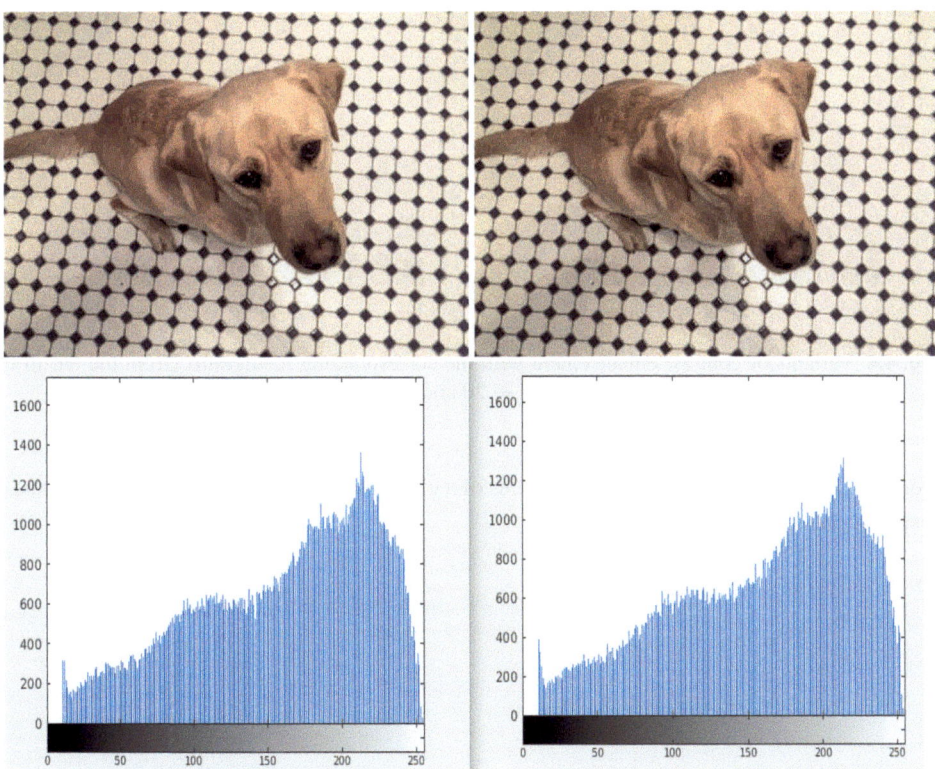

Fig. 4.3 Variants of contrast enhancement with the corresponding histograms (from the left to the right): $c = 10$ and $window = 5$, $c = 10$ and $window = 7$

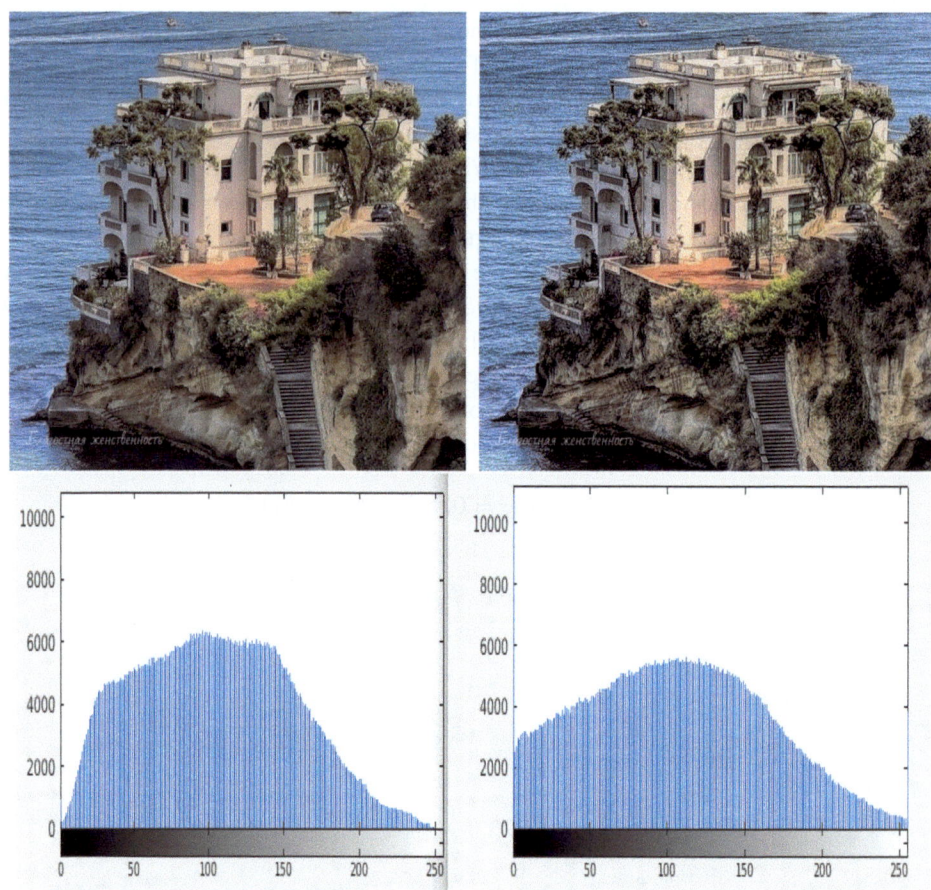

Fig. 4.4 Variants of contrast enhancement with the corresponding histograms (from the left to the right): original image, restored image with $c = 20$ and $window = 15$

Here, I_R, I_G, and I_N stand for the intensities of a given image in R,G and B spectral channels, respectively.

4.5 Numerical Results

Fig. 4.5 Variants of contrast enhancement with the corresponding histograms (from the left to the right): original and restored with $c = 10$ and $window = 7$

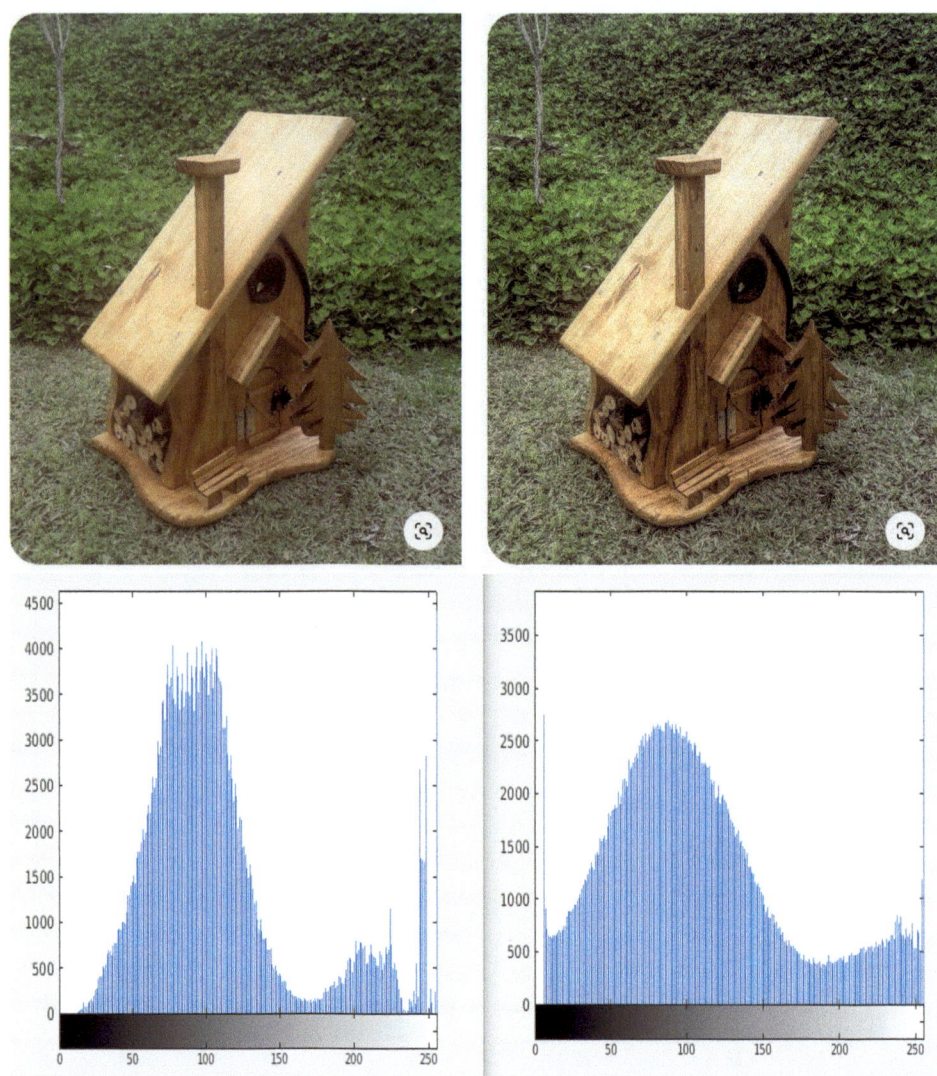

Fig. 4.6 Variants of contrast enhancement with the corresponding histograms (from the left to the right): original image, restored image with $c = 10$ and $window = 5$

4.5 Numerical Results

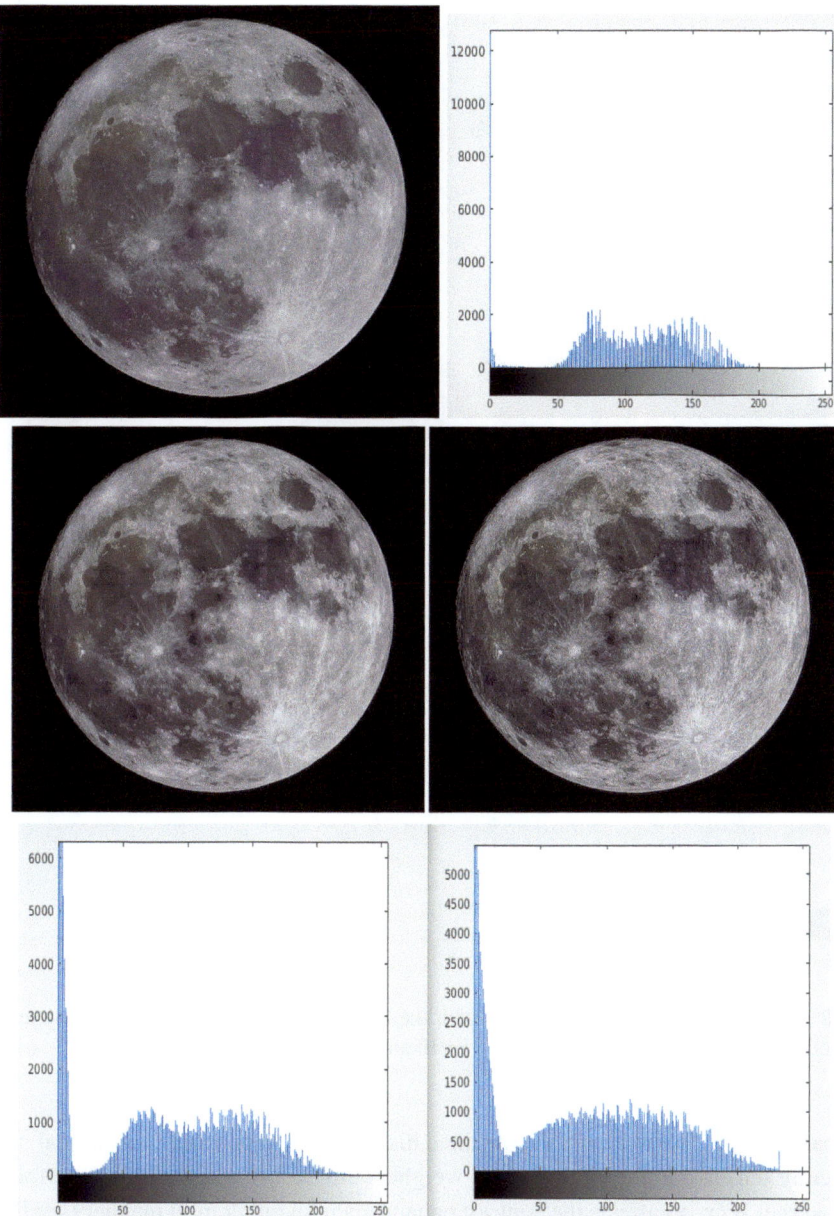

Fig. 4.7 Influence of the contrast enhancement scale on the result (from the left to the right): original, $c = 10$ and $window = 5$, $c = 20$ and $window = 5$

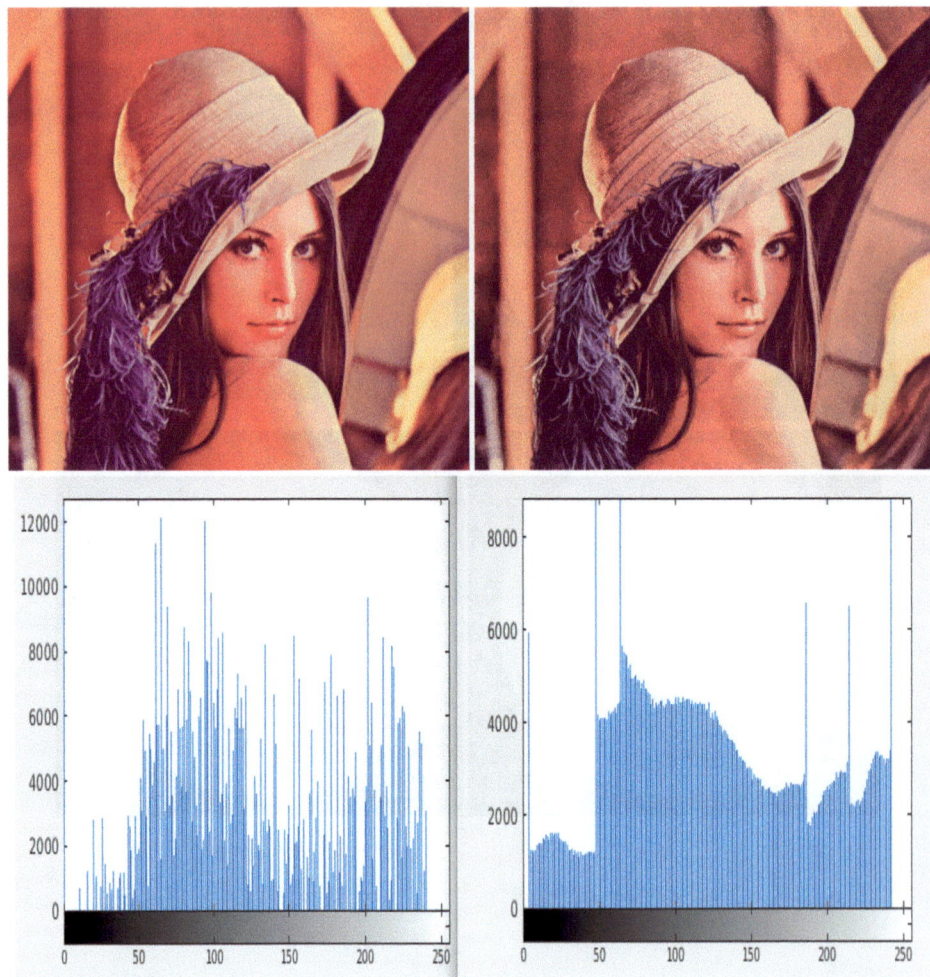

Fig. 4.8 Variants of contrast enhancement with the corresponding histograms (from the left to the right): original image, restored image with $c = 10$ and $window = 5$

In particular, as follows from the obtained histograms, the proposed variational model is sufficiently sensitive to the choice of the weight coefficient c, whereas the size of $window$ for the kernel $W(x, y)$ affects the contrast enhancement in rather mild manner (see Figs. 4.2 and 4.3).

4.5 Numerical Results

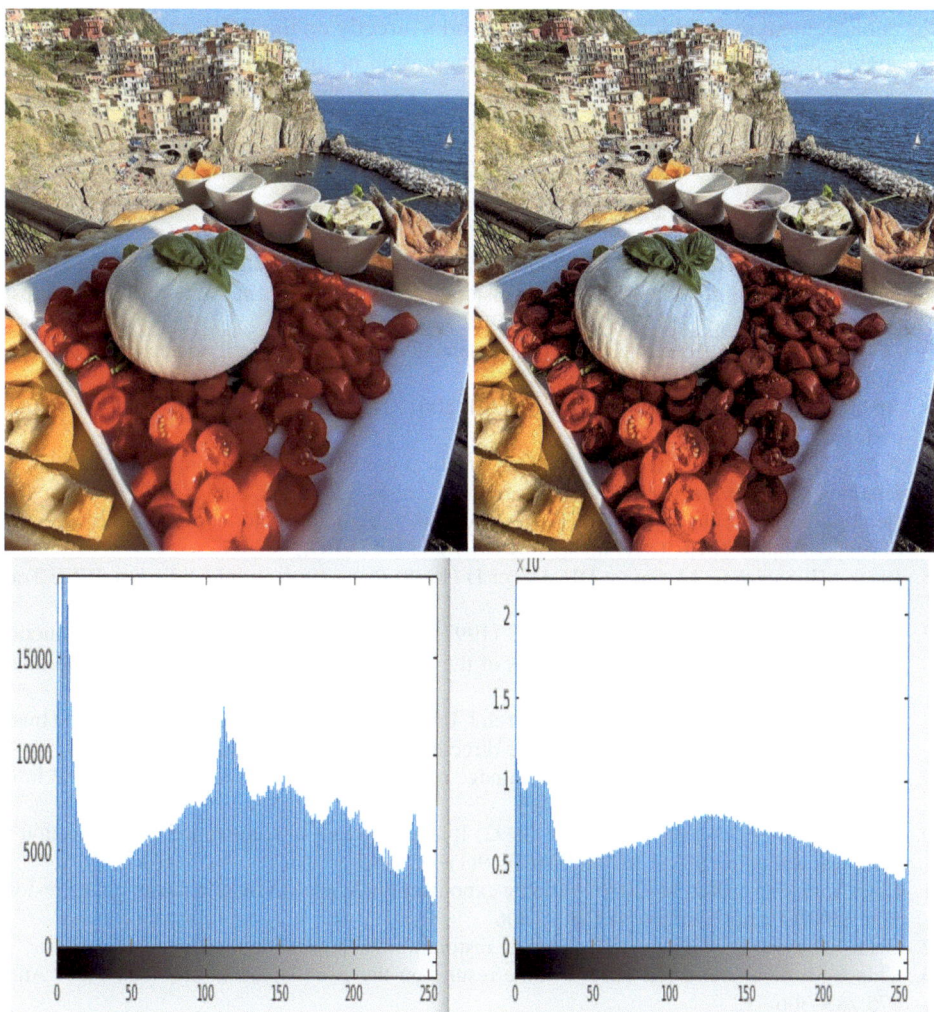

Fig. 4.9 Variants of contrast enhancement with the corresponding histograms (from the left to the right): original image, restored image with $c = 10$ and $window = 5$

4.6 Conclusions

We introduced a variational model with nonstandard growth condition for the restoration and contrast enhancement of multi-band images. We show that increasing the average local contrast measure improves the perceived contrast of the image. We have obtained sufficient conditions for the convergence of the minimization algorithm. The contrast scale and level in our model are adjustable, so that the proposed approach can be considered as fully adaptive.

Our enhancement method for color images works directly on the RGB images without any pre- and/or post-processing. The automatic adaptation of the parameters to the content of the considered image could be a future direction of research.

References

1. Alaa H, Alaa NE, Bouchriti A, Charkaou A (2022) An improved nonlinear anisotropic PDE with $p(x)$-growth conditions applied to image restoration and enhancement. Authorea (2022). https://doi.org/10.22541/au.165717367.72990650/v1
2. Alvarez L, Lions PL, Morel JM (1992) Image selective smoothing and edge detection by nonlinear diffusion. SIAM J Numer Anal 29:845–866
3. Ambrosio L, Caselles V, Masnou S, Morel JM (2001) The connected components of sets of finite perimeter. Eur J Math 3:39–92
4. Aubert G, Kornprobst P (eds) (2006) Mathematical problems in image processing: partial differential equations and the calculus of variations, vol 147. Springer, New York
5. Bertalmío M, Caselles V, Provenzi E, Rizzi A (2007) Perceptual color correction through variational techniques. IEEE Trans Image Process 16(4):1058–1072
6. Black MJ, Sapiro G, Marimont DH, Heger D (1998) Robust anisotropic diffusion. IEEE Trans Image Process 7(3):421–432
7. Blomgren P, Chan T, Mulet P, Wong C (1997) Total variation image restoration: numerical methods and extensions. In: Proceedings of the 1997 IEEE international conference on image processing, vol 42, pp 384–387
8. Bungert L, Coomes DA, Ehrhardt MJ, Rasch J, Reisenhofer R, Schönlieb CB (2018) Blind image fusion for hyperspectral imaging with the directional total variation. Inverse Probl 34(4)
9. Bungert L, Ehrhardt MJ (2020) Robust image reconstruction with misaligned structural information. IEEE Access 8:222944–222955
10. Catté F, Lions PL, Morel JM, Coll T (1992) Image selective smoothing and edge detection by nonlinear diffusion. SIAM J Numer Anal 29(1):182–193
11. Chen Y, Levine S, Rao M (2006) Variable exponent, linear growth functionals in image restoration. SIAM J Appl Math 66(4):1383–1406
12. Chen Y, Levine S, Stanich J (2014) Image restoration via nonstandard diffusion. Figshare
13. Chipot M, de Oliveira HB (2019) Some results on the $p(u)$-laplacian problem. Math Annal 375:283–306
14. D'Apice C, De Maio U, Kogut PI (2008) Gap phenomenon in homogenization of parabolic optimal control problem. IMA J Math Control Inf 25:461–480
15. D'Apice C, De Maio U, Kogut PI (2009) Boundary velocity suboptimal control of incompressible flow in cylindrically perforated domain. Disc Contin Dyn Syst—B 11(2):283–314
16. D'Apice C, De Maio U, Kogut PI (2020) An indirect approach to the existence of quasi-optimal controls in coefficients for multi-dimensional thermistor problem. In: Sadovnichiy VA, Zgurovsky M (eds) Contemporary approaches and methods in fundamental mathematics and mechanics. Springer, New York, pp 489–522
17. D'Apice C, Kogut PI, Kupenko O, Manzo R (2023) On variational problem with nonstandard growth functional and its applications to image processing. J Math Imaging Vis 65(3):472–491
18. D'Apice C, Kogut PI, Manzo R (2023) On coupled two-level variational problem in Sobolev-Orlicz space. Differ Integral Equ 36(7–8):621–660
19. D'Apice C, Kogut PI, Manzo R (2023) Qualitative analysis of an optimal sparse control problem for quasi-linear parabolic equation with variable order of nonlinearity. J Optim Diff Eqs Their Appl 31(2):125–173

20. D'Apice C, Kogut PI, Manzo R (2023) A two-level variational algorithm in the Sobolev-Orlicz space to predict daily surface reflectance at LANDSAT high spatial resolution and MODIS temporal frequency. J Comput Appl Math 434:1–23
21. D'Apice C, Kogut PI, Manzo R, Parisi A (2024) Variational model with nonstandard growth condition in image restoration and contrast enhancement. Commun Appl Math Comput. https://doi.org/10.1007/s42967-024-00382-1
22. D'Apice C, Kogut PI, Manzo R, Uvarov M (2023) Variational model with nonstandard growth conditions for restoration of satellite optical images using synthetic aperture radar. Eur J Appl Math 34(1):77–105
23. Dautray R, Lion JL (eds) (1985) Mathematical analysis and numerical methods for science and technology, vol 5. Springer, Berlin, Heidelberg
24. Diening L, Harjulehto P, Hästö P, Růžička M (eds) Lebesgue and Sobolev spaces with variable exponents. Springer, New York (2011)
25. Evans LC, Gariepy RF (eds) (1992) Measure theory and fine properties of functions. CRC Press, Boca Raton
26. Jia Z, Ng MK, Wang W (2019) Color image restoration by saturation-value total variation. SIAM J Imaging Sci 12(2):1000–9722
27. Karami F, Meskine D, Sadik K (2019) A new nonlocal model for the restoration of textured images. J Appl Anal Comput 9(6):2070–2095
28. Kogut P (2019) On optimal and quasi-optimal controls in coefficients for multi-dimensional thermistor problem with mixed dirichlet-neumann boundary conditions. Control Cybern 48(1):31–68
29. Kogut PI (1996) Variational S-convergence of minimization problems. Part I. Definitions and basic properties. Problemy Upravleniya i Informatiki (Avtomatika) 5:29–42
30. Kogut PI (1997) S-convergence of the conditional optimization problems and its variational properties. Problemy Upravleniya i Informatiki (Avtomatika) 4:64–79
31. Kogut PI (2014) On approximation of an optimal boundary control problem for linear elliptic equation with unbounded coefficients. Disc Contin Dyn Syst—A 34(5):2105–2133
32. Kogut PI, Kohut Y, Manzo R (2022) Existence result and approximation of an optimal control problem for the Perona-Malik equation. Ric di Mat 1–18. https://doi.org/10.1007/s11587-022-00730-4
33. Kogut PI, Kupenko OP (2019) Approximation methods in optimization of nonlinear systems. In: De Gruyter series in nonlinear nalysis and applications, vol 32. Walter de Gruyter GmbH, Berlin
34. Kogut PI, Leugering G (2001) On S-homogenization of an optimal control problem with control and state constraint. Z Fur Anal Ihre Anwend 20(2):395–429
35. Kohr H (2017) Total variation regularization with variable Lebesgue priors. arXiv: 1702.08807
36. Lieu LH, Vese LA (2008) Image restoration and decomposition via bounded total variation and negative Hilbert-Sobolev spaces. Appl Math Optim 58:167–193
37. Meyer Y (2002) Oscillating patterns in image processing and nonlinear evolution equations. In: Univ Lecture Ser 2. AMS, Providence, RI
38. Osher S, Rudin LI (1990) Feature-oriented image enhancement using shock filters. SIAM J Numer Anal 27(4):458–474
39. Piella G (2009) Image fusion for enhanced visualization: a variational approach. Int J Comput Vis 83(1):1–11
40. Pierre F, Aujol JF, Bugeau A, Steidl G, Ta V (2017) Variational contrast enhancement of Gray-Scale and RGB Images. J Math Imaging Vis 57:99–116
41. Prasath V, Urbano JM, Vorotnikov D (2015) Analysis of adaptive forward-backward diffusion flows with applications in image processing. Inverse Probl. 31:1–30
42. Ring, W.: Structural properties of solutions to Total Variation regularization problems. ESAIM: Math. Model. Numer. Anal. **34**, 799–810 (2000)

43. Schönlieb CB (2009) Total variation minimization with an h^{-1} constraint (2009). Research Gate Publication
44. Sugimura D, Mikami T, Yamashita H, Hamamoto T (2015) Enhancing color images of extremely low light scenes based on RGB/NIR images acquisition with different exposure times. IEEE Trans Image Process 24(11):3586–3597
45. Wunderli T (2010) On time flows of minimizers of general convex functionals of linear growth with variable exponent in BV space and stability of pseudosolutions. J Math Anal Appl 364(2):59–591
46. Zhikov V (2008) Solvability of the three-dimensional thermistor problem. Proc Steklov Inst Math 281:98–111
47. Zhikov V (2011) On variational problems and nonlinear elliptic equations with nonstandard growth conditions. J Math Sci 175(5):463–570

Variational Approach to Simultaneous Fusion and Denoising of the Color Images with Different Spatial Resolution

Abstract

The main purpose of in this chapter is to describe a robust approach for the simultaneous fusion and denoising of non-smooth multispectral images defined on grids with different resolution using for that a special extremal problem with nonstandard growth of the energy functional. In fact, we use the L^1-norm of the noise in the minimization function and a special form of anisotropic diffusion tensor for the regularization term. Following this approach, we increase the noise robustness of the proposed model albeit it makes such variational problem completely non-smooth, non-convex, and, hence, significantly more difficult from a minimization point of view. The principle characteristic feature of the proposed model is that we consider the energy functional with nonstandard growth for each spectral channel separately. The second point that should be emphasized is the fact that we do not predefine the variable exponents a priori using for that the original noisy images, but instead we associate these characteristics with each feasible solution.

The synthesis of several source images of the same scene into a single image that would contain much more visual information (see, for instance, [6, 8, 39]) is an important issue appearing in various fields such as remote sensing, medical diagnosis, defect inspection, and military surveillance. Since the observed source images are inevitably corrupted by noise, they can be blurred, and arguably, are geometrically. A very promising approach to image quality enhancement is to fuse several sources with different degradations together in order to extract as much useful information as possible.

A significant part of the existing fusion methods (the so-called pixel-level methods) is based on the estimation of the value for each point in the fused image through some feature selection rule [27]. In particular, several methods have been developed such as spatial domain fusion methods [38], transform domain fusion methods [34], variational methods based on fusing the gradient information [45], or their combinations [35]. In [30], the authors proposed a new variational model by fusing the first- and second-order gradient information from the source images. However, this approach has originally been aimed at the fusion of images without visible noise corruptions.

Regarding the fusion methods of the noisy source images, apparently, [40] was one of the first paper dedicated to this problem. The authors proposed a weighted variational method based on the total variation (TV) regularization and with some regularization parameter in the objective functional that trades off the fit to the noisy source images and the smoothness from TV. So, the TV regularization term was added to the proposed model to reduce the influence from the noise.

Another approach has been introduced in [33], where the authors considered fractional-order derivatives as regularization in the variational model for image fusion and denoising. Their goal is to obtain a fused image of high quality, preserving sharp edges while maintaining smoothness in homogeneous regions, even when the source images are corrupted by noise. To achieve this, the authors of [33] aim to match the fractional-order gradient of the fused image with a target fractional-order gradient, using either L^2-norm or L^1-norm. However, selecting the appropriate target fractional-order gradient is a challenging task, and the practical implementation of this approach becomes complicated as a result.

Recent papers [18, 19, 32] also deserve mention, where the authors address the contrast enhancement, multimodal image fusion, and denoising problem using different techniques, such as a Retinex-based variational model, a Siamese convolutional neural network, and quaternion-based dictionary learning with saturation-value Total Variation regularization.

In this chapter we consider a constrained minimization problem with a special objective functional. The main feature of this functional is the fact that it contains a spatially variable exponent characterizing the growth conditions of the objective functional and it can be seen as a replacement for the 1-norm in TV regularization. The idea of using a spatially varying exponent in a TV-like regularization method for image denoising dates back as early as 1997 [5] and put into practice in 2006 [9]. Both papers as well as some subsequent articles try to tackle variants of the problem

$$J(u) = \mathcal{D}(u) + \lambda \int_\Omega |\nabla u(x)|^{p(\nabla u(x))}\, dx \longrightarrow \inf, \qquad (5.1)$$

where the exponent depends directly on the image, e.g.,

$$p(\nabla u) = 1 + \frac{a^2}{a^2 + |\nabla G_\sigma * u|^2}.$$

Here, $(G_\sigma * v)(x)$ determines the convolution of function v with the 2-dimensional Gaussian filter kernel G_σ

$$G_\sigma(x) = \frac{1}{\left(\sqrt{2\pi}\sigma\right)^2} \exp\left(-\frac{|x|^2}{2\sigma^2}\right), \quad x \in \mathbb{R}^2,\ \sigma > 0 \text{ is a fixed parameter}, \qquad (5.2)$$

$$(\nabla G_\sigma * u)(x) = \int_{\mathbb{R}^2} \nabla G_\sigma(x-y)\widetilde{u}(y)\,dy, \quad \forall x \in \Omega, \tag{5.3}$$

$\widetilde{u}$ is zero-extension of u outside Ω, $|\xi|$ stands for the Euclidean norm of $\xi \in \mathbb{R}^2$ given by the rule $|\xi| = \sqrt{(\xi, \xi)}$,

It has been demonstrated that this model possesses some favorable properties, particularly when edge preservation and effective noise suppression are primary goals in image reconstruction.

Furthermore, this model has been introduces specifically to address the issue of staircasing [36], which refers to the regularizer's inclination towards piecewise constant functions. The appearance of the staircasing effect is a notable drawback of the classical TV model. However, the non-convex model (5.1) did not gain significant attention for a long period due to its high numerical complexity and the absence of a rigorous mathematical substantiation of its consistency. Only partial solutions to this problem have been derived for a smoothed version of the integrand, using a weak notion of solution (see, for instance, [41]).

A recently developed alternative variant is the TV-like method [26], which computes the variable exponent p in an offline step and keeps it as a fixed parameter in the final optimization problem This approach allows the exponent to vary based on spatial location, enabling users to locally select whether to preserve edges or smooth intensity variations. However, there are only two natural types of imaging problems where this approach can be applied:

- single-channel imaging where first the exponent is computed from the given data and then is applied as prior in the subsequent minimization problem;
- dual-channel imaging where the secondary channel provides the exponent map that is used for regularization of the primary channel.

Thus, this circumstance imposes significant limitations from practical point of view, especially in the case of multi-spectral satellite noisy images, where different channels can differ drastically (for instance, red and infrared channels).

The main purpose of this chapter is to describe a robust approach for the simultaneous fusion and denoising of non-smooth multispectral images defined on grids with different resolution using for that the energy functional with nonstandard growth. In fact, we use the L^1-norm of the noise in the minimization function and a special form of anisotropic diffusion tensor for the regularization term. By following this approach, we aim to increase the noise robustness of the proposed model albeit it makes such variational problem completely non-smooth, non-convex, and, hence, significantly more difficult from a minimization point of view. The key results of this chapter has been obtained in cooperation with C. D'Apice, R. Manzo, and C. Pipino (see [15]).

The principle characteristic feature of the proposed model is that we consider the energy functional with nonstandard growth for each spectral channel separately. Moreover, the edge information for fusion of two images with different resolution is mainly accumulated in the variable exponents $p_1(x)$, $p_2(x)$, ..., $p_m(x)$. The second point that should be emphasized is the fact that we do not predefine the variable exponents $p_i(x)$ a priori using for that the original noisy images, but instead we associate these characteristics with each feasible solution.

5.1 Preliminaries

Let $\Omega \subset \mathbb{R}^2$ be a bounded connected open set with a sufficiently smooth boundary $\partial\Omega$ and nonzero Lebesgue measure. In majority cases Ω can be interpreted as a rectangle domain. Let G_H and G_L be two sample grids on Ω such that $G_H = \widehat{G}_H \cap \Omega$ and $G_H = \widehat{G}_H \cap \Omega$, where

$$\widehat{G}_H = \left\{(x_i, y_j) \middle| \begin{array}{l} x_1 = x_H, \; x_i = x_1 + \Delta_{H,x}(i-1), \; i = 1, \ldots, N_x, \\ y_1 = y_H, \; y_j = y_1 + \Delta_{H,y}(j-1), \; j = 1, \ldots, N_y, \end{array}\right\},$$

$$\widehat{G}_L = \left\{(x_i, y_j) \middle| \begin{array}{l} x_1 = x_L, \; x_i = x_1 + \Delta_{L,x}(i-1), \; i = 1, \ldots, M_x, \\ y_1 = y_L, \; y_j = y_1 + \Delta_{L,y}(j-1), \; j = 1, \ldots, M_y, \end{array}\right\},$$

with some fixed points (x_H, y_H) and (x_L, y_L). Hereinafter, it is assumed that $N_x \gg M_x$ and $N_y \gg M_y$.

Let $S : G_H \to \mathbb{R}^m$ and $M : G_L \to \mathbb{R}^m$, $m \geq 1$, be a couple of multispectral images, containing the same scene albeit they are defined on grids with different resolution. The principle point is that the image with low resolution $M : G_L \to \mathbb{R}^m$ contains some extra objects which are invisible or absent in the image $S : G_H \to \mathbb{R}^m$. It is assumed that:

(i) Each of the given images $S : G_H \to \mathbb{R}^m$ and $M : G_L \to \mathbb{R}^m$ can be corrupted by some additive Gaussian noise with zero mean.
(ii) All spectral channels of the image $M = [M_1, M_2, \ldots, M_m]$ have similar spectral characteristics to the corresponding channels of the image $S = [S_1, S_2, \ldots, S_m]$, respectively;
(iii) The images $M : G_L \to \mathbb{R}^m$ and $S : G_H \to \mathbb{R}^m$ are rigidly co-registered. This means that the image M, after arguably some affine transformation and the image S after the resampling to the grid with low resolution G_L, could be successfully matched with the exception of the zone where there are new objects.

In practice, the co-registration procedure is usually applied not to the original images directly, but rather to their spectral energies $Y_M : G_L \to \mathbb{R}$ and $Y_S : G_H \to \mathbb{R}$, where the last ones should be previously resampled to the grid of the low resolution G_L. Here,

$$Y_M(z) := \alpha_1 M_1(z) + \alpha_2 M_2(z) + \cdots + \alpha_m M_m(z), \quad \forall z = (x, y) \in G_L,$$
$$Y_S(z) := \alpha_1 S_1(z) + \alpha_2 S_2(z) + \cdots + \alpha_m S_m(z), \quad \forall z = (x, y) \in G_H$$

with appropriate weight coefficients α_i, $i = 1, \ldots, m$.

5.1 Preliminaries

Our main purpose is to present a robust approach for the simultaneous denoising and fusion of non-smooth multi-spectral images defined on grids with different resolution. With that in mind, we use a special form of anisotropic diffusion tensor for the regularization term and the L^1-norms for the fidelity terms. Namely, we deal with the following family of optimization problems:

$$J_i(v) = \int_\Omega |\nabla v(x)|^{\mathfrak{F}(v(x))}\,dx + \lambda \int_\Omega |\nabla v(x) - \nabla S_i(x)|\,dx$$
$$+ \mu \int_\Omega |T_S v(x) - S_i(x)|\,dx$$
$$+ \frac{1-\mu}{2} T_M\left([(G_\sigma * v)(\cdot) - M_i(\cdot)]^2\right) \longrightarrow \inf, \quad (5.4)$$

subject to the constraints

$$v \in W^{1,\mathfrak{F}(v(\cdot))}(\Omega), \quad 1 \leq \gamma_{i,0} \leq v(x) \leq \gamma_{i,1} \quad \text{a.e. in } \Omega, \quad (5.5)$$

where $i = 1, \ldots, m$, $S_i \in L^1(\Omega)$ and $M_i \in L^1(\Omega)$ are a particular spectral channel of the original noisy images $S = [S_1, S_2, \ldots, S_m]^T \in L^1(\Omega; \mathbb{R}^m)$ and $M = [M_1, M_2, \ldots, M_m]^T \in L^1(\Omega; \mathbb{R}^m)$, respectively, $\lambda > 0$ and $\mu \in (0, 1)$ are the tuning parameters, $W^{1,\mathfrak{F}(v(\cdot))}(\Omega)$ stands for the so-called Sobolev-Orlicz space associated with a feasible solution v, $T_S \in \mathcal{L}(L^1(\Omega))$ and $T_M \in \mathcal{L}(L^1(\Omega), \mathbb{R})$ are bounded linear operators with unbounded inverse,

$$\mathfrak{F}(v(x)) = 1 + g\left(|(\nabla G_\sigma * v)(x)|\right) \quad \text{in } \Omega,$$

and $g : [0, \infty) \to (0, \infty)$ is a continuous monotone decreasing function such that $g(0) = 1$ and $g(t) > 0$ for all $t > 0$ with $\lim_{t \to +\infty} g(t) = 0$.

In particular, if we set $p(x) := 1 + g\left(|(\nabla G_\sigma * v)(x)|\right)$, where the edge-stopping function $g(s)$ is taken in the form of the Cauchy law

$$g(t) = \frac{1}{1 + (t/a)^2} \quad \text{with an appropriate } a > 0,$$

it implies that $p(x) \approx 1$ in places in Ω where edges or discontinuities are present in the image $v(x)$, and $p(x) \approx 2$ in places where $v(x)$ is smooth or contains homogeneous features.

We define the parameters $\gamma_{i,0}$, $\gamma_{i,1}$, and the operator $T_M \in \mathcal{L}(L^1(\Omega), \mathbb{R})$, as follows:

$$\gamma_{i,0} = \min_{(x,y) \in G_H} S_i(x, y), \quad \gamma_{i,1} = \max_{(x,y) \in G_H} S_i(x, y), \quad T_M = \sum_{(x_i, y_j) \in S_L} \delta_{(x_i, y_j)}, \quad (5.6)$$

where $\delta_{(x_i, y_j)}$ is the Dirac's delta at the point (x_i, y_j) of the sample grid G_L.

It is worth to emphasize that, in contrast to the quadratic data-fitting term in the well-known model, introduced by Rudin et al. [37], we take the fidelity terms in L^1-norm just in order to increase the noise robustness of the model (5.4) albeit it makes such variational problem completely non-smooth and, hence, significantly more difficult from a minimization point of view.

Thus, the problem of simultaneous fusion and denoising of multi-spectral images with different spatial resolution consists in generation of a new multi-spectral image $I^0 = \left[I_1^0, I_2^0, \cdots, I_m^0\right]^t : G_H \to \mathbb{R}^m$, which would be well defined on the entire grid G_H, such that

$$J_i(I_i^0) = \inf_{v \in \Xi_i} J_i(v), \quad \forall i = 1, \ldots, m, \tag{5.7}$$

where

$$\Xi_i = \left\{u \in W^{1,\mathfrak{F}(v(\cdot))}(\Omega) \, : \, 1 \leq \gamma_{i,0} \leq u(x) \leq \gamma_{i,1} \text{ a.e. in } \Omega\right\} \tag{5.8}$$

stands for the set of feasible solutions to the minimization problem (5.4).

So, the main characteristic feature of the model (5.4) is the energy functional with non-standard growth where the main information for the simultaneous fusion and denoising of images S and M is accumulated in the variable exponents $[\mathfrak{F}(v_1(x)), \ldots, \mathfrak{F}(v_m(x))]$. However, in contrast to [1, 9, 10, 29], we do not predefine the variable exponents $p(x)$ a priori using for that the original noisy images S or/and M, but instead we associate this characteristic with each feasible solution. As a result, we admit that each feasible solution to this problem lives in the corresponding "individual" functional space. Formally it means that we look for the true image $I^0 = \left[I_1^0, I_2^0, \cdots, I_m^0\right]^t$ such that

$$I^0 \in W^{1,\mathfrak{F}(I_1^0(\cdot))}(\Omega) \times W^{1,\mathfrak{F}(I_2^0(\cdot))}(\Omega) \times \cdots \times W^{1,\mathfrak{F}(I_m^0(\cdot))}(\Omega).$$

As follows from the definition of Sobolev-Orlicz space $W^{1,\mathfrak{F}(I_i^0(\cdot))}(\Omega)$, its regularity is completely determined by the exponent $\mathfrak{F}(I_i^0(\cdot))$ which depends on i-th spectral channel of the true image I^0 and, hence, is unknown a priori. Moreover, the exponents $\{\mathfrak{F}(I_1^0(\cdot)), \mathfrak{F}(I_2^0(\cdot)), \ldots, \mathfrak{F}(I_m^0(\cdot))\}$ may significantly differ from channel to channel. In particular, some pixels, which are the local minimum points in the red channel, become local maximum points in near-infrared channel and wise versa. Moreover, the different feasible solutions $v \neq u$ to the above problem live in different functional spaces: we have $v \in W^{1,\mathfrak{F}(v(\cdot))}(\Omega)$ whereas $u \in W^{1,\mathfrak{F}(u(\cdot))}(\Omega)$. As a consequence, any minimizing sequence to this problem is, in fact, a sequence living in the scale of variable spaces. As a result, the notions such as convergence concept, compactness, density and others should be specified for the case of variable Sobolev-Orlicz spaces.

Thus, in spite of the fact that in the literature there are many approaches to the study of variational problems in abstract functional spaces, the above mentioned circumstances make the problem (5.4) rather challenging (see [7, 9, 10, 12, 13, 21] for recent studies in this field).

5.2 Existence Result

Our main intention in this section is to show that constrained minimization problem (5.4)–(5.5) is consistent and admits at least one solution. Because of the specific form of the energy functional $J_i(v)$, the structure and main topological properties of the set of feasible solution to minimization problem (5.4)–(5.5) are challenging issues. So, the study of these issues is the main subject of this section (we can refer to [5, 7, 9, 14, 16] for some specific details that can appear in this case).

We begin with the following key assumptions:

(A1) The true intensities I_i^0 of all spectral channels for the retrieved image $I^0 = \left[I_1^0, \ldots, I_m^0\right]^t$ are subjected to the constraints $\gamma_{i,0} \leq I_i^0(x) \leq \gamma_{i,1}$ a.e. in Ω, where the thresholds $\gamma_{i,0}$ and $\gamma_{i,1}$ are defined in (5.6).

(A2) There exist a couple of vector value functions $\widetilde{S} \in W^{1,1}(\Omega; \mathbb{R}^m)$ and $\widetilde{M} \in C(\overline{\Omega}; \mathbb{R}^m)$ such that the grids G_H and G_L are the sets of Lebesgue point of $\widetilde{S}$ and $\widetilde{M}$, respectively, and

$$\widetilde{S}\big|_{G_H} = S, \quad \widetilde{M}\big|_{G_L} = M. \tag{5.9}$$

Remark 5.1 Let us mention that in the case of digital images, the only accessible information is a sampled and quantized version of $I : \Omega \to \mathbb{R}^m$, i.e., $I(x_i, y_j)$, where $\{(x_i, y_j) \in \Omega\}$ is a set of discrete points and for each spectral channel $k = 1, \ldots, m$, $I_k(x_i, y_j)$ belongs in fact to a discrete set of values, $0, 1, \ldots, 255$ in the mostly cases. Due to the Shannon's theory, it is plausible to assume that I_k is recoverable at any point $(x, y) \in \Omega$ from the samples $I_k(x_i, y_j)$. So, in view of assumption (A2), we may assume that the images S and M are known in a continuous domain Ω and, therefore, the objective functional (5.4) should be interpreted as follows

$$J_i(v) = \int_\Omega |\nabla v(x)|^{\mathfrak{F}(v(x))}\, dx + \lambda \int_\Omega |\nabla v(x) - \nabla \widetilde{S}_i(x)|\, dx$$
$$+ \mu \int_\Omega |T_S v(x) - \widetilde{S}_i(x)|\, dx + \frac{1-\mu}{2} T_M\left(|(G_\sigma * v)(\cdot) - \widetilde{M}_i(\cdot)|^2\right).$$

However, in practice, such reconstruction is not a trivial problem.

We say that a function $I^0 = \left[I_1^0, \ldots, I_m^0\right]^t : \Omega \to \mathbb{R}^m$ is the result of simultaneous fusion and denoising of the noise contaminated images $S : G_H \to \mathbb{R}^m$ and $M : G_L \to \mathbb{R}^m$ if for given regularization parameters $\lambda > 0$, $\mu \in (0, 1)$, and a given linear blur operator $T_S \in \mathcal{L}(L^1(\Omega))$, each spectral component I_i^0 is the solution of the corresponding constrained minimization problem (5.7)–(5.8), i.e., for each $i = 1, \ldots, m$,

$$I_i^0 \in \Xi_i \quad \text{and} \quad J_i\left(I_i^0\right) = \inf_{v \in \Xi_i} J_i(v).$$

Hereinafter, we associate with each spectral channel v_i of an arbitrary image $v = [v_1, v_2, \ldots, v_m]^t : \Omega \to \mathbb{R}^m$ the so-called texture index $p_i : \Omega \to \mathbb{R}$ following the rule

$$p_i(x) := \mathfrak{F}(v_i(x)) = 1 + g\left(|(\nabla G_\sigma * v_i)(x)|\right), \quad \forall x \in \Omega, \ \forall i = 1, \ldots, m, \quad (5.10)$$

where $g:[0, \infty) \to (0, \infty)$ is the edge-stopping function that we take in the form of the Cauchy law $g(t) = \frac{1}{1+(t/a)^2}$.

As follows from representation (5.10) and smoothness of the Gaussian filter kernel G_σ, we have the following estimates

$$|(\nabla G_\sigma * v)(x)| \leq \int_\Omega |\nabla G_\sigma(x-y)||v(y)|\,dy$$

$$\leq \|G_\sigma\|_{C^1(\overline{\Omega-\Omega})}\|v\|_{L^1(\Omega)} \leq \|G_\sigma\|_{C^1(\overline{\Omega-\Omega})}|\Omega|\gamma_{1,i}, \quad \forall x \in \Omega,$$

$$\mathfrak{F}(v(x)) = 1 + \frac{a^2}{a^2 + (|(\nabla G_\sigma * v)(x)|)^2}$$

$$\geq 1 + \frac{a^2}{a^2 + \|G_\sigma\|^2_{C^1(\overline{\Omega-\Omega})}\|v\|^2_{L^1(\Omega)}} \geq 1 + \delta, \quad \forall x \in \Omega,$$

$$\mathfrak{F}(v(x)) \leq 2 \quad \text{in } \Omega,$$

where

$$\delta = \frac{a^2}{a^2 + \|G_\sigma\|^2_{C^1(\overline{\Omega-\Omega})}|\Omega|^2 \max_{1 \leq i \leq m} \gamma^2_{1,i}} \ll 1, \quad (5.11)$$

$$\|G_\sigma\|_{C^1(\overline{\Omega-\Omega})} = \max_{\substack{z=x-y \\ x\in\overline{\Omega}, y\in\overline{\Omega}}} \left[|G_\sigma(z)| + |\nabla G_\sigma(z)|\right] \frac{e^{-1}}{2\pi\sigma^2}\left[1 + \frac{1}{\sigma^2}\operatorname{diam}\Omega\right]. \quad (5.12)$$

Hence,

$$\alpha \leq \mathfrak{F}(v(x)) \leq \beta \quad \text{in } \Omega, \text{ where } \alpha := 1+\delta \text{ and } \beta := 2.$$

The following results play a crucial role in the sequel (for the proof, we refer to [14]).

Lemma 5.2 *Let $\{v_k\}_{k\in\mathbb{N}} \subset L^\infty(\Omega)$ be a sequence of measurable non-negative functions such that $\gamma_{i,0} \leq v_k(x) \leq \gamma_{i,1}$ a.e. in Ω and $v_k(x) \to v(x)$ weakly-$*$ in $L^\infty(\Omega)$ for some $v \in L^\infty(\Omega)$, and each element of this sequence is extended by zero outside of Ω. Let $\{p_k = 1 + g(|(\nabla G_\sigma * v_k)|)\}_{k\in\mathbb{N}}$ be the corresponding sequence of texture indices. Then*

$$p_k(\cdot) \to p(\cdot) = 1 + g\left(|(\nabla G_\sigma * v)(\cdot)|\right) \quad \text{uniformly in } \overline{\Omega} \text{ as } k \to \infty,$$

$$\alpha := 1 + \delta \leq p_k(x) \leq \beta := 2, \quad \forall x \in \Omega, \ \forall k \in \mathbb{N}. \quad (5.13)$$

5.2 Existence Result

Proposition 5.3 *Let* $\{p_k = 1 + g\left(|(\nabla G_\sigma * v_k)|\right)\}_{k \in \mathbb{N}}$ *be a sequence of texture indices such that*

$$p_k(\cdot) \to p(\cdot) = 1 + g\left(|(\nabla G_\sigma * v)(\cdot)|\right) \quad \text{uniformly in } \overline{\Omega} \text{ as } k \to \infty$$

and conditions (5.13) *hold true. If a bounded sequence* $\left\{f_k \in L^{p_k(\cdot)}(\Omega)\right\}_{k \in \mathbb{N}}$ *converges weakly in* $L^{1+\delta}(\Omega)$ *to* f, *then* $f \in L^{p(\cdot)}(\Omega)$, $f_k \rightharpoonup f$ *in variable* $L^{p_k(\cdot)}(\Omega)$, *and*

$$\liminf_{k \to \infty} \int_\Omega |f_k(x)|^{p_k(x)} \, dx \geq \int_\Omega |f(x)|^{p(x)} \, dx. \tag{5.14}$$

Following, in some technical aspects, recent studies [12–14, 21], we can give the following existence result.

Theorem 5.4 *For each* $i = 1, \ldots, m$ *and given* $\mu \in (0, 1)$, $\lambda > 0$, $S \in L^1(\Omega; \mathbb{R}^m)$, $M : G_L \to \mathbb{R}^m$, *and* $T_S \in \mathcal{L}(L^1(\Omega))$, *the minimization problem* (5.7)–(5.8) *admits at least one solution* $I_i^0 \in \Xi_i$.

Proof Since $\Xi_i \neq \emptyset$ and $0 \leq J_i(v) < +\infty$ for all $v \in \Xi_i$, it follows that there exists a non-negative value $\zeta \geq 0$ such that $\zeta = \inf_{v \in \Xi_i} J_i(v)$. Let $\{v_k\}_{k \in \mathbb{N}} \subset \Xi_i$ be a minimizing sequence to the problem (5.7)–(5.8), i.e.

$$v_k \in \Xi_i, \ \forall k \in \mathbb{N}, \quad \text{and} \quad \lim_{k \to \infty} J_i(v_k) = \zeta.$$

So, without loss of generality, we can suppose that $J_i(v_k) \leq \zeta + 1$ for all $k \in \mathbb{N}$.

Utilizing the fact that $v_k \in \Xi_i, \forall k \in \mathbb{N}$ and, therefore, $v_k(x) \leq \gamma_{1,i}$ for almost all $x \in \Omega$, we see that

$$\|v_k\|_{L^1(\Omega)} \leq \gamma_{1,i} |\Omega|, \quad \forall k \in \mathbb{N}.$$

Then setting $p_k(x) = 1 + g\left(|(\nabla G_\sigma * v_k)(x)|\right)$ in Ω and arguing as in the proof of Lemma 5.2, it can be shown that $p_k \in C^{0,1}(\overline{\Omega})$ and

$$\alpha := 1 + \delta \leq p_k(x) \leq \beta := 2, \quad \forall x \in \Omega, \ \forall k \in \mathbb{N}, \tag{5.15}$$

where δ is defined by the rule (5.11). From this, we deduce that

$$\int_\Omega |v_k(x)|^\alpha \, dx \leq \int_\Omega \gamma_{1,i}^\alpha \, dx \leq \gamma_{1,i}^\alpha |\Omega|, \quad \forall k \in \mathbb{N},$$

$$\int_\Omega |\nabla v_k(x)|^{p_k(x)} \, dx \leq \zeta + 1, \quad \forall k \in \mathbb{N}, \tag{5.16}$$

with $\alpha = 1 + \delta$.

Taking this fact into account, we infer from (5.16), (5.15), and (5.5) that

$$\|v_k\|_{W^{1,\alpha}(\Omega)} = \left(\int_\Omega \left[|v_k(x)|^\alpha + |\nabla v_k(x)|^\alpha\right] dx\right)^{1/\alpha}$$

$$\leq (1+|\Omega|)^{1/\alpha} \left(\int_\Omega \left[|v_k(x)|^{p_k(x)} + |\nabla v_k(x)|^{p_k(x)} \right] dx + 2 \right)^{1/\alpha}$$

$$\stackrel{by\ (5.16)}{\leq} (1+|\Omega|)^{1/\alpha} \left(\gamma_{1,i}^2 |\Omega| + \zeta + 3 \right)^{1/\alpha}$$

uniformly with respect to $k \in \mathbb{N}$. Therefore, there exist a subsequence of $\{v_k\}_{k\in\mathbb{N}}$, still denoted by the same index, and a function $I_i^0 \in W^{1,\alpha}(\Omega)$ such that

$$v_k \to I_i^0 \text{ strongly in } L^q(\Omega) \text{ for all } q \in [1, \alpha^*),$$
$$v_k \rightharpoonup I_i^0 \text{ weakly in } W^{1,\alpha}(\Omega) \text{ as } k \to \infty, \qquad (5.17)$$

where, by Sobolev embedding theorem, $\alpha^* = \frac{2\alpha}{2-\alpha} = \frac{2+2\delta}{1-\delta} > 2+\delta$.

Moreover, passing to a subsequence if necessary, we have (see Proposition 5.3 and Lemma 5.2):

$$v_k(x) \to I_i^0(x) \text{ a.e. in } \Omega \qquad (5.18)$$
$$v_k \rightharpoonup I_i^0 \text{ weakly in } L^{p_k(\cdot)}(\Omega),$$
$$\nabla v_k \rightharpoonup \nabla I_i^0 \text{ weakly in } L^{p_k(\cdot)}(\Omega; \mathbb{R}^N),$$
$$p_k(\cdot) \to p_i^0(\cdot) = 1 + g\left(\left|(\nabla G_\sigma * I_i^0)(\cdot)\right|\right) \text{ uniformly in } \overline{\Omega} \text{ as } k \to \infty,$$

where $I_i^0 \in W^{1,p^0(\cdot)}(\Omega)$.

Since $\gamma_{0,i} \leq v_k(x) \leq \gamma_{1,i}$ a.a. in Ω for all $k \in \mathbb{N}$, it follows from (5.18) that the limit function I_i^0 is also subjected to the same restriction. Thus, I_i^0 is a feasible solution to minimization problem (5.7)–(5.8).

Let us show that I_i^0 is a minimizer of this problem. With that in mind we note that due to the obvious inequality

$$|T_S(v_k(x)) - \widetilde{S}_i(x)| \leq \left(\|T_S\|_{\mathcal{L}(L^1(\Omega))} \gamma_{1,i} + |\widetilde{S}_i(x)| \right),$$

we have: the sequence $\left\{ T_S(v_k(x)) - \widetilde{S}_i(x) \right\}_{k\in\mathbb{N}}$ is bounded in $L^1(\Omega)$, equi-integrable in Ω, and because of (5.18), it strongly converges in $L^1(\Omega)$ to $T_S(I_i^0) - \widetilde{S}_i$. Hence,

$$\liminf_{k\to\infty} \int_\Omega |T_S(v_k(x)) - \widetilde{S}_i(x)| \, dx = \int_\Omega \left| T_S(I_i^0(x)) - \widetilde{S}_i(x) \right| dx. \qquad (5.19)$$

In view of the piecewise convergence (5.18), we have a similar relation for the last term in (5.4)

$$\liminf_{k\to\infty} T_M\left(|(G_\sigma * v_k)(\cdot) - \widetilde{M}_i(\cdot)|^2\right) = T_M\left(|(G_\sigma * I_i^0)(\cdot) - \widetilde{M}_i(\cdot)|^2\right).$$

It remains to notice that due to the properties (5.16), (5.17), the sequence $\left\{ |\nabla v_k| \in L^{p_k(\cdot)}(\Omega) \right\}_{k\in\mathbb{N}}$ is bounded and weakly convergent to $|\nabla I_i^0|$ in $L^\alpha(\Omega)$. Hence, by Proposition 5.3, the following lower semicontinuous properties

$$\liminf_{k\to\infty} \int_\Omega |\nabla v_k(x)|^{p_k(x)}\,dx \geq \int_\Omega |\nabla I_i^0(x)|^{p_i^0(x)}\,dx, \qquad (5.20)$$

$$\liminf_{k\to\infty} \int_\Omega |\nabla v_k(x) - \nabla \widetilde{S}_i(x)|\,dx \geq \int_\Omega |\nabla I_i^0(x) - \nabla \widetilde{S}_i(x)|\,dx \qquad (5.21)$$

hold true.

As a result, utilizing relations (5.19)–(5.21), we finally obtain

$$\zeta = \inf_{v\in \Xi_i} J_i(v) = \lim_{k\to\infty} J_i(v_k) = \liminf_{k\to\infty} J_i(v_k) \geq J_i(I_i^0).$$

Thus, I_i^0 is a minimizer to the problem (5.7)–(5.8), whereas its uniqueness remains an open question.

5.3 On Relaxation of the Minimization Problem (5.7)–(5.8)

It is clear that because of the specific choice of the exponent

$$\mathfrak{F}(v(x)) = 1 + g\left(|(\nabla G_\sigma * v)(x)|\right) \text{ in } \Omega,$$

constrained minimization problem (5.7)–(5.8) is not trivial in its practical implementation. Moreover, in this case the objective functional $J_i(v)$ is not convex. Even if we represent the minimization problem (5.7)–(5.8) in the form

$$\text{Find } (I_i^0, p_i^0) \in \Lambda_i \text{ such that } F_i(I_i^0, p_i^0) = \inf_{(v,p)\in\Lambda_i} F_i(v,p), \qquad (5.22)$$

where

$$F_i(v,p) = \int_\Omega |\nabla v(x)|^{p(x)}\,dx + \lambda \int_\Omega |\nabla v(x) - \nabla \widetilde{S}_i(x)|\,dx$$
$$+ \mu \int_\Omega |T_S v(x) - \widetilde{S}_i(x)|\,dx$$
$$+ \frac{1-\mu}{2} T_M\left(|(G_\sigma * v)(\cdot) - \widetilde{M}_i(\cdot)|^2\right), \qquad (5.23)$$

$$\Lambda_i = \Big\{(u,p) \in W^{1,\mathfrak{F}(v(\cdot))}(\Omega) \times C^{0,1}(\overline{\Omega}) \;\Big|$$
$$\begin{array}{l} 1 \leq \gamma_{i,0} \leq u(x) \leq \gamma_{i,1} \text{ a.e. in } \Omega \\ p(x) = 1 + g\left(|(\nabla G_\sigma * u)(x)|\right) \text{ in } \Omega \end{array}\Big\} \qquad (5.24)$$

the main difficulty in its study comes from the state constraints

$$p(x) = 1 + g\left(|(\nabla G_\sigma * v)(x)|\right)$$

with the non-convex right-hand side. This motivates us to pass to some relaxation scheme of variational problem (5.22)–(5.24). It will be shown in the sequel that using this approach, the non-convexity can be negligible in practice and that reliable solutions can be computed using a variety of different optimization algorithms.

As the main step of this procedure we propose to consider the function $p(\cdot) := \mathfrak{F}(v(\cdot))$ as a fictitious control subjected to some special constraints and interpret the fulfillment of equality $\mathfrak{F}(v(x)) = 1 + g\left(|(\nabla G_\sigma * v)(x)|\right)$ with some accuracy in Ω. To do so, we notice that if $v \in \Xi_i$ is a feasible solution to the problem (5.7)–(5.8) then $\mathfrak{F}(v(\cdot))$ is subjected to the two-side inequality (5.15) with $\delta \in (0, 1)$ given by (5.11). Keeping this in mind and following in some aspects the standard penalty method [42, Chapter 2] (see also [22–25, 28]), we consider the following family of approximating problems:

$$\text{Minimize } J_{i,\varepsilon}(v, p) = \int_\Omega |\nabla v(x)|^{p(x)} \, dx + \lambda \int_\Omega |\nabla v(x) - \nabla \widetilde{S}_i(x)| \, dx$$

$$+ \mu \int_\Omega |T_S v(x) - \widetilde{S}_i(x)| \, dx + \frac{1-\mu}{2} T_M \left(|(G_\sigma * v)(\cdot) - \widetilde{M}_i(\cdot)|^2\right)$$

$$+ \frac{1}{\varepsilon} \int_\Omega |p(x) - 1 - g\left(|(\nabla G_\sigma * v)(x)|\right)|^2 \, dx \qquad (5.25)$$

subject to the constraints $(v, p) \in \Xi_{i,\varepsilon}$, where

$$\Xi_{i,\varepsilon} = \left\{ (v, p) \, \middle| \, \begin{array}{l} v \in W^{1,\alpha}(\Omega), \ p \in \mathfrak{S}_{ad}, \ J_{i,\varepsilon}(v, p) < +\infty, \\ 0 \leq \gamma_{i,0} \leq v(x) \leq \gamma_{i,1} \ a.e. \ \text{in} \ \Omega, \end{array} \right\} \qquad (5.26)$$

$$\mathfrak{S}_{ad} = \left\{ h \in C(\Omega) \, \middle| \, \begin{array}{l} |h(x) - h(y)| \leq C|x - y|, \ \forall x, y \in \Omega, \\ 1 < \alpha \leq h(\cdot) \leq \beta \ \text{in} \ \overline{\Omega}. \end{array} \right\} \qquad (5.27)$$

Here, $\alpha = 1 + \delta$, $\delta > 0$ is given by (5.11), $\beta = 2$, and

$$C := \frac{2\|G_\sigma\|_{C^1(\overline{\Omega - \Omega})} \gamma_{1,i}^2 |\Omega| C_G}{a^2} \qquad (5.28)$$

with a positive constant C_G coming from the inequality

$$\int_\Omega |\nabla G_\sigma(x - z) - \nabla G_\sigma(y - z)| \, dz \leq C_G |x - y|, \quad \forall x, y \in \Omega.$$

To justify the choice (5.28) for the constant C, we make use of the following observation. If we assume for a moment that $p(x) = 1 + g\left(|(\nabla G_\sigma * v)|\right)$ for some $v \in \Xi_i$, then the following chain of estimates holds true

$$|p(x) - p(y)| \leq a^2 \left| \frac{|(\nabla G_\sigma * v)(x)|^2 - |(\nabla G_\sigma * v)(y)|^2}{\left(a^2 + |(\nabla G_\sigma * v)(x)|^2\right)\left(a^2 + |(\nabla G_\sigma * v)(y)|^2\right)} \right|$$

$$\leq \frac{2\|G_\sigma\|_{C^1(\overline{\Omega - \Omega})} \|v\|_{L^1(\Omega)}}{a^2} \left| |(\nabla G_\sigma * v)(x)| - |(\nabla G_\sigma * v)(y)| \right|$$

5.3 On Relaxation of the Minimization Problem (5.7)–(5.8)

$$\leq \frac{2\|G_\sigma\|_{C^1(\overline{\Omega-\Omega})}\gamma_1^2|\Omega|}{a^2}\int_\Omega |\nabla G_\sigma(x-z) - \nabla G_\sigma(y-z)|\,dz,$$

$$\forall x, y \in \Omega \text{ with } \gamma_1 = \|v\|_{L^\infty(\Omega)} \leq \gamma_{i,1}.$$

Then taking into account the smoothness of the function $\nabla G_\sigma(\cdot)$, we deduce: there exists a positive constant $C_G > 0$ independent of k such that

$$|p(x) - p(y)| \leq \frac{2\|G_\sigma\|_{C^1(\overline{\Omega-\Omega})}\gamma_{1,i}^2|\Omega|C_G}{a^2}|x-y|, \ \forall x, y \in \Omega.$$

Hereinafter, we assume that the parameter ε varies within a strictly decreasing sequence of positive real numbers which converges to 0. So, when we write $\varepsilon > 0$, we consider only the elements of this sequence.

Definition 5.5 We say that a pair (v, p) is quasi-feasible to minimization problem (5.22)–(5.24) if $(v, p) \in \Xi_{i,\varepsilon}$ for some $\varepsilon > 0$ small enough. We also say that $(u^0_{i,\varepsilon}, p^0_{i,\varepsilon}) \in W^{1,p^0_\varepsilon(\cdot)}(\Omega) \times C^{0,1}(\overline{\Omega})$ is a quasi-optimal solution to the problem (5.22)–(5.24) if

$$(u^0_{i,\varepsilon}, p^0_{i,\varepsilon}) \in \Xi_{i,\varepsilon} \text{ and } J_{i,\varepsilon}(u^0_{i,\varepsilon}, p^0_{i,\varepsilon}) = \inf_{(v,p) \in \Xi_{i,\varepsilon}} J_{i,\varepsilon}(v, p).$$

Remark 5.6 It is clear that condition $p \in \mathfrak{S}_{ad}$ together with the fact that $\mathfrak{S}_{ad}$ is a compact subset in $C(\overline{\Omega})$ implies: every cluster point of a sequence $\{p_k\}_{k \in \mathbb{N}} \subset \mathfrak{S}_{ad}$ with respect to the uniform topology is a regular exponent, i.e. it is an exponent satisfying the log-Hölder continuity condition [44]. In this case the set $C^\infty_0(\mathbb{R}^2)$ is dense in $W^{1,p(\cdot)}(\Omega)$ [11] and this fact plays a crucial role in the study of minimization problem (5.25).

The principle point in the statement of approximated problem (5.25) is the fact that we pass from the state constrained optimization problem (5.22) with the variable exponent $p(x) = \mathfrak{F}(v(x))$ strongly depending on the function of interest v to its approximation where we eliminate the equality constraint $p(x) = \mathfrak{F}(v(x))$ for the state $v(x)$ and the exponent $p(x)$ and allow such pairs run freely in their respective sets of feasibility.

We begin with the following existence result.

Theorem 5.7 *For each $i = 1, \ldots, M$, every positive value $\varepsilon > 0$, and given $\mu > 0$, $\lambda > 0$, $\widetilde{S}_i, \widetilde{M}_i \in L^1(\Omega)$, and $T_S \in \mathcal{L}(L^1(\Omega))$, the minimization problem (5.25) has at least one solution.*

Proof Since the set $\Xi_{i,\varepsilon}$ is nonempty, we can assert the existence of a minimizing sequence $\{(u_k, p_k)\}_{k \in \mathbb{N}} \subset \Xi_{i,\varepsilon}$. Then arguing as in the proof of Theorem 5.4, we deduce the boundedness of the sequence $\{u_k\}_{k \in \mathbb{N}}$ in $W^{1,p_k(\cdot)}(\Omega)$ and, hence, the existence of a subsequence, still denoted in the same way, such that $u_k \rightharpoonup u_\varepsilon^0$ in $W^{1,\alpha}(\Omega)$ and in variable $W^{1,p_k(\cdot)}(\Omega)$. As for the sequence $\{p_k\}_{k\in\mathbb{N}}$, we see that

$$\{p_k(\cdot)\} \subset \mathfrak{S} = \left\{ h \in C^{0,1}(\overline{\Omega}) \left| \begin{array}{l} |h(x) - h(y)| \leq C|x-y|, \ \forall x,y \in \Omega, \\ 1 < \alpha \leq h(\cdot) \leq \beta \text{ in } \overline{\Omega}, \end{array} \right. \right\}$$

and $\max_{x\in\overline{\Omega}}|p_k(x)| \leq \beta$. Since each element of the sequence $\{p_k\}_{k\in\mathbb{N}}$ has the same modulus of continuity, it follows that this sequence is uniformly bounded and equi-continuous. Hence, by Arzelà–Ascoli Theorem the sequence $\{p_k\}_{k\in\mathbb{N}}$ is relatively compact with respect to the norm topology of $C(\overline{\Omega})$. Since the set $\mathfrak{S}$ is closed with respect to the uniform convergence, it follows that

$$p_k(\cdot) \to p_\varepsilon^0(\cdot) \text{ uniformly in } \overline{\Omega} \text{ as } k \to \infty \text{ and, therefore, } p_\varepsilon^0 \in \mathfrak{S}_{ad}.$$

Thus, we can suppose that for a given minimizing sequence there exists a subsequence of $\{(u_k, p_k)\}_{k\in\mathbb{N}}$ in $W^{1,p_k(\cdot)}(\Omega) \times C^{0,1}(\Omega)$, still denoted in the same way, and a pair $(u_\varepsilon^0, p_\varepsilon^0)$ such that $p_k \to p_\varepsilon^0$ in $C(\overline{\Omega})$, $u_k \rightharpoonup u_\varepsilon^0$ in $W^{1,\alpha}(\Omega)$ and in variable $W^{1,p_k(\cdot)}(\Omega)$. Then, by the Sobolev embedding theorem, we deduce that $u_k \to u_\varepsilon^0$ strongly in $L^q(\Omega)$ for all $q \in [1, \frac{2\alpha}{2-\alpha})$, and, therefore, we can suppose that $u_k(x) \to u_\varepsilon^0(x)$ almost everywhere in Ω as $k \to \infty$. As a result, we have

$$\gamma_{0,i} \leq u_\varepsilon^0(x) \leq \gamma_{1,i} \quad \text{and} \quad \alpha \leq p_\varepsilon^0(x) \leq \beta \text{ a.a. in } \Omega,$$

$$\lim_{k\to\infty} \int_\Omega |T_S(u_k(x)) - \widetilde{S}_i(x)| \, dx = \int_\Omega |T_i(u_\varepsilon^0(x)) - \widetilde{S}_i(x)| \, dx,$$

$$\lim_{k\to\infty} T_M \left(|(G_\sigma * u_k)(\cdot) - \widetilde{M}_i(\cdot)|^2 \right) = T_M \left(|(G_\sigma * u_\varepsilon^0)(\cdot) - \widetilde{M}_i(\cdot)|^2 \right),$$

$$\liminf_{k\to\infty} \int_\Omega |\nabla u_k(x)|^{p_k(x)} \, dx \stackrel{\text{by (5.14)}}{\geq} \int_\Omega |\nabla u_\varepsilon^0(x)|^{p_\varepsilon^0(x)} \, dx.$$

Thus, $(u_\varepsilon^0, p_\varepsilon^0) \in \Xi_{i,\varepsilon}$. It remains to notice that

$$\left| p_k - 1 - g\left(|(\nabla G_\sigma * u_k)(x)|\right) \right|^2 \to \left| p_\varepsilon^0 - 1 - g\left(|(\nabla G_\sigma * u_\varepsilon^0)(x)|\right) \right|^2$$

in $C(\overline{\Omega})$, and the Lebesgue dominated convergence theorem implies

$$\lim_{k\to\infty} \int_\Omega \left| p_k - 1 - g\left(|(\nabla G_\sigma * u_k)(x)|\right) \right|^2 dx$$
$$= \int_\Omega \left| p_\varepsilon^0 - 1 - g\left(|(\nabla G_\sigma * u_\varepsilon^0)(x)|\right) \right|^2 dx.$$

Utilizing the above mentioned properties, we finally obtain

$$J_{i,\varepsilon}(u_\varepsilon^0, p_\varepsilon^0) \leq \liminf_{k\to\infty} J_{i,\varepsilon}(u_k, p_k) = \inf_{(v,p)\in\Xi_i} J_{i,\varepsilon}(v, p).$$

Thus, $(u_\varepsilon^0, p_\varepsilon^0) \in \Xi_{i,\varepsilon}$ is an optimal pair to the problem (5.25).

5.3 On Relaxation of the Minimization Problem (5.7)–(5.8)

Taking this existence result into account, we pass to the study of approximation properties of the problems (5.25). Namely, we establish the convergence of minima of (5.25) to minima of (5.22)–(5.24) as ε tends to zero. In other words, we show that some optimal solutions to (5.22)–(5.24) can be approximated by the quasi-optimal solutions of this problem.

Theorem 5.8 *Let $\{(u_\varepsilon^0, p_\varepsilon^0) \in \Xi_{i,\varepsilon}\}_{\varepsilon > 0}$ be a sequence of minimizers to the problem (5.25). Then there exists a subsequence of $\{(u_\varepsilon^0, p_\varepsilon^0)\}_{\varepsilon > 0}$, still denoted by the same index ε, such that*

$$p_\varepsilon^0 \to p^0 \text{ in } C(\overline{\Omega}) \text{ as } \varepsilon \to 0, \tag{5.29}$$

$$u_\varepsilon^0 \rightharpoonup u^0 \text{ in } W^{1,\alpha}(\Omega) \text{ as } \varepsilon \to 0, \tag{5.30}$$

$$u_\varepsilon^0 \rightharpoonup u^0 \text{ in } W^{1, p_\varepsilon^0(\cdot)}(\Omega), \ u^0 \in W^{1, p^0(\cdot)}(\Omega), \tag{5.31}$$

$$p^0(x) = 1 + g\left(\left|(\nabla G_\sigma * u^0)(x)\right|\right) \text{ in } \Omega, \tag{5.32}$$

$$J_i(u^0) = \inf_{v \in \Xi_i} J_i(v)$$

$$= \lim_{\varepsilon \to 0} \inf_{(u,p) \in \Xi_{i,\varepsilon}} J_{i,\varepsilon}(v, p) = \lim_{\varepsilon \to 0} J_{i,\varepsilon}(u_\varepsilon^0, p_\varepsilon^0), \tag{5.33}$$

and $u^0 \in \Xi_i$.

Proof Let $u^* \in \Xi_i$ be an arbitrary feasible solution to the original problem (5.7)–(5.8). We set $p^* = \mathfrak{F}(u^*(\cdot))$ in Ω. Then $u^* \in W^{1,\alpha}(\Omega)$, $p^* \in \mathfrak{S}_{ad}$, $J_{i,\varepsilon}(u^*, p^*) = J_i(u^*) < +\infty$, and, as a consequence, $(u^*, p^*) \in \Xi_{i,\varepsilon}$ for each $\varepsilon > 0$.

Since $J_{i,\varepsilon}(u_\varepsilon^0, p_\varepsilon^0) \leq J_{i,\varepsilon}(u^*, p^*) = J_i(u^*) =: C^*$, it follows from (5.25) that

$$\sup_{\varepsilon > 0} \int_\Omega |\nabla u_\varepsilon^0(x)|^{p_\varepsilon^0(x)} \, dx \leq C^*, \tag{5.34}$$

$$\int_\Omega \left|p_\varepsilon^0(x) - 1 - g\left(\left|(\nabla G_\sigma * u_\varepsilon^0)(x)\right|\right)\right|^2 dx \leq \varepsilon C^*, \quad \forall \varepsilon > 0. \tag{5.35}$$

Since $\{p_\varepsilon^0 \in C^{0,1}(\overline{\Omega})\}$ is a bounded sequence in $C(\overline{\Omega})$ with the same modulus of continuity, it follows, by Arzelà–Ascoli Theorem, that this sequence is relatively compact with respect to the norm topology of $C(\overline{\Omega})$. Without loss of generality, we can suppose that there exists a function $p^0 \in C(\overline{\Omega})$ such that assertion (5.29) is valid. Moreover, as follows from definition of the set $\mathfrak{S}_{ad}$, the limit function p^0 is subjected to the pointwise constraints

$$\alpha := 1 + \delta \leq p^0(x) \leq \beta := 2, \quad \forall x \in \Omega.$$

Arguing in a similar manner, we can infer from (5.34) and the two-side inequality

$$0 \leq \gamma_{0,i} \leq u_\varepsilon^0(x) \leq \gamma_{1,i} \quad \text{a.a. in } \Omega, \ \forall \varepsilon > 0 \tag{5.36}$$

that the sequence $\{u_\varepsilon^0\}$ is relatively compact with respect to the weak topology of $W^{1,\alpha}(\Omega)$. Indeed, taking into account (5.36) and observing that

$$\sup_{\varepsilon>0} \int_\Omega |u_\varepsilon^0(x)|^{p_\varepsilon^0(x)} \, dx \overset{\text{by (5.36)}}{\leq} +\infty,$$

we see that $u_\varepsilon^0 \in W^{1,p_\varepsilon^0(\cdot)}(\Omega)$ for all $\varepsilon > 0$ and the sequence $\{u_\varepsilon^0\}$ is bounded in variable space $W^{1,p_\varepsilon^0(\cdot)}(\Omega)$. Hence, this sequence is bounded in $W^{1,\alpha}(\Omega)$. Therefore, in view of completeness of $W^{1,\alpha}(\Omega)$, there exists a function $u^0 \in W^{1,\alpha}(\Omega)$ such that, up to a subsequence, property (5.30) holds true. As a result, Proposition 5.3 and Sobolev embedding theorem lead us to the conclusion:

$$\begin{aligned} u_\varepsilon^0 &\rightharpoonup u^0 \text{ in } W^{1,p_\varepsilon^0(\cdot)}(\Omega), \ u^0 \in W^{1,p^0(\cdot)}(\Omega), \\ u_\varepsilon^0 &\to u^0 \text{ strongly in } L^q(\Omega) \text{ for all } q \in [1, \alpha^*), \end{aligned} \quad (5.37)$$

where $\alpha^* = \frac{2\alpha}{2-\alpha}$. So, we can suppose that $u_\varepsilon^0(x) \to u^0(x)$ a.e. in Ω. Then passing to the limit in (5.36) as $\varepsilon \to 0$, we see that the limit function u^0 is also subjected to the point-wise constraints

$$0 \leq \gamma_{0,i} \leq u^0(x) \leq \gamma_{1,i} \quad \text{a.a. in } \Omega. \quad (5.38)$$

Moreover, utilizing the estimate (5.35) and properties (5.29)–(5.30), we get

$$\lim_{\varepsilon \to 0} \int_\Omega \left| p_\varepsilon^0(x) - 1 - g\left(\left|(\nabla G_\sigma * u_\varepsilon^0)(x)\right|\right)\right|^2 dx$$
$$= \int_\Omega \left| p^0(x) - 1 - g\left(\left|(\nabla G_\sigma * u^0)(x)\right|\right)\right|^2 dx = 0.$$

Hence, $p^0(x) = 1 + g\left(\left|(\nabla G_\sigma * u^0)(x)\right|\right)$ in Ω. Thus, $u^0 \in W^{1,\mathfrak{F}(u^0(\cdot))}(\Omega)$. Combining this fact with (5.38), we see that the limit function u^0 is a feasible solution to minimization problem (5.7)–(5.8).

Let us show that this function is optimal to the problem (5.7)–(5.8). Since

$$\lim_{\varepsilon \to 0} \int_\Omega |T_S(u_\varepsilon^0(x)) - \widetilde{S}_i(x)| \, dx \overset{\text{by (5.37)}}{=} \int_\Omega |T_S(u^0(x)) - \widetilde{S}_i(x)| \, dx,$$

$$\lim_{\varepsilon \to 0} T_M\left(|(G_\sigma * u_\varepsilon^0)(\cdot) - \widetilde{M}_i(\cdot)|^2\right) \overset{\text{by (5.37)}}{=} T_M\left(|(G_\sigma * u^0)(\cdot) - \widetilde{M}_i(\cdot)|^2\right),$$

it follows from Proposition 5.3 that

$$\liminf_{\varepsilon \to 0} J_{i,\varepsilon}(u_\varepsilon^0, p_\varepsilon^0) \geq J_i(u^0). \quad (5.39)$$

Then, assuming the converse—namely, there is a function $\widehat{u} \in \Xi_i$ such that $J_i(\widehat{u}) < J_i(u^0)$, we get:

5.3 On Relaxation of the Minimization Problem (5.7)–(5.8)

$$(\widehat{u}, \widehat{p}) \in \Xi_{i,\varepsilon} \quad \forall \varepsilon > 0 \text{ with } \widehat{p} := \mathfrak{F}(\widehat{u}(\cdot)),$$
$$J_i(\widehat{u}) \equiv J_{i,\varepsilon}(\widehat{u}, \widehat{p}) \geq \inf_{(v,p) \in \Xi_{i,\varepsilon}} J_{i,\varepsilon}(v, p) = J_{i,\varepsilon}(u_\varepsilon^0, p_\varepsilon^0).$$

Hence,

$$J_i(\widehat{u}) \geq \limsup_{\varepsilon \to 0} J_{i,\varepsilon}(u_\varepsilon^0, p_\varepsilon^0) \geq \liminf_{\varepsilon \to 0} J_{i,\varepsilon}(u_\varepsilon^0, p_\varepsilon^0) \stackrel{\text{by (5.39)}}{\geq} J_i(u^0), \quad (5.40)$$

and we come into contradiction with the initial assumptions. Thus, u^0 is a solution of the original problem (5.7)–(5.8). In order to establish the equality (5.33), it is enough, instead of $(\widehat{u}, \widehat{p})$, to take (u^0, p^0) in (5.40).

Since Theorem 5.8 does not give an answer whether the entire set of solutions to the problem (5.7)–(5.8) can be attained in such a way, the following result sheds some light on this matter.

Corollary 5.9 *Let* $u^0 \in \Xi_i$ *be a minimizer to optimization problem (5.7)–(5.8) such that there is a closed neighborhood* $\mathcal{U}(u^0)$ *of* u^0 *in the norm topology of* $L^\alpha(\Omega)$ *satisfying*

$$J_i(u^0) < J_i(v) \quad \forall v \in \Xi_i \cap \mathcal{U}(u^0).$$

Then there exists a sequence of local minima $\{(u_\varepsilon^0, p_\varepsilon^0)\}_{\varepsilon>0}$ *of problems (5.25) such that*

$$(u_\varepsilon^0, p_\varepsilon^0) \to (u^0, \mathfrak{F}(u^0(\cdot))) \quad \text{in the sense of Theorem 5.8.}$$

Proof By the strict local optimality of u^0, we have that it is the unique solution of the problem

$$\min_{v \in \Xi_i, v \in \mathcal{U}(u^0)} J_i(v). \quad (5.41)$$

For every $\varepsilon > 0$ let us consider the following optimization problems

$$\min_{(v,p) \in \Xi_{i,\varepsilon}, v \in \mathcal{U}(u^0)} J_{i,\varepsilon}(v, p). \quad (5.42)$$

Since the set $\{(v, p) \in \Xi_{i,\varepsilon}, v \in \mathcal{U}(u^0)\}$ is nonempty, it follows that the problem (5.42) has at least one solution $(u_\varepsilon^0, p_\varepsilon^0)$ for every $\varepsilon > 0$. Now, arguing as in the proof of Theorem 5.8, we deduce that $(u_\varepsilon^0, p_\varepsilon^0) \to (\widetilde{u}^0, \widetilde{p}^0)$ in the sense of convergences (5.29)–(5.33), and $\widetilde{u}^0$ is a solution of (5.41). Since u^0 is the unique solution of (5.41), we infer that $u^0 = \widetilde{u}^0$ and, therefore, $(u_\varepsilon^0, p_\varepsilon^0) \to (u^0, \mathfrak{F}(u^0(\cdot)))$ in the sense of Theorem 5.8. This implies the existence of $\varepsilon^0 > 0$ such that u_ε^0 belongs to the interior of $\mathcal{U}(u^0)$ for every $\varepsilon \leq \varepsilon^0$. Consequently, $(u_\varepsilon^0, p_\varepsilon^0)$ is a local minimum of (5.25) for every $\varepsilon \leq \varepsilon^0$. This concludes the proof.

5.4 Proximal Alternating Minimization Algorithm and Its Modification

In this section we discuss an algorithm that will attempt to numerically compute the solutions to the state constrained minimization problem (5.22)–(5.24). As follows from Theorem 5.8, some optimal solutions to (5.22)–(5.24) can be obtained as cluster points of the quasi-optimal solutions to this problem. From practical point of view, it means that we can focus on the mathematical model of approximating problem (5.25)–(5.26), with $\varepsilon > 0$ small enough, which models the solution that we are after. For a concise presentation, we cast problem (5.25)–(5.26) in the form

$$(v^*, p^*) \in \underset{(v,p)\in\Xi_{i,\varepsilon}}{\operatorname{Argmin}} J_{i,\varepsilon}(v, p). \tag{5.43}$$

Since the objective functional $J_{i,\varepsilon}(v, p)$ is neither convex in the joint variables (v, p) nor bi-convex (i.e., convex in each of the variables v and p), an abstract algorithm for finding solution of (5.43) is the proximal alternating minimization algorithm [2]. Given the initial pair $(v_0, p_0) \in \Lambda_i \subset \Xi_{i,\varepsilon}$, where

$$v_0(x) = \widetilde{M}_i(x) \quad \text{and} \quad p_0(x) = 1 + g\left(|(\nabla G_\sigma * v_0)(x)|\right) \quad \text{in } \Omega$$

and the step sizes $\tau_k^u, \tau_k^q > 0$, the next iterations can be computed by the update scheme

$$(v_k, p_k) \longrightarrow (v_{k+1}, p_k) \longrightarrow (v_{k+1}, p_{k+1}), \tag{5.44}$$

$$v_{k+1} \in \underset{\substack{u\in W^{1,\alpha}(\Omega) \\ \gamma_{i,0}\le u(x)\le \gamma_{i,1}}}{\operatorname{Argmin}} \left\{ \frac{1}{2\tau_k^u} \|u - v_k\|_{L^2(\Omega)}^2 + J_{i,\varepsilon}(u, p_k) \right\}, \tag{5.45}$$

$$p_{k+1} \in \underset{q\in\mathfrak{S}_{ad}}{\operatorname{Argmin}} \left\{ \frac{1}{2\tau_k^q} \|q - p_k\|_{L^2(\Omega)}^2 + J_{i,\varepsilon}(v_{k+1}, q) \right\}. \tag{5.46}$$

It is well known that under reasonably mild conditions on the regularity of $J_{i,\varepsilon}$ (which are obviously satisfied in our case, see [2] for the details), the proximal alternating minimization algorithm monotonously decreases the objective functional and its iterates converge to a critical point of $J_{i,\varepsilon}$. However, as it was mentioned in [2], very few general results ensure that the sequence $\{(v_k, p_k)\}_{k\in\mathbb{N}}$ converges to a global minimizer of (5.22)–(5.24), even for strictly convex functions. Meanwhile, exploiting the fact that minimization problem (5.46) with ε small enough admits a unique minimizer p_{k+1} at each step of iteration, we see that

$$p_{k+1}(x) = 1 + g\left(|(\nabla G_\sigma * v_{k+1})(x)|\right) \quad \text{in } \Omega. \tag{5.47}$$

In fact, it means that due to the equality (5.47), we can alleviate this approach. Indeed, in view of the representation (5.47), we can specify the above mentioned iteration procedure as follows

5.4 Proximal Alternating Minimization Algorithm and Its Modification

$$(v_k, p_k) \longrightarrow (v_{k+1}, p_{k+1}), \tag{5.48}$$

$$v_{k+1} \in \underset{\substack{u \in W^{1,\alpha}(\Omega) \\ \gamma_{i,0} \leq u(x) \leq \gamma_{i,1}}}{\text{Argmin}} \left\{ \frac{1}{2\tau_k^u} \|u - v_k\|_{L^2(\Omega)}^2 + J_{i,\varepsilon}(u, p_k) \right\}, \tag{5.49}$$

$$p_{k+1}(x) = 1 + g\left(|(\nabla G_\sigma * v_{k+1})(x)|\right) \text{ in } \Omega, \tag{5.50}$$

provided the parameter $\varepsilon > 0$ is chosen small enough. However, as follows from the structure of the penalized objective functional $J_{i,\varepsilon}$, we still deal with a non-convex optimization problem in (5.49).

In view of this, the main idea, we are going to push forward in this section, is to represent the iteration procedure (5.48)–(5.50) as follows

$$(v_k, p_k) \longrightarrow (v_{k+1}, p_{k+1}), \tag{5.51}$$

$$v_{k+1} \in \underset{u \in \mathcal{B}_{i,p_k(\cdot)}}{\text{Argmin}} \left\{ \frac{1}{2\tau_k^u} \|u - v_k\|_{L^2(\Omega)}^2 + F_i(u, p_k) \right\}, \tag{5.52}$$

$$p_{k+1}(x) = 1 + g\left(|(\nabla G_\sigma * v_{k+1})(x)|\right) \text{ in } \Omega, \tag{5.53}$$

where the cost functional F_i is defined in (5.23) and

$$\mathcal{B}_{i,p(\cdot)} = \left\{ v \in W^{1,p(\cdot)}(\Omega) : 1 \leq \gamma_{i,0} \leq v(x) \leq \gamma_{i,1} \text{ a.e. in } \Omega \right\}.$$

The main benefit of this modification is to pass to convex optimization problems at each step of iteration. Then arguing as in the proof of Theorem 5.4 and using convexity arguments, it can be shown that, for each $p_k(\cdot) \in \mathfrak{S}_{ad}$, there exists a unique element $v_{k+1} \in \mathcal{B}_{i,p_k(\cdot)}$ such that $v_{k+1} = \underset{u \in \mathcal{B}_{i,p_k(\cdot)}}{\text{Argmin}} \left\{ \frac{1}{2\tau_k^u} \|u - v_k\|_{L^2(\Omega)}^2 + F_i(u, p_k) \right\}$. This fact reflects the principle difference between optimization problems (5.52) and (5.22), where the problem (5.52) can be viewed as a minimization of the growth energy functional (5.23) with "the frozen exponent" $p_k(x)$. Thus, the sequence $\{v_k\}_{k \in \mathbb{N}}$ can be defined in a unique way. Moreover, the iteration procedure (5.48)–(5.50) possesses the following property.

Proposition 5.10 *For any sequences of stepsizes $\{\tau_k^u\}_{k \in \mathbb{N}}$, $\{\tau_k^q\}_{k \in \mathbb{N}}$ such that $\tau_k^u, \tau_k^q \in (r_-, +\infty)$ for all $k \in \mathbb{N}$ with some positive r_-, the numerical sequence $\{F_i(v_k, p_k)\}_{k \in \mathbb{N}}$ does not increase and the estimates*

$$F_i(v_{k+1}, p_{k+1}) + \frac{1}{2\tau_k^q} \|p_{k+1} - p_k\|_{L^2(\Omega)}^2 + \frac{1}{2\tau_k^u} \|v_{k+1} - v_k\|_{L^2(\Omega)}^2$$

$$\leq F_i(v_k, p_k), \quad \forall k \in \mathbb{N}, \tag{5.54}$$

$$\sum_{k=1}^{\infty} \left[\|v_k - v_{k-1}\|_{L^2(\Omega)}^2 + \|p_k - p_{k-1}\|_{L^2(\Omega)}^2 \right] < +\infty \tag{5.55}$$

hold true.

Proof To begin with, we notice that the equality (5.53) can be rewritten in an equivalent form as follows

$$p_{k+1} \in \underset{q \in \mathfrak{S}_{ad}}{\text{Argmin}} \left\{ \frac{1}{2\tau_k^q} \|q - p_k\|_{L^2(\Omega)}^2 + F_i(v_{k+1}, q) \right\} \qquad (5.56)$$

provided the stepsizes τ_k^q is greater than a fixed positive parameter which can be chosen arbitrarily large. In this case the algorithm (5.48)–(5.50) is very close to a coordinate descent method. Then

$$F_i(v_{k+1}, p_k) + \frac{1}{2\tau_k^u} \|v_{k+1} - v_k\|_{L^2(\Omega)}^2 \overset{\text{by (5.52)}}{\leq} F_i(v_k, p_k),$$

$$F_i(v_{k+1}, p_{k+1}) + \frac{1}{2\tau_k^q} \|p_{k+1} - p_k\|_{L^2(\Omega)}^2 \overset{\text{by (5.56)}}{\leq} F_i(v_{k+1}, p_k).$$

Hence, an elementary induction

$$F_i(v_{k+1}, p_{k+1}) + \frac{1}{2\tau_k^q} \|p_{k+1} - p_k\|_{L^2(\Omega)}^2 + \frac{1}{2\tau_k^u} \|v_{k+1} - v_k\|_{L^2(\Omega)}^2$$

$$\leq F_i(v_{k+1}, p_k) + \frac{1}{2\tau_k^u} \|v_{k+1} - v_k\|_{L^2(\Omega)}^2 \leq F_i(v_k, p_k), \quad \forall k \in \mathbb{N}$$

ensures that estimate (5.54) is valid.

As for estimate (5.55), it is enough to observe that $F_i(v, p) \geq 0$ for all feasible pairs (v, p). Hence, (5.55) immediately follows from (5.54). As a consequence, we have,

$$\lim_{k \to \infty} \|v_k - v_{k-1}\|_{L^2(\Omega)} = \lim_{k \to \infty} \|p_k - p_{k-1}\|_{L^2(\Omega)} = 0.$$

We say that a function $\widehat{u}_i$ is a weak solution to the original problem (5.7)–(5.8) if

$$\widehat{u}_i = \underset{v \in \mathcal{B}_{\widehat{p}_i}}{\text{Argmin}}\, F_i(v, \widehat{p}_i(\cdot)), \quad \widehat{u}_i \in \mathcal{B}_{i, \widehat{p}_i(\cdot)},$$
$$\widehat{p}_i(x) = 1 + g\left(|(\nabla G_\sigma * \widehat{u}_i)(x)|\right), \quad \forall x \in \Omega. \qquad (5.57)$$

Remark 5.11 The relation between a weak solution and a solution to the problem (5.7)–(5.8) is rather intricate. Since the uniqueness of solutions to (5.7)–(5.8) is a questionable option, it follows that, in principle, these definitions can describe the different functions in Ξ_i. As immediately follows from (5.57), a weak solution is a merely feasible one to the original problem. However, if the problem (5.22) admits a unique solution $(u_i^0, p_i^0) \in \Lambda_i$, then (5.57) implies that the function u_i^0 can be considered as a weak solution.

Before proceeding further, we note that, for given $i = 1, \ldots, m$, the sequence of exponents $\{p_k\}_{k \in \mathbb{N}}$ is compact with respect to the strong topology of $C(\overline{\Omega})$. Our next goal is to

5.4 Proximal Alternating Minimization Algorithm and Its Modification

establish the existence of a weak solution to the original problem (5.7)–(5.8) and show that it can be attained by the iterative algorithm (5.51)–(5.53). To do so, we begin with some technical results.

Lemma 5.12 *For each $i = 1, \ldots, m$ and given $\mu \in (0, 1)$, $\lambda > 0$, $\widetilde{S} \in L^1(\Omega; \mathbb{R}^m)$, $\widetilde{M} : G_L \to \mathbb{R}^m$, and $T_S \in \mathcal{L}(L^1(\Omega))$, the sequence of minimizers $\{v_k \in W^{1, p_k(\cdot)}(\Omega)\}_{k \in \mathbb{N}}$ of (5.52) is compact with respect to the weak topology of $W^{1,\alpha}(\Omega)$.*

Proof Let us show that the sequence of minimizers $\{v_k\}_{k \in \mathbb{N}}$ of (5.52) is bounded in the following sense

$$\limsup_{k \to \infty} \int_\Omega |v_k(x)|^{p_k(x)} \, dx < +\infty.$$

Let $\widehat{u} \in C^1(\overline{\Omega})$ be an arbitrary function such that $\gamma_{0,i} \leq \widehat{u}(x) \leq \gamma_{1,i}$ in Ω. Since

$$F_i(v_k, p_k) \leq F_i(\widehat{u}, p_k) + \frac{1}{2\tau_{k-1}^u} \|\widehat{u} - v_{k-1}\|^2_{L^2(\Omega)}, \quad \forall k = 1, 2, \ldots$$

and

$$\int_\Omega |\nabla \widehat{u}(x)|^{p_k(x)} \, dx \leq \int_\Omega \left(1 + \|\widehat{u}\|_{C^1(\overline{\Omega})}\right)^{p_k(x)} \, dx$$
$$\leq |\Omega| \left(1 + \|\widehat{u}\|_{C^1(\overline{\Omega})}\right)^2,$$

$$\int_\Omega |\nabla \widehat{u}(x) - \nabla \widetilde{S}_i(x)| \, dx \leq \int_\Omega \left[\|\widehat{u}\|_{C^1(\overline{\Omega})} + |\nabla \widetilde{S}_i(x)|\right] dx$$
$$\leq |\Omega| \|\widehat{u}\|_{C^1(\overline{\Omega})} + \|\widetilde{S}_i\|_{W^{1,1}(\Omega)},$$

$$\int_\Omega |T_S \widehat{u}(x) - \widetilde{S}_i(x)| \, dx \leq \int_\Omega \left[\|T_S\|_{\mathcal{L}(L^1(\Omega))} \gamma_{1,i} + |\widetilde{S}_i(x)|\right] dx$$
$$\leq |\Omega| \|T_S\|_{\mathcal{L}(L^1(\Omega))} \gamma_{1,i} + \|\widetilde{S}_i\|_{L^1(\Omega)},$$

$$T_M \left(|(G_\sigma * \widehat{u}(\cdot)) - \widetilde{M}_i(\cdot)|^2\right) \leq \left(\|G_\sigma * \widehat{u}\|_{C(\overline{\Omega})} + \|\widetilde{M}_i\|_{C(\overline{\Omega})}\right)^2$$
$$\leq \left(\frac{1}{\left(\sqrt{2\pi}\sigma\right)^2} \|\widehat{u}\|_{C(\overline{\Omega})} + \|\widetilde{M}_i\|_{C(\overline{\Omega})}\right)^2,$$

$$\|\widehat{u} - v_{k-1}\|^2_{L^2(\Omega)} \leq 4\gamma_{1,i}^2 |\Omega|,$$

it follows that

$$\sup_{k \in \mathbb{N}} F_i(v_k, p_k) \leq \sup_{k \in \mathbb{N}} \left[F_i(\widehat{u}, p_k) + \frac{1}{2\tau_{k-1}^u} \|\widehat{u} - v_{k-1}\|^2_{L^2(\Omega)}\right] \leq \widehat{C}$$

with some appropriate constant $\widehat{C} > 0$.

From this and definition of the set $\mathcal{B}_{i,p_k(\cdot)}$, we deduce

$$\int_\Omega |v_k(x)|^\alpha \, dx \le \gamma_{1,i}^\alpha |\Omega|, \quad \forall k \in \mathbb{N},$$

$$\int_\Omega |\nabla v_k(x)|^{p_k(x)} \, dx \le \widehat{C}, \quad \forall k \in \mathbb{N}.$$

Since (see [11, 17, 43] for the details)

$$\|f\|_{L^{p(\cdot)}(\Omega)}^\alpha - 1 \le \int_\Omega |f(x)|^{p(x)} \, dx \le \|f\|_{L^{p(\cdot)}(\Omega)}^\beta + 1, \quad \forall f \in L^{p(\cdot)}(\Omega), \tag{5.58}$$

it follows that the sequence $\{v_k\}_{k\in\mathbb{N}}$ is bounded in $W^{1,\alpha}(\Omega)$. So, its weak compactness is a direct consequence of the reflexivity of $W^{1,\alpha}(\Omega)$.

We notice that boundedness of the sequence $\{v_k\}_{k\in\mathbb{N}}$ in $W^{1,\alpha}(\Omega)$ and compactness of the embedding $W^{1,\alpha}(\Omega) \hookrightarrow L^q(\Omega)$ for $q \in \left[1, \frac{2\alpha}{2-\alpha}\right)$ imply the existence of an element $u^* \in W^{1,\alpha}(\Omega)$ such that, up to a subsequence,

$$v_k(x) \to u^*(x) \text{ a.e. in } \Omega, \tag{5.59}$$

$$v_k \to u^* \text{ in } L^q(\Omega), \text{ and } \nabla v_k \rightharpoonup \nabla u^* \text{ in } L^\alpha(\Omega; \mathbb{R}^2). \tag{5.60}$$

Then using (5.59) and passing to the limit in two-side inequality $\gamma_{0,i} \le v_k(x) \le \gamma_{1,i}$, we obtain

$$\gamma_{0,i} \le u^*(x) \le \gamma_{1,i} \quad \text{for a.a. } x \in \Omega.$$

Utilizing this fact together with the pointwise convergence (5.59), by the Lebesgue dominated convergence theorem, we have

$$\lim_{k\to\infty} p_k(x) = \lim_{k\to\infty} \mathfrak{F}(v_k(x)) = 1 + \frac{a^2}{a^2 + \left(\lim_{k\to\infty} |(\nabla G_\sigma * v_k)(x)|\right)^2}$$

$$= 1 + \frac{a^2}{a^2 + \left(\left|\left(\nabla G_\sigma * \lim_{k\to\infty} v_k\right)(x)\right|\right)^2} = \mathfrak{F}(u^*(x)), \quad \forall x \in \Omega. \tag{5.61}$$

Since, by Arzelà–Ascoli theorem, the set $\left\{p_k = 1 + g\left(|(\nabla G_\sigma * v_k)(x)|\right)\right\}_{k\in\mathbb{N}}$ is compact with respect to the norm topology of $C(\overline{\Omega})$, it follows from (5.61) (see also the proof of Lemma 5.2) that

$$p_k \to p^* = \mathfrak{F}(u^*(x)) \text{ strongly in } C(\overline{\Omega}) \text{ as } k \to \infty, \text{ and } p^* \in \mathfrak{S}_{ad}. \tag{5.62}$$

Then properties (5.59)–(5.62) and Proposition 5.3 imply:

$$u^* \in \mathcal{B}_{i,p^*(\cdot)} = \left\{ u \in W^{1,p^*(\cdot)}(\Omega) \, : \, 1 \le \gamma_{i,0} \le u(x) \le \gamma_{i,1} \text{ a.e. in } \Omega \right\}.$$

5.4 Proximal Alternating Minimization Algorithm and Its Modification

Thus, the iterative procedure (5.48)–(5.50) has a cluster point $(u^*, p^*) \in \mathcal{B}_{i, p^*(\cdot)} \times \mathfrak{S}_{ad}$ with respect to the convergence (5.59)–(5.60), (5.62).

We are now in a position to state the main result of this section.

Theorem 5.13 *Let $\mu \in (0, 1)$, $\lambda > 0$, $\widetilde{S} \in L^1(\Omega; \mathbb{R}^m)$, $\widetilde{M} : G_L \to \mathbb{R}^m$, and $T_S \in \mathcal{L}(L^1(\Omega))$ be given data. Let $\{\tau_k^u\}_{k \in \mathbb{N}}$ be a monotonically increasing sequence of positive stepsizes such that $\tau_k^u \to \infty$ as $k \to \infty$. Then, for each $i = 1, \ldots, m$, the sequence $\{(v_k, p_k)\}_{k \in \mathbb{N}}$, coming from the iteration procedure (5.48)–(5.50), possesses the following asymptotic properties:*

$$v_k(x) \to \widetilde{u}_i(x) \text{ a.e. in } \Omega, \tag{5.63}$$

$$v_k \to \widetilde{u}_i \text{ in } L^q(\Omega), \, \forall q \in \left[1, \frac{2\alpha}{2-\alpha}\right), \, \nabla v_k \rightharpoonup \nabla \widetilde{u}_i \text{ in } L^\alpha(\Omega; \mathbb{R}^2), \tag{5.64}$$

$$p_k \to \widetilde{p}_i = \mathfrak{F}(\widetilde{u}_i) \text{ strongly in } C(\overline{\Omega}) \text{ as } k \to \infty, \tag{5.65}$$

where $\widetilde{u}_i$ is a weak solution to the problem (5.7)–(5.8), that is,

$$\widetilde{u}_i \in \mathcal{B}_{i, \widetilde{p}_i(\cdot)}, \quad \widetilde{u}_i = \underset{v \in \mathcal{B}_{i, \widetilde{p}_i(\cdot)}}{\text{Argmin}} \, F_i(v, \widetilde{p}_i),$$

and, in addition, the following variational property holds true

$$F_i(v_k, p_k) \geq F_i(v_{k+1}, p_{k+1}), \quad \forall k \in \mathbb{N}, \tag{5.66}$$

$$\lim_{k \to \infty} F_i(v_k, p_k(\cdot)) = \lim_{k \to \infty} \left[\inf_{v \in \mathcal{B}_{i, p_k(\cdot)}} F_i(v, p_k) \right]$$
$$= \inf_{v \in \mathcal{B}_{i, \widetilde{p}_i(\cdot)}} F_i(v, \widetilde{p}_i(\cdot)) = J_i(\widetilde{u}_i). \tag{5.67}$$

Proof "Lemma 5.12, the sequence $\{(v_k, p_k)\}_{k \in \mathbb{N}}$ is compact with respect to the convergence (5.63)–(5.65). Let $(\widetilde{u}_i, \widetilde{p}_i)$ be its cluster point. In order to show that the function $\widetilde{u}_i$ is a weak solution to the problem (5.7)–(5.8), we assume the converse—namely, there is another function $z \in \mathcal{B}_{i, \widetilde{p}_i(\cdot)}$ such that

$$F_i(z, \widetilde{p}_i) = \inf_{v \in \mathcal{B}_{i, \widetilde{p}_i(\cdot)}} F_i(v, \widetilde{p}_i) < F_i(\widetilde{u}_i, \widetilde{p}_i) \equiv J_i(\widetilde{u}_i). \tag{5.68}$$

Using the procedure of the standard direct smoothing, we set

$$u_\varepsilon(x) = \frac{1}{\varepsilon^2} \int_{\mathbb{R}^2} K\left(\frac{x-s}{\varepsilon}\right) \widetilde{z}(s) \, ds,$$

where $\varepsilon > 0$ is a small parameter, K is a positive compactly supported smooth function with properties

$$K \in C_0^\infty(\mathbb{R}^2), \quad \int_{\mathbb{R}^2} K(x)\,dx = 1, \quad \text{and} \quad K(x) = K(-x),$$

and $\tilde{z}$ is zero extension of z outside of Ω.

Since $z \in W^{1,\tilde{p}(\cdot)}(\Omega)$ and $\tilde{p}(x) \geq \alpha = 1 + \delta$ in Ω, it follows that $z \in W^{1,\alpha}(\Omega)$. Then

$$u_\varepsilon \in C_0^\infty(\mathbb{R}^2) \text{ for each } \varepsilon > 0,$$
$$u_\varepsilon \to z \text{ in } L^\alpha(\Omega), \quad \nabla u_\varepsilon \to \nabla z \text{ in } L^\alpha(\Omega;\mathbb{R}^2) \tag{5.69}$$

by the classical properties of smoothing operators. From this we deduce that

$$u_\varepsilon(x) \to z(x) \text{ a.e. in } \Omega. \tag{5.70}$$

Moreover, taking into account the estimates

$$u_\varepsilon(x) = \int_{\mathbb{R}^2} K(y)\,\tilde{z}(x - \varepsilon y)\,dy \leq \gamma_{1,i} \int_{\mathbb{R}^2} K(y)\,dy = \gamma_{1,i},$$
$$u_\varepsilon(x) \geq \int_{y \in \varepsilon^{-1}(x-\Omega)} K(y)\,\tilde{z}(x - \varepsilon y)\,dy \geq \gamma_{0,i} \int_{y \in \varepsilon^{-1}(x-\Omega)} K(y)\,dy \geq \gamma_{0,i},$$

we see that each element u_ε is subjected to the pointwise constraints

$$\gamma_{0,i} \leq u_\varepsilon(x) \leq \gamma_{1,i} \text{ a.e. in } \Omega, \quad \forall \varepsilon > 0.$$

Since, for each $\varepsilon > 0$, $u_\varepsilon \in W^{1,p_k(\cdot)}(\Omega)$ for all $k \in \mathbb{N}$, it follows that $u_\varepsilon \in \mathcal{B}_{i,p_k(\cdot)}$, i.e., each element of the sequence $\{u_\varepsilon\}_{\varepsilon>0}$ is a feasible solution to all approximating problems

$$\inf_{v \in \mathcal{B}_{i,p_k(\cdot)}} \left\{ \frac{1}{2\tau_k^u} \|v - v_k\|_{L^2(\Omega)}^2 + F_i(v, p_k) \right\}, \quad k \in \mathbb{N}. \tag{5.71}$$

Hence,

$$F_i(v_{k+1}, p_k) + \frac{1}{2\tau_k^u} \|v_{k+1} - v_k\|_{L^2(\Omega)}^2 \stackrel{\text{by (5.52)}}{\leq} F_i(u_\varepsilon, p_k) + \frac{1}{2\tau_k^u} \|u_\varepsilon - v_k\|_{L^2(\Omega)}^2,$$
$$F_i(v_{k+1}, p_{k+1}) + \frac{1}{2\tau_k^q} \|p_{k+1} - p_k\|_{L^2(\Omega)}^2 \stackrel{\text{by (5.56)}}{\leq} F_i(v_{k+1}, p_k)$$

for all $\varepsilon > 0$ and $k = 0, 1, \ldots$. From this we deduce that

$$F_i(v_{k+1}, p_{k+1}) \leq F_i(u_\varepsilon, p_k)$$
$$+ \frac{1}{2\tau_k^u} \|u_\varepsilon - v_k\|_{L^2(\Omega)}^2, \quad \forall \varepsilon > 0, \forall k = 0, 1, \ldots \tag{5.72}$$

Further, we notice that

$$\liminf_{k \to \infty} F_i(v_k, p_k) \geq F_i(\tilde{u}_i, \tilde{p}_i) \tag{5.73}$$

by Proposition 5.3 and Fatou's lemma. Since

$$|\nabla u_\varepsilon(x)|^{p_k(x)} \to |\nabla u_\varepsilon(x)|^{\widetilde{p}_i(x)} \quad \text{uniformly in } \Omega \text{ as } k \to \infty,$$

it follows from the Lebesgue dominated convergence theorem that the objective functional $F_i(u_\varepsilon, \cdot)$ is continuous with respect to the norm convergence in $C(\overline{\Omega})$, i.e.

$$\lim_{k \to \infty} F_i(u_\varepsilon, p_k) = F_i(u_\varepsilon, \widetilde{p}_i), \quad \forall \varepsilon > 0. \tag{5.74}$$

As a result, passing to the limit in (5.72) and utilizing properties (5.73)–(5.74), we obtain

$$\lim_{k \to \infty} \frac{1}{2\tau_k^u} \|u_\varepsilon - v_k\|_{L^2(\Omega)}^2 = \frac{1}{2 \lim_{k \to \infty} \tau_k^u} \|u_\varepsilon - \widetilde{u}_i\|_{L^2(\Omega)}^2 = 0.$$

Therefore,

$$F_i(\widetilde{u}_i, \widetilde{p}_i) \leq F_i(u_\varepsilon, \widetilde{p}_i) = \int_\Omega |\nabla u_\varepsilon(x)|^{\widetilde{p}_i(x)} \, dx$$
$$+ \lambda \int_\Omega |\nabla u_\varepsilon(x) - \nabla \widetilde{S}_i(x)| \, dx + \mu \int_\Omega |T_S u_\varepsilon(x) - \widetilde{S}_i(x)| \, dx$$
$$+ \frac{1-\mu}{2} T_M \left(|(G_\sigma * u_\varepsilon)(\cdot) - \widetilde{M}_i(\cdot)|^2 \right), \quad \forall \varepsilon > 0. \tag{5.75}$$

Taking into account the pointwise convergence (see (5.70) and property (5.69))

$$|\nabla u_\varepsilon(x)|^{\widetilde{p}_i(x)} \to |\nabla z(x)|^{\widetilde{p}_i(x)}, \quad \text{a.e. in } \Omega,$$
$$|T_S u_\varepsilon(x) - \widetilde{S}_i(x)| \to |T_S z(x) - \widetilde{S}_i(x)|, \quad \text{a.e. in } \Omega,$$
$$\int_\Omega |\nabla u_\varepsilon(x) - \nabla \widetilde{S}_i(x)| \, dx \to \int_\Omega |\nabla z(x) - \nabla \widetilde{S}_i(x)| \, dx,$$
$$(G_\sigma * u_\varepsilon)(x) - \widetilde{M}_i(x) \to (G_\sigma * z)(x) - \widetilde{M}_i(x), \quad \text{in } \Omega$$

as $\varepsilon \to 0$, and the fact that, for ε small enough,

$$|\nabla u_\varepsilon|^{\widetilde{p}_i(\cdot)} \leq (1 + |\nabla z|)^{\widetilde{p}_i(\cdot)} \in L^1(\Omega),$$
$$|T_S u_\varepsilon(\cdot) - \widetilde{S}_i(\cdot)| \leq [\|T_S\| (1 + |z(\cdot)|) + |\widetilde{S}_i(\cdot)|] \in L^1(\Omega),$$

we can pass to the limit in (5.75) as $\varepsilon \to 0$ by the Lebesgue dominated convergence theorem. This yields

$$F_i(\widetilde{u}_i, \widetilde{p}_i(\cdot)) \leq \lim_{\varepsilon \to 0} F_i(u_\varepsilon, \widetilde{p}_i(\cdot)) = F_i(z, \widetilde{p}_i(\cdot)).$$

Combining this inequality with (5.75) and (5.68), we finally get

$$F_i(z, \widetilde{p}_i) = \inf_{v \in \mathcal{B}_{i, \widetilde{p}_i(\cdot)}} F_i(v, \widetilde{p}_i) < F_i(\widetilde{u}_i, \widetilde{p}_i) \leq F_i(z, \widetilde{p}_i),$$

that leads us into conflict with the initial assumption. Thus,

$$J_i(\widetilde{u}_i) = F_i(\widetilde{u}_i, \widetilde{p}_i(\cdot)) = \inf_{v \in \mathcal{B}_{i,\widetilde{p}_i(\cdot)}} F_i(v, \widetilde{p}_i(\cdot)) \qquad (5.76)$$

and, therefore, $\widetilde{u}_i$ is a weak solution to the original problem (5.7)–(5.8). As for the variational property (5.67) and property (5.66), they immediately follow from (5.76), (5.74), and Proposition 5.10.

5.5 Optimality Conditions

To characterize the solution $u^{0,p(\cdot)} \in \mathcal{B}_{i,p(\cdot)}$ of the approximating optimization problem $\langle \inf_{v \in \mathcal{B}_{i,p(\cdot)}} F_i(v, p(\cdot)) \rangle$, we check whether the objective functional $\mathcal{J}_i(v, p)$

$$\begin{aligned}F_i(v, p) &= \int_\Omega |\nabla v(x)|^{p(x)}\, dx + \lambda \int_\Omega |\nabla v(x) - \nabla \widetilde{S}_i(x)|\, dx \\ &\quad + \mu \int_\Omega |T_S v(x) - \widetilde{S}_i(x)|\, dx \\ &\quad + \frac{1-\mu}{2} T_M \left(|(G_\sigma * v)(\cdot) - \widetilde{M}_i(\cdot)|^2 \right) \end{aligned} \qquad (5.77)$$

is Gâteaux differentiable with respect to v. Namely, let us show that

$$\begin{aligned}&\lim_{t \to 0} \frac{F_i(u^{0,p(\cdot)} + tv, p) - F_i(u^{0,p(\cdot)}, p)}{t} \\ &= \int_\Omega p(x) \left(|\nabla u^{0,p(\cdot)}(x)|^{p(x)-2} \nabla u^{0,p(\cdot)}(x), \nabla v(x) \right) dx \\ &\quad + \lambda \int_\Omega \frac{(\nabla u^{0,p(\cdot)}(x) - \nabla \widetilde{S}_i(x), \nabla v)}{|\nabla u^{0,p(\cdot)}(x) - \nabla \widetilde{S}_i(x)|}\, dx + \mu \int_\Omega \frac{T_S\left(u^{0,p(\cdot)}\right)}{|T_S\left(u^{0,p(\cdot)}\right) - \widetilde{S}_i|} T_S(v)\, dx \\ &\quad + (1-\mu) T_M \left(\left[\left(G_\sigma * u^{0,p(\cdot)}\right) - \widetilde{M}_i \right] G_\sigma * v \right), \quad \forall v \in W^{1,p(\cdot)}(\Omega). \end{aligned} \qquad (5.78)$$

To this end, we note that

$$\frac{|\nabla u^{0,p(\cdot)}(x) + t\nabla v(x)|^{p(x)} - |\nabla u^{0,p(\cdot)}(x)|^{p(x)}}{p(x)t}$$

$$\to \left(|\nabla u^{0,p(\cdot)}(x)|^{p(x)-2} \nabla u^{0,p(\cdot)}(x), \nabla v(x) \right)$$

as $t \to 0$ almost everywhere in Ω. Indeed, by convexity,

$$|\xi|^p - |\eta|^p \leq 2p \left(|\xi| p - 1 + |\eta|^{p-1} \right) |\xi - \eta|,$$

it follows that

5.5 Optimality Conditions

$$\left| \frac{1}{p(x)t} \left(|\nabla u^{0,p(\cdot)}(x) + t\nabla v(x)|^{p(x)} - |\nabla u^{0,p(\cdot)}(x)|^{p(x)} \right) \right|$$
$$\leq 2 \left(\|\nabla u^{0,p(\cdot)}(x) + t\nabla v(x)\|^{p(x)-1} + \|\nabla u^{0,p(\cdot)}(x)\|^{p(x)-1} \right) \|\nabla v(x)\|$$
$$\leq \operatorname{const} \left(|\nabla u^{0,p(\cdot)}(x)|^{p(x)-1} + |\nabla v(x)|^{p(x)-1} \right) |\nabla v(x)|. \tag{5.79}$$

Taking into account that

$$\int_\Omega |\nabla u^{0,p(\cdot)}(x)|^{p(x)-1} |\nabla v(x)| \, dx$$
$$\leq 2 \| |\nabla u^{0,p(\cdot)}(x)|^{p(x)-1} \|_{L^{p'(\cdot)}(\Omega)} \| |\nabla v(x)| \|_{L^{p(\cdot)}(\Omega)}$$
$$\leq 2 \| |\nabla u^{0,p(\cdot)}(x)|^{p(x)-1} \|_{L^{p'(\cdot)}(\Omega,\mathbb{R}^2)} \| \nabla v(x) \|_{L^{p(\cdot)}(\Omega,\mathbb{R}^2)},$$

and $\int_\Omega |\nabla v(x)|^{p(x)} \, dx \stackrel{\text{by (5.58)}}{\leq} \|\nabla v\|^2_{L^{p(\cdot)}(\Omega,\mathbb{R}^2)} + 1$, we see that the right hand side of inequality (5.79) is an $L^1(\Omega)$-function. Therefore,

$$\int_\Omega \frac{|\nabla u^{0,p(\cdot)}(x) + t\nabla v(x)|^{p(x)} - |\nabla u^{0,p(\cdot)}(x)|^{p(x)}}{t} \, dx$$
$$\to \int_\Omega p(x) \left(|\nabla u^{0,p(\cdot)}(x)|^{p(x)-2} \nabla u^{0,p(\cdot)}(x), \nabla v(x) \right) dx \text{ as } t \to 0$$

by the Lebesgue dominated convergence theorem.

Utilizing similar arguments to the rest of terms in (5.77), we deduce that the representation (5.78) for the Gâteaux differential of $F_i(\cdot, p(\cdot))$ at the point $u^{0,p(\cdot)} \in \mathcal{B}_{i,p(\cdot)}$ is valid.

Thus, in order to derive some optimality conditions for the minimizing element $v_{k+1} \in \mathcal{B}_{i,p_k(\cdot)}$ to the problem (5.71), we note that $\mathcal{B}_{i,p_k(\cdot)}$ is a nonempty convex subset of $W^{1,p_k(\cdot)}(\Omega)$ and the objective functional

$$\left\{ \frac{1}{2\tau_k^u} \| \cdot - v_k \|^2_{L^2(\Omega)} + F_i(\cdot, p_k) \right\} : \mathcal{B}_{i,p_k(\cdot)} \to \mathbb{R}$$

is strictly convex. Hence, the well known classical result (see [31, Theorem 1.1.3]) and representation (5.78) lead us to the following conclusion.

Theorem 5.14 *Let $p_k(\cdot) \in \mathfrak{S}$ be an exponent given by the iterative rule (5.53). Let $i \in \{1, \ldots, m\}$ be the number of a fixed spectral channel. Then the unique minimizer $v_{k+1} \in \mathcal{B}_{i,p_k(\cdot)}$ to the approximating problem (5.71) is characterized by*

$$\int_\Omega \left(p_k(x) |\nabla v_{k+1}(x)|^{p_k(x)-2} \nabla v_{k+1}(x), \nabla v(x) - \nabla v_{k+1}(x) \right) dx$$
$$+ \lambda \int_\Omega \frac{(\nabla v_{k+1}(x) - \nabla \widetilde{S}_i(x), \nabla v - \nabla v_{k+1}(x))}{|\nabla v_{k+1}(x) - \nabla \widetilde{S}_i(x)|} \, dx$$

$$+ \mu \int_\Omega \frac{T_S(v_{k+1})}{|T_S(v_{k+1}) - \widetilde{S}_i|} T_S(v - v_{k+1}) \, dx$$
$$+ (1 - \mu) T_M \left(\left[(G_\sigma * v_{k+1}) - \widetilde{M}_i \right] G_\sigma * (v - v_{k+1}) \right)$$
$$+ \frac{1}{\tau_k^u} (v_{k+1} - v_k, v - v_{k+1})_{L^2(\Omega)} \geq 0, \quad \forall v \in \mathcal{B}_{i, p_k(\cdot)}. \quad (5.80)$$

5.6 Numerical Scheme and Settings

In order to illustrate the proposed algorithm for the simultaneous fusion and denoising of color images with different spacial resolution, we conduct the numerical simulations setting $T_S = Id$ for each spectral channels $i = R, G, B$ and extending the set of feasible solutions $\mathcal{B}_{i, p_k(\cdot)}$ to the form $\mathcal{B}_{i, p_k(\cdot)} = W^{1, p_k(\cdot)}(\Omega)$. In other words, we have dropped the two-side constraints $\gamma_{i,0} \leq u(x) \leq \gamma_{i,1}$ from the sets $\mathcal{B}_{i, p_k(\cdot)}$, and instead we control the fulfilment of this two-side constraints at each step of the numerical approximations. We also use the L^1-norm for the fidelity terms. As a result, it allows us to rewrite the variational problem (5.80) in the form of the following boundary value problem

$$-\operatorname{div}\left(p_k |\nabla v_{k+1}|^{p_k(\cdot) - 2} \nabla v_{k+1}\right)$$
$$= \lambda \operatorname{div}\left(\frac{\nabla v_{k+1} - \nabla \widetilde{S}_i}{|\nabla v_{k+1} - \nabla \widetilde{S}_i|}\right) - \mu T_S^* \left(\frac{T_S(v_{k+1})}{|T_S(v_{k+1}) - \widetilde{S}_i|}\right)$$
$$- (1 - \mu) \sum_{(x_i, y_j) \in S_L} \delta_{(x_i, y_j)} \left[(G_\sigma * v_{k+1}) - \widetilde{M}_i \right]$$
$$- \frac{1}{\tau_k^u} (v_{k+1} - v_k), \quad \text{in } \Omega, \quad (5.81)$$

$$\frac{\partial v_{k+1}}{\partial n} = 0 \quad \text{on } \partial \Omega \quad (5.82)$$

with $p_k(\cdot)$ defined in (5.53), and $k = 1, 2, \ldots$.

Since, in practical implementations, it is reasonable to define the solution of the problem (5.81)–(5.82) using a "gradient descent" strategy, we can start with some initial image u_{k+1}^* and pass to the following initial-boundary value problem for quasi-linear parabolic equation with Nuemann boundary conditions

$$\frac{\partial v_{k+1}}{\partial t} = \operatorname{div}\left(p_k |\nabla v_{k+1}|^{p_k(\cdot) - 2} \nabla v_{k+1}\right) + \lambda \operatorname{div}\left(\frac{\nabla v_{k+1} - \nabla \widetilde{S}_i}{|\nabla v_{k+1} - \nabla \widetilde{S}_i|}\right)$$
$$- \mu T_S^* \left(\frac{T_S(v_{k+1})}{|T_S(v_{k+1}) - \widetilde{S}_i|}\right)$$
$$- (1 - \mu) \sum_{(x_i, y_j) \in S_L} \delta_{(x_i, y_j)} \left[(G_\sigma * v_{k+1}) - \widetilde{M}_i \right]$$

5.6 Numerical Scheme and Settings

$$-\frac{1}{\tau_k^u}(v_{k+1} - v_k), \quad \text{in } (0, T) \times \Omega, \tag{5.83}$$

$$\frac{\partial v_{k+1}}{\partial n} = 0 \quad \text{on } (0, T) \times \partial\Omega, \tag{5.84}$$

$$v_{k+1}(0, \cdot) = u_{k+1}^*(\cdot), \quad k = 0, 1, \ldots, \quad v_0(0, \cdot) = \widetilde{S}_i(\cdot), \quad \text{in } \Omega. \tag{5.85}$$

There are numerous approaches to solve quasi-linear partial differential equations (see the references [3, 20] for various techniques). Since we are dealing with pixels in image processing, finite differences approaches and an explicit scheme of the forward Euler method are arguably the best options. Let Δt be a time step size. Then setting

$$t = n\Delta t, \quad n = 0, 1, 2, \ldots, \quad x = l \, (1 \leq l \leq N_x), \quad y = j \, (1 \leq j \leq N_y),$$

where (x, y) stands for image pixel and $N_x \times N_y$ is the original image size at the grid G_H, we define the following discrete notations

$$\Delta_{\pm}^x v_{lj}^n = \pm \left(v_{l\pm 1, j}^n - v_{lj}^n \right), \quad \Delta_{\pm}^y v_{lj}^n = \pm \left(v_{l, j\pm 1}^n - v_{lj}^n \right),$$

$$m(a, b) = \text{minmod}(a, b) = \frac{\text{sgn } a + \text{sgn } b}{2} \min(|a|, |b|),$$

where $v_{l,j}^n$ denotes the approximation of $v_{k+1}(n\Delta t, l, j)$. Then the numerical approximation of the principle components of the boundary value problem (5.83)–(5.85) takes the form

$$\left(\frac{\partial v_{k+1}}{\partial t} \right)_{l,j}^n \approx \frac{v_{l,j}^{n+1} - v_{l,j}^n}{\Delta t},$$

$$\left(\text{div} \left(p_k |\nabla v_{k+1}|^{p_k(\cdot) - 2} \nabla v_{k+1} \right) \right)_{l,j}^n \approx \Delta_-^x \left(P_{l,j}^n \right) + \Delta_-^y \left(Q_{l,j}^n \right),$$

$$P_{l,j}^n = \frac{p_{l,j}^n}{\left(\sqrt{\varepsilon^2 + \left(\Delta_+^x v_{l,j}^n \right)^2 + \left(m \left(\Delta_+^y v_{l,j}^n, \Delta_-^y v_{l,j}^n \right) \right)^2} \right)^{2 - p_{l,j}^n}} \Delta_+^x v_{l,j}^n,$$

$$Q_{l,j}^n = \frac{p_{l,j}^n}{\left(\sqrt{\varepsilon^2 + \left(\Delta_+^y v_{l,j}^n \right)^2 + \left(m \left(\Delta_+^x v_{l,j}^n, \Delta_-^x v_{l,j}^n \right) \right)^2} \right)^{2 - p_{l,j}^n}} \Delta_+^y v_{l,j}^n,$$

$$\left(\text{div} \frac{\nabla v_{k+1} - \nabla \widetilde{S}_i}{|\nabla v_{k+1} - \nabla \widetilde{S}_i|} \right)_{l,j}^n \approx \Delta_-^x \left(R_{l,j}^n \right) + \Delta_-^y \left(W_{l,j}^n \right),$$

$$R_{l,j}^n = \frac{\Delta_+^x v_{l,j}^n - \Delta_+^x (\widetilde{S}_i)_{l,j}^n}{\sqrt{\varepsilon^2 + \left(\Delta_+^x v_{l,j}^n - \Delta_+^x (\widetilde{S}_i)_{l,j}^n \right)^2 + A_1^2}},$$

$$W_{l,j}^n = \frac{\Delta_+^y v_{l,j}^n - \Delta_+^y (\widetilde{S}_i)_{l,j}^n}{\sqrt{\varepsilon^2 + \left(\Delta_+^y v_{l,j}^n - \Delta_+^y (\widetilde{S}_i)_{l,j}^n \right)^2 + B_1^2}},$$

$$A_1 = m\left(\Delta_+^y v_{l,j}^n - \Delta_+^y (\widetilde{S}_i)_{l,j}^n, \Delta_-^y v_{l,j}^n - \Delta_-^y (\widetilde{S}_i)_{l,j}^n\right),$$

$$B_1 = m\left(\Delta_+^x v_{l,j}^n - \Delta_+^x (\widetilde{S}_i)_{l,j}^n, \Delta_-^x v_{l,j}^n - \Delta_-^x (\widetilde{S}_i)_{l,j}^n\right),$$

where

$$p_{l,j}^n = 1 + \frac{a^2}{a^2 + [(|(\nabla G_\sigma * v_k)(x)|)^2]_{l,j}^n},$$

$$(|(\nabla G_\sigma * v_k)(x)|)_{l,j}^n = \sum_{k_1=-5}^{5} \sum_{k_2=-5}^{5} G_\sigma(k_1, k_2) v_{l-k_1, j-k_2}^n.$$

As a result, utilizing the formulas given above and associating each step $k = 1, 2, \ldots$ of the iterative procedure (5.51)–(5.53) with the corresponding time step $n\Delta t$ in the numerical approximation of the parabolic problem (5.83)–(5.85), we arrive at the following numerical scheme associated with the initial boundary problem (5.83)–(5.85):

$$\begin{aligned}
v_{l,j}^{n+1} &= v_{l,j}^n + \Delta_-^x \left[P_{l,j}^n\right] \Delta t + \Delta_-^y \left[Q_{l,j}^n\right] \Delta t \\
&+ \lambda \Delta_-^x \left[R_{l,j}^n\right] \Delta t + \lambda \Delta_-^y \left[W_{l,j}^n\right] \Delta t \\
&+ \mu \frac{v_{l,j}^n}{\sqrt{\varepsilon^2 + \left(v_{l,j}^n\right)^2 + \left((\widetilde{S})_{l,j}^n\right)^2}} \\
&+ (1-\mu) \begin{cases} \left[v_{l,j}^n - \widetilde{M}_{l,j}\right] & \text{if } (l,j) \in S_L, \\ 0 & \text{otherwise} \end{cases}
\end{aligned} \quad (5.86)$$

$$\forall l = 1, \ldots, N_x, \quad \forall j = 1, \ldots, N_y, \quad \forall n = 0, 1, \ldots$$

with the initial conditions

$$v_{l,j}^0 = (\widetilde{S}_i)_{l,j}, \quad \forall l = 1, \ldots, N_x, \quad \forall j = 1, \ldots, N_y$$

and boundary conditions

$$v_{0,j}^n = v_{1,j}^n, \quad v_{N_x,j}^n = v_{N_x-1,j}^n, \quad v_{l,0}^n = v_{l,1}^n, \quad v_{l,N_y}^n = v_{l,N_y-1}^n, \quad (5.87)$$

$$\forall l = 1, \ldots, N_x, \quad \forall j = 1, \ldots, N_y.$$

To conclude this section, we note that the step size Δt should be small enough in order to guarantee the stability of the numerical scheme (5.86)–(5.87). As for the stopping condition

$$v_{l,j}^{n+1} \approx v_{l,j}^n \quad \text{for all } l \text{ and } j$$

it can be formalized as follows

$$\max_{1 \leq l \leq N_x} \max_{1 \leq j \leq N_y} \left|v_{l,j}^{n+1} - v_{l,j}^n\right| \leq \varepsilon.$$

5.7 Numerical Results

For numerical simulations in this section, we set: $\sigma = 0.5$, $\varepsilon = 0.001$, $\tau_k^u = 100 * k$, $\lambda = 0.01$, $\mu = 0.1$. As for the noise estimator $a > 0$, we use the choice of Black et al. [4], i.e.

$$a = \frac{1.4826}{\sqrt{2}} MAD(\nabla \widetilde{S}_i),$$

where MAD denotes the median absolute deviation of the corresponding spectral channel $S_i : G_H \to \mathbb{R}$ of original image $S : G_H \to \mathbb{R}^m$ that can bee computed as

$$MAD(\nabla \widetilde{S}_i) = \text{median}\left[\left|\nabla \widetilde{S}_i - \text{median}\left(|\nabla \widetilde{S}_i|\right)\right|\right]$$

and median $\left(|\nabla \widetilde{S}_i|\right)$ represents the median over the band $S_i : G_H \to \mathbb{R}$ to the gradient amplitude. To guarantee the stability of the proposed algorithm, we make use of the following condition

$$2\left[\frac{1}{\varepsilon} + \lambda\right] \Delta t < 1,$$

where ε comes from the approximation formulae for $P_{l,j}^n$ and $Q_{l,j}^n$, and we set $\varepsilon = 0.001$.

In order to illustrate the proposed approach we have used three images $S^I : G_H^I \to \mathbb{R}^3$ (Dog), $S^{II} : G_H^{II} \to \mathbb{R}^1$ (Barbara), and $S^{III} : G_H^{III} \to \mathbb{R}^3$ (Christmas Tree) with the resolutions in pixels $G_H^I = 342 \times 458$, $G_H^{II} = 512 \times 512$, and $G_H^{III} = 1200 \times 800$, respectively. Each of these images has been previously corrupted by the additive zero-mean Gaussian white noise with variance 0.01 (see Figs. 5.1, 5.2 and 5.3).

As for the images of the same scenes with low resolution and with some extra objects, we have considered two collections. The first one is defined on the grids $G_L^I = 114 \times 152$,

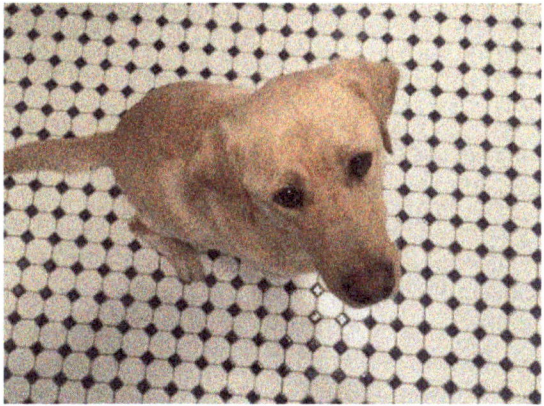

Fig. 5.1 Noisy image $S^I : G_H^I \to \mathbb{R}^3$ (Dog) defined on the grid $G_H^I = 342 \times 458$

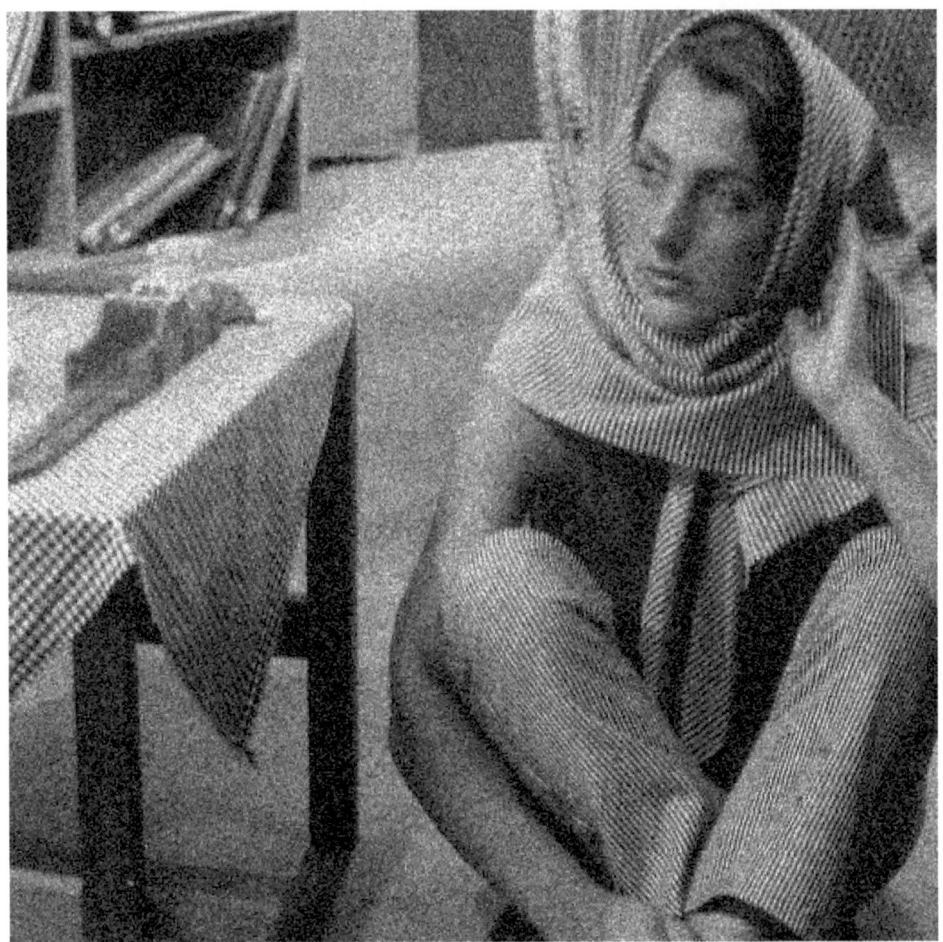

Fig. 5.2 Noisy image $S^{II} : G_H^{II} \to \mathbb{R}^1$ (Barbara) defined on the grid $G_H^{II} = 512 \times 512$

$G_L^{II} = 170 \times 170$, and $G_L^{III} = 400 \times 266$, respectively, and the second one has the resolution $G_L^I = 68 \times 91$, $G_L^{II} = 102 \times 102$, and $G_L^{III} = 240 \times 160$, respectively (see Figs. 5.4 and 5.5).

Then following the proximal alternating minimization algorithm described in Sect. 5.4, we realize the fusion procedure of given images with simultaneous denoising procedure. Obtained results are depicted in Figs. 5.6, 5.7 and 5.8.

It is worth to emphasize that the proposed algorithm is rather sensitive to the choice of parameter μ (see Fig. 5.9 for illustration). As for the running time of processing, it takes for the Matlab-realization about 30, 95, and 280 s for the images depicted in Figs. 5.6, 5.7 and 5.8, respectively.

5.7 Numerical Results

Fig. 5.3 Noisy image $S^{III} : G_H^{III} \to \mathbb{R}^3$ (Christmas Tree) defined on the grid $G_H^{III} = 1200 \times 800$

Fig. 5.4 Images with extra objects and which are defined on the grids with low resolution ($G_L^I = 114 \times 152$, $G_L^{II} = 170 \times 170$, and $G_L^{III} = 400 \times 266$), respectively

Fig. 5.5 Images with extra objects and which are defined on the grids with low resolution ($G_L^I = 68 \times 91$, $G_L^{II} = 102 \times 102$, and $G_L^{III} = 240 \times 160$), respectively

5.7 Numerical Results

Fig. 5.6 Result of Simultaneous Fusion and Denoising of $S^I : G_H^I \to \mathbb{R}^3$ with $M^I : (114 \times 152) \to \mathbb{R}^3$ (left) and $S^I : G_H^I \to \mathbb{R}^3$ with $M^I : (68 \times 91) \to \mathbb{R}^3$ (right)

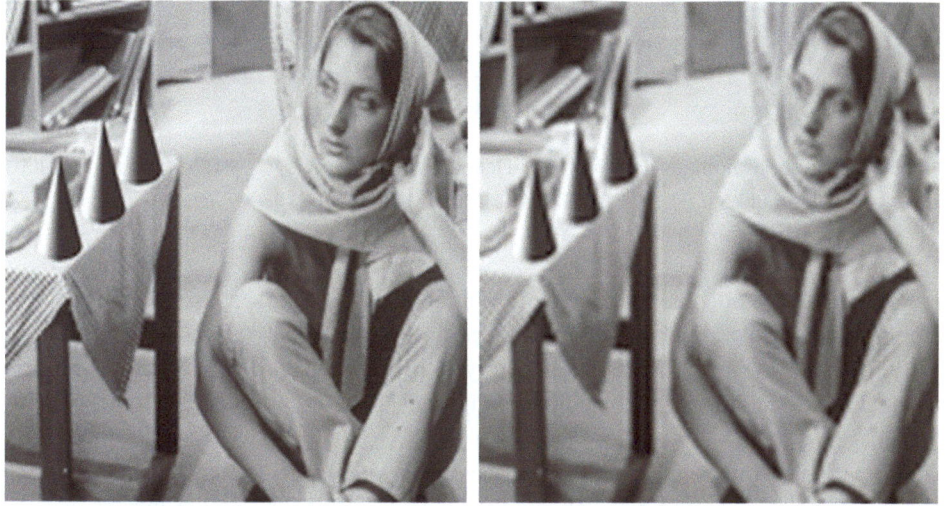

Fig. 5.7 Result of Simultaneous Fusion and Denoising of $S^{II} : G_H^{II} \to \mathbb{R}^3$ with $M^{II} : (170 \times 170) \to \mathbb{R}^3$ (up) and $S^{II} : G_H^{II} \to \mathbb{R}^3$ with $M^{II} : (102 \times 102) \to \mathbb{R}^3$ (bottom)

The next portion of numerical simulations shows that the proposed technique can be successfully applied to the well-known spatial increasing resolution problem of MODIS-like multi-spectral satellite images via their fusion with the Lansat-like imagery at higher resolution. As input data we have used a MODIS (the Moderate Resolution Imaging Spectro-

Fig. 5.8 Result of Simultaneous Fusion and Denoising of $S^{III} : G_H^{III} \to \mathbb{R}^3$ with $M^{III} : (400 \times 266) \to \mathbb{R}^3$ (left) and $S^{III} : G_H^{III} \to \mathbb{R}^3$ with $M^{III} : (240 \times 160) \to \mathbb{R}^3$ (right)

radiometer) image of some region with resolution $350m/pixel$ (see Fig. 5.10). This region represents a typical agricultural area with medium sides fields of various shapes.

We also have the image of the same territory with resolution $25m/pixel$ that was delivered from Landsat satellite at higher resolution. Figure 5.11 shows the RGB spectral channels of this image.

Figure 5.12 displays the result of image fusion corresponding to the data given by Figs. 5.10 and 5.11. In order to validate the obtained result for satellite images, we have provided the following calculations.

- Closednees of the means $\rho_2 = |\text{Mean } I - \text{Mean } L| = 0$;
- Closedness of the variances $\rho_3 = 100 \frac{|\text{Var } I - \text{Var } L|}{\text{Var } L} \approx 6\%$;
- ERGAS metric

$$ERGAS = 100 \frac{h}{l} \sqrt{\frac{1}{3} \sum_{k=1}^{3} \left(\frac{\text{RMSE(k)}}{\mu_0(k)} \right)^2} = 2.24,$$

where h/l is the ratio between the size of the high spatial resolution image pixel and the size of the pixel in the MODIS-like image.

5.7 Numerical Results

Fig. 5.9 Data Fusion of $S^{III} : G_H^{III} \to \mathbb{R}^3$ with $M^{III} : (400 \times 266) \to \mathbb{R}^3$ with a semi-transparency effect ($\mu = 0.8$ left) and $\mu = 0.4$ (right)

Fig. 5.10 The MODIS image with resolution $350 m/pixel$

Fig. 5.11 The Landsat image with resolution $25m/pixel$

It is worth to notice that in view of the suggestions of Prof. L. Wald, if the ERGAS value is less than 3, the spectral quality of an image is satisfactory.

Fig. 5.12 The retrieved image at high resolution $25m/pixel$ as a result of simultaneous fusion and denoising of the MODIS and Landsat images

References

1. Alaa H, Alaa N, Bouchriti A, Charkaoui A (2022) An improved nonlinear anisotropic PDE with $p(x)$-growth conditions applied to image restoration and enhancement. Authorea (2022). https://doi.org/10.22541/au.165717367.72990650/v1
2. Attouch H, Bolte J, Redont P, Soubeyran A (2013) Proximal alternating minimization and projection methods for nonconvex problems. an approach based on the kurdyka-lojasiewicz inequality. arXiv:0801.1780v3
3. Aubert G, Kornprobst P (eds) (2006) Mathematical problems in image processing: partial differential equations and the calculus of variations, vol 147. Springer, New York
4. Black MJ, Sapiro G, Marimont DH, Heger D (1998) Robust anisotropic diffusion. IEEE Trans Image Process 7(3):421–432

5. Blomgren P, Chan T, Mulet P, Wong C (1997) Total variation image restoration: Numerical methods and extensions. In: Proceedings of the 1997 IEEE international conference on image processing, vol 42, pp 384–387
6. Blum R, Xue Z, Zhang Z (2006) An overview of image fusion. In: Multi-sensor image fusion and its applications, vol Chapter 1. CRC Press, Boca Raton
7. Bungert L, Coomes DA, Ehrhardt MJ, Rasch J, Reisenhofer R, Schönlieb CB (2018) Blind image fusion for hyperspectral imaging with the directional total variation. Inverse Probl 34(4)
8. Burgos N, Cardoso MJ, Modat M, Pedemonte S, Dickson J, Barnes A, Duncan JS, Atkinson D, Arridge SR, Hutton BF, Ourselin S (2014) Attenuation correction synthesis for hybrid pet-mr scanners. IEEE Trans Med Imaging 33(12):147–154
9. Chen Y, Levine S, Rao M (2006) Variable exponent, linear growth functionals in image restoration. SIAM J Appl Math 66(4):1383–1406
10. Chen Y, Levine S, Stanich J (2014) Image restoration via nonstandard diffusion. Figshare
11. Cruz-Uribe D, Fiorenza A (eds) (2013) Variable Lebesgue spaces: foundations and harmonic analysis. Birkhauser, New York
12. D'Apice C, De Maio U, Kogut P (2018) Thermistor problem: Multi-dimensional modelling, optimization, and approximation. In: Proceedings of the 32nd European conference on modelling and simulation, May 22nd–May 25th Wilhelmshaven Germany, pp 348–356
13. D'Apice C, De Maio U, Kogut PI (2020) An indirect approach to the existence of quasi-optimal controls in coefficients for multi-dimensional thermistor problem. In: Sadovnichiy VA, Zgurovsky M (eds) Contemporary approaches and methods in fundamental mathematics and mechanics. Springer, New York, pp 489–522
14. D'Apice C, Kogut P, Kupenko O, Manzo R (2023) On variational problem with nonstandard growth functional and its applications to image processing. J Math Imaging Vis 65:472–491
15. D'Apice C, Kogut PI, Manzo R, Pipino C (2024) Variational approach to simultaneous fusion and denoising of color images with different spacial resolution. Commun Math Sci 22(4):1099–1132
16. D'Apice C, Kogut PI, Manzo R, Uvarov M (2023) Variational model with nonstandard growth conditions for restoration of satellite optical images using synthetic aperture radar. Eur J Appl Math 34(1):77–105
17. Diening L, Harjulehto P, Hästö P, Růžička M (eds) (2011) Lebesgue and Sobolev spaces with variable exponents. Springer, New York
18. Goyal S, Singh V, Rani A, Yadav N (2022) Multimodal image fusion and denoising in NSCT domain using CNN and FOTGV. Biomed Signal Proc Control 71(Article ID 103214):3769–3781
19. Huang C, Ng M, Wu T, Zeng T (2021) Quaternion-based dictionary learning and saturation-value total variation regularization for color image restoration. IEEE Trans Multimedia 24:3769–3781
20. Karami F, Meskine D, Sadik K (2019) A new nonlocal model for the restoration of textured images. J Appl Anal Comput 9(6):2070–2095
21. Kogut P (2019) On optimal and quasi-optimal controls in coefficients for multi-dimensional thermistor problem with mixed dirichlet-neumann boundary conditions. Control Cybern 48(1):31–68
22. Kogut P, Kohut Y, Manzo R (2022) Fictitious controls and approximation of an optimal control problem for perona-malik equation. J Optim Diff Equ Their Appl (JODEA) 30(1):42–70
23. Kogut P, Kupenko O, Manzo R (2020) On regularization of an optimal control problem for ill-posed nonlinear elliptic equations. Abst Appl Anal 2020(Article ID 7418707):15 p
24. Kogut P, Manzo R (2013) On vector-valued approximation of state constrained optimal control problems for nonlinear hyperbolic conservation laws. J Dyn Control Syst 19(3):381–404
25. Kogut PI, Kupenko OP (2019) Approximation methods in optimization of nonlinear systems. In: De Gruyter series in nonlinear nalysis and applications, vol 32. Walter de Gruyter GmbH, Berlin
26. Kohr H (2017) Total variation regularization with variable Lebesgue priors. arXiv: 1702.08807

27. Kumar M, Dass S (2009) A total variation-based algorithm for pixel-level image fusion. IEEE Trans Image Process 18(9):2137–2143
28. Kupenko O, Manzo R (2016) Approximation of an optimal control problem in the coefficient for variational inequality with anisotropic p-laplacian. Nonlinear Differ. Equs. Appl. 23(3):Article 35
29. Li F, Li Z, Pi L (2010) Variable exponent functionals in image restoration. Appl Math Comput 216:870–882
30. Li F, Zeng T (2016) Variational image fusion with first and second-order gradient information. J Comput Math 34(2):200–222
31. Lions J (ed) (1971) Optimal control of systems governed by partial differential equations. Springer, Berlin
32. Ma Q, Wang Y, Zeng T (2022) Retinex-based variational framework for low-light image enhancement and denoising. IEEE Trans Multimedia. https://doi.org/10.1109/TMM.2022.3194993
33. Mei J, Dong Y, Huang T (2001) Simultaneous image fusion and denoising by using fractional-order gradient information. J Comput Appl Math 351:212–227
34. Naidu V, Raol J (2008) Pixel-level image fusion using wavelets and principal component analysis. Def Sci J 58(3):338–348
35. Piella G (2009) Image fusion for enhanced visualization: a variational approach. Int J Comput Vis 83(1):1–11
36. Ring W (2000) Structural properties of solutions to Total Variation regularization problems. ESAIM: Math Model Numer Anal 34:799–810
37. Rudin L, Osher S, Fatemi E (1992) Nonlinear total variation based noise removal algorithms. Physica 60(D):259–268
38. Shah P, Merchant S, Desai U (2011) An efficient spatial domain fusion scheme for multifocus images using statistical properties of neighborhood. IEEE Int Conf Multimedia Expo 18(9):1–6
39. Šroubek F, Flusser J (2006) Fusion of blurred images. In: Multi-sensor image fusion and its applications. CRC Press, Boca Raton
40. Wang W, Shui P, Feng X (2008) Variational models for fusion and denoising of multifocus images. IEEE Signal Process Lett 15:65–68
41. Wunderli T (2010) On time flows of minimizers of general convex functionals of linear growth with variable exponent in BV space and stability of pseudosolutions. J Math Anal Appl 364(2):59–591
42. Zaslavski A (ed) (2010) Optimization on metric and normed spaces, springer optimization and its applications, vol 44. Springer, New York
43. Zhikov V (2008) Solvability of the three-dimensional thermistor problem. Proc Steklov Inst Math 281:98–111
44. Zhikov V (2011) On variational problems and nonlinear elliptic equations with nonstandard growth conditions. J Math Sci 175(5):463–570
45. Zhou Z, Li S, Wang B (2014) Multi-scale weighted gradient-based fusion for multi-focus images. Inf Fusion 20:60–72

MIX
Papier aus verantwortungsvollen Quellen
Paper from responsible sources
FSC® C105338

If you have any concerns about our products,
you can contact us on
ProductSafety@springernature.com

In case Publisher is established outside the EU,
the EU authorized representative is:
**Springer Nature Customer Service Center GmbH
Europaplatz 3, 69115 Heidelberg, Germany**

Printed by Libri Plureos GmbH
in Hamburg, Germany